Practical Mass Spectrometry
A Contemporary Introduction

Practical Mass Spectrometry

A Contemporary Introduction

Edited by

Brian S. Middleditch

University of Houston
Houston, Texas

PLENUM PRESS • NEW YORK AND LONDON

Library of Congress Cataloging in Publication Data

Main entry under title: CHEMISTRY

Practical mass spectrometry.

  Includes bibliographies and index.
  1. Mass spectrometry. I. Middleditch, Brian S.
QD96.M3P72   545'.33   79-351
ISBN 0-306-40230-0

© 1979 Plenum Press, New York
A Division of Plenum Publishing Corporation
227 West 17th Street, New York, N.Y. 10011

Printed in the United States of America

Contributors

Thomas Aczel, Exxon Research and Engineering Company, Baytown, Texas 77520

Charles J. W. Brooks, Department of Chemistry, University of Glasgow, Glasgow G12 8QQ, Scotland

Edward C. M. Chen, School of Sciences and Technologies, University of Houston at Clear Lake City, Houston, Texas 77058

Charles G. Edmonds, Department of Chemistry, University of Glasgow, Glasgow G12 8QQ, Scotland. *Present address:* Divisaõ Química, Instituto Nacional de Pesquisas da Amazônia, Manaus, Amazonas, Brasil

Stephen R. Heller, U. S. Environmental Protection Agency, MIDSD, PM-218, Washington, D. C. 20460

H. D. M. Jager, Atomic Weapons Research Establishment, Aldermaston, Reading, Berkshire RG7 4PR, England

David C. Maxwell, Atomic Weapons Research Establishment, Aldermaston, Reading, Berkshire RG7 4PR, England

Andrew McCormick, Atomic Weapons Research Establishment, Aldermaston, Reading, Berkshire RG7 4PR, England

Brian S. Middleditch, Department of Biophysical Sciences, University of Houston, Houston, Texas 77004

Daryl Nooner, Department of Biophysical Sciences, University of Houston, Houston, Texas 77004. *Present address*: Spectrix Corporation, Houston, Texas 77054

J. Oró, Department of Biophysical Sciences, University of Houston, Houston, Texas 77004

Henry Shanfield, Department of Chemistry, University of Houston, Houston, Texas 77004

Richard N. Stillwell, Institute for Lipid Research, Baylor College of Medicine, Houston, Texas 77030

William J. A. VandenHeuvel, Merck, Sharp and Dohme Research Laboratories, Rahway, New Jersey 07065

Anthony G. Zacchei, Merck Institute for Therapeutic Research, West Point, Pennsylvania 19486

Albert Zlatkis, Department of Chemistry, University of Houston, Houston, Texas 77004

Preface

It has been estimated that more than 80% of the world's scientists who have ever lived are still alive today. It would not be unreasonable to suggest that more than 95% of those who have ever used a mass spectrometer are not only alive but are still actively employed. Most have never had any formal training in the subject since, with a few notable exceptions, universities have only recently begun to offer courses in mass spectrometry.

We have written this book for the student of modern mass spectrometry: it is for the novice who wished to know what the instruments can do and how the techniques can be applied. There are other books on the market which delve into the history of mass spectrometry and go deeply into the mathematical theory and instrumentation. There are yet more books which guide one through the art of interpreting spectra. We have deliberately avoided these topics so that the reader is confronted only with the basic principles and is allowed a taste of the applications.

One of the best methods of developing a useful textbook is to teach a course based upon its content. This is what we did. We met in Houston in 1976 to teach a course on "Perspectives in Mass Spectrometry" and to coordinate our writing. The authors of five of the chapters met again in St. Louis in 1978 to teach a course on combined gas chromatography–mass spectrometry. The final product is a series of chapters covering the major aspects of the subject, often written by those who developed them. Some fairly heavy editing was required to maintain uniformity in style, but the characteristic styles of the individual authors have been largely preserved.

The first chapter is for the true neophyte. In the second chapter, rather than discuss all of the various components which can be assembled to produce a working instrument, discussion has been limited to three commercial machines. The LKB 9000 is similar to the LKB 9000S referred to in several of the other chapters. Likewise, the Hewlett-Packard 5992 is one of the latest of the quadrupole instruments, a descendent of the H.-P.

5390 referred to in Chapter 4. The reader is introduced to gas chromatography–mass spectrometry (GC–MS) in Chapter 3, and a more detailed discussion of selective ion monitoring follows in the next chapter. The special problems associated with the analysis of volatile samples are reviewed in Chapter 5, and automatic data processing is discussed in Chapter 6. Since those who are new to mass spectrometry will initially, at least, identify compounds by comparison with spectra in reference collections, the next three chapters review the tools that are available to accomplish this. The final four chapters provide a sampling of the applications for which mass spectrometry is most suited.

The exercises provided in most chapters are designed to allow the student to evaluate his comprehension of the subject matter. Answers to problems are given at the end of the book. We have also provided lists of selected review articles and books for further reading.

In addition to the 16 authors of the various chapters, many other individuals and companies have made minor, but valuable, contributions. Some of these are acknowledged in the text. Among the others are Kratos-AEI for suggesting that we undertake this task and for underwriting our first course, and the following: *Chapter 2*—LKB Instruments, Hewlett-Packard, and Kratos-AEI for providing information concerning their instruments and some of the illustrations; *Chapter 3*—the Medical Research Council and Science Research Council for support of much of the work described, colleagues and co-workers in many laboratories for assistance and support, and special thanks to E.C. Horning for research facilities at Baylor College of Medicine, Houston, Texas; *Chapter 5*—the National Aeronautics and Space Administration, Life Sciences Directorate, Johnson Space Center, Houston, Texas (Contract NAS 9-14534); *Chapter 6*—the National Institutes of Health (GM-13901) and the Robert A. Welch Foundation (Q-125); *Chapter 7*—J.A. McCloskey for collaborating in the production of the "Guide to Collections of Mass Spectral Data," upon which this chapter is based; *Chapter 8*—Her Majesty's Stationery Office for permission to reprint excerpts from the *Mass Spectrometry Bulletin*; *Chapter 9*—the following persons, who have worked on and contributed to the Mass Spectral Search System (MSSS): K. Biemann (MIT); A. Bridy, H.M. Fales, R.J. Feldmann, and G.W.A. Milne (NIH); R.S. Heller (University of Maryland); D.C. Maxwell and A. McCormick (Mass Spectrometry Data Centre); W.L. Budde and J.M. McGuire (Environmental Protection Agency); and F.W. McLafferty (Cornell University); *Chapter 10*—colleagues at the U.S. Environmental Protection Agency whose research is described, including R.G. Webb (Power Plant Cooling Water); L.H. Keith (Industrialized Shipping Channel); J.W. Eichelberger, W.M. Middleton, and W.L. Budde (Municipal Drinking Water); and R.G. Weber and C.E. Taylor (Landfill Leachate); *Chapter 11*—the following colleagues who have contributed to the research described: B. Arison, J. Carlin, M. Christy, S. Date, B. Ellsworth, V. Gruber, K. Hooke, H.B. Hucker, T. Jacob, C. Porter, R.E. Rhodes, A. Rosegay, J. Smith, D.J. Tocco, T. Tyler, S. Vickers, R. Walker, L.L. Weidner, T.I. Wishousky, D. Wolf, F. Wolf, and O. Woltersdorf; *Chapter 13*—present and former students and colleagues working in cooperation with J. Oró whose research forms the basis of this article, the National Aeronautics and Space Administration for research support, and authors and publishers for permission to use copyrighted figures and tables.

I would like to thank Ellis Rosenberg and Robert Golden of Plenum Press, who skillfully and efficiently expedited the book through its various stages of production.

I owe a special debt to my wife, Tin Tin, and my daughter, Courtney, for their patience while I was working on the manuscript.

University of Houston   Brian S. Middleditch

Contents

3.   Combined Gas Chromatography–Mass Spectrometry

Charles J. W. Brooks and Charles G. Edmonds

4.   Selective Ion Monitoring

Edward C. M. Chen

5. Concentration Techniques for Volatile Samples

Albert Zlatkis and Henry Shanfield

6. Automatic Data Processing

Richard N. Stillwell

7. Collections of Mass Spectral Data

Brian S. Middleditch

8. The Mass Spectrometry Data Centre

H.D.M. Jager, David C. Maxwell, and Andrew McCormick

9. The Mass Spectral Search System
Stephen R. Heller

10. Environmental Applications of Mass Spectrometry
Stephen R. Heller

11. Applications of Mass Spectrometry in the Pharmaceutical Industry
Anthony G. Zacchei and William J. A. VandenHeuvel

12. Applications of Mass Spectrometry in the Petrochemical Industry
Thomas Aczel

13. Cosmochemical and Geochemical Applications of Mass Spectrometry

J. Oró and Daryl Nooner

Principles of Mass Spectrometry

Brian S. Middleditch

1. History

The term "mass spectrometry" did not appear in the scientific literature until around 1920, when F. W. Aston reported for the first time the precise atomic weights of neon-20 and neon-22. This was a decade later than the discovery of these, the first, isotopes by Sir J. J. Thomson, confirming an earlier suggestion by F. Soddy that such entities should exist. These far-reaching discoveries resulted in the award of Nobel prizes, and led to the establishment of a vast new field of science which was eventually to affect each and every one of us. While the discovery of isotopes led to the development of some of the most horrendous weapons known to man, it has indisputably enriched our lives in many ways.

The history of mass spectrometry for almost half a century was mainly limited to one of instrumental refinements required for the cataloging of isotopes of the other elements. Thus the early development of mass spectrometry was in the realm of the physical chemists.

This situation was to change rather abruptly during the Second World War. Until that time the petroleum industry was content to distill its crude oil to obtain products such as gasolines, kerosines, solvents, lubricating oils, waxes, and tars. Little regard was paid to the actual chemical compositions of these fractions, and those of no commercial value were sometimes discarded. With the establishment of the plastics industry,

Brian S. Middleditch • Department of Biophysical Sciences, University of Houston, Houston, Texas 77004

there was a sudden demand for large quantities of the appropriate feed-stocks which could not be met simply by distilling crude petroleum. Thus the petrochemical industry was born, with crude petroleum fractions being chemically converted to more useful raw materials.

The petrochemists needed more precise knowledge of, and control over, the chemical compositions of their products. A change in the composition of the feedstock for a plastics plant could easily modify the physical characteristics of the product. Fortunately, it was not difficult to modify the mass spectrometer to enable it to perform analyses of these volatile petrochemicals. It was even possible to perform both qualitative and quantitative analyses of simple mixtures.

The scope of mass spectrometry was further extended in the 1950s by the development of techniques for the analysis of compounds of higher molecular weights, including those such as steroids which were formerly regarded as nonvolatile. Further impetus was added in the 1960s by the invention of efficient devices for coupling gas chromatographs and mass spectrometers. The past decade has witnessed a proliferation of innovations in instrumentation and techniques. Many of these developments are described in succeeding chapters, often by those who were responsible for their introduction.

2. The Electron Impact Mass Spectrum

Mass spectrometry is very different from most other forms of spectrometry. In electron spin resonance, nuclear magnetic resonance, infrared, raman, visible, ultraviolet, and Mössbauer spectrometry, for example, one irradiates the sample and determines the effect of the sample upon the radiation. Thus (Fig. 1) a spectrophotometer is used to determine the types and positions of functional groups in molecules. The mass spectroscopist, on the other hand, observes the effects of a source of ionizing energy upon the sample molecule. Usually, ionization yields a detectable molecular ion from which the molecular weight of the sample can be determined directly. If sufficient energy is imparted to the molecule during its ioniza-

INFRARED SPECTROPHOTOMETRY

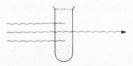

Ultraviolet/visible spectrophotometry
Nuclear magnetic resonance
Electron spin resonance

FIG. 1. Principle of spectrophotometry.

tion, bonds will be cleaved to yield fragment ions. These ions, together with the molecular ion, comprise the mass spectrum. Since various functional groups direct specific bond cleavages, the mass spectrum can be "interpreted" to obtain information concerning the structure of the molecule.

2.1. The Molecular Ion

2.1.1. Production

To produce a molecular ion, the molecule should be exposed to a source of energy sufficient to effect its ionization. Most instruments used for organic mass spectrometry contain an electron-impact ion source. Electrons emitted from a heated filament are attracted toward an anode, or trap. Sample molecules drifting into the beam may be struck by an electron. This may impart sufficient energy to remove an electron from the sample molecule, resulting in the production of a molecular ion:

$$M + \epsilon^- \rightarrow M^{+\cdot} + 2\epsilon^-$$

It is, perhaps, less confusing to consider the bombarding electron as a source of energy:

$$M + energy \rightarrow M^{+\cdot} + \epsilon^-$$

The electron removed from the sample molecule may originate from:

  (i) a lone pair of electrons on a heteroatom such as oxygen, sulfur, or nitrogen,
 (ii) a pair of electrons in the π orbital of a multiple bond, or
(iii) a pair of electrons in the σ orbital of a single bond.

In each case, a positive charge is imparted and an unpaired electron remains, so the molecular ion is represented as $M^{+\cdot}$.

2.1.2. Mass-to-Charge Ratio

Only charged species appear in the mass spectrum. The instrument cannot determine the mass of an ion, only its mass-to-charge ratio (m/e). Under normal operating conditions most ions are singly charged, so the m/e value of an ion is the sum of the atomic weights of its components. Thus the m/e value of an ion is loosely referred to as the mass of the ion. Under certain circumstances an ion may have a multiple charge: a doubly charged ion will have an m/e value half that of the corresponding singly charged ion.

2.1.3. Isotopes

Most elements have more than one naturally occurring isotope (Table 1), so molecules of any compound can have a variety of compositions. Hydrogen chloride, for example, will comprise $^{1}H^{35}Cl$, $^{2}H^{35}Cl$, $^{1}H^{37}Cl$, and $^{2}H^{37}Cl$ with molecular weights of 36 to 39. The natural abundance of deuterium (0.012%) is small enough to ignore in this example so the species observed in the spectrum (Fig. 2) are $^{1}H^{35}Cl$ (*m/e* 36) and $^{1}H^{37}Cl$ (*m/e* 38), in a ratio of approximately 3:1. Similarly, the major molecular ions of hydrogen bromide appear in its spectrum (Fig. 3) at *m/e* 80 ($^{1}H^{79}Br$) and *m/e* 82 ($^{1}H^{81}Br$) in a ratio of approximately 1:1.

The major molecular ions in the spectrum of methylene chloride (Fig. 4) are $^{12}C^{1}H_2^{35}Cl_2$ (*m/e* 84), $^{12}C^{1}H_2^{35}Cl^{37}Cl$ (*m/e* 86), and $^{12}C^{1}H_2^{37}Cl_2$ (*m/e* 88). It should be noted that the relative amounts of these species are in a binomial distribution ($a^2:2ab:b^2$, where a and b represent the relative abundances of the two isotopes) so that the abundances of the molecular

TABLE 1.   Relative Abundances
of Naturally Occurring Isotopes

Element	Isotope	Abundance
Hydrogen	1	99.98
	2	0.012
Carbon	12	98.892
	13	1.108
Nitrogen	14	99.635
	15	0.365
Oxygen	16	99.76
	17	0.037
	18	0.204
Fluorine	19	100
Silicon	28	92.2
	29	4.7
	30	3.1
Phosphorus	31	100
Sulfur	32	95.0
	33	0.76
	34	4.22
	36	0.014
Chlorine	35	75.5
	37	24.5
Bromine	79	50.5
	81	49.5
Iodine	127	100

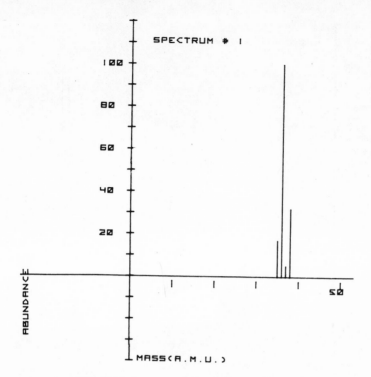

FIG. 2. Mass spectrum of hydrogen chloride.

ions are, approximately,

$$(3 \times 3) : (2 \times 3 \times 1) : (1 \times 1), \quad \text{or}$$
$$9 \quad : \quad 6 \quad : \quad 1$$

Similarly, the relative abundances of the major molecular ions of methylene bromide, $^{12}C^1H_2{}^{79}Br_2$ (*m/e* 172), $^{12}C^1H_2{}^{79}Br^{81}Br$ (*m/e* 174), and $^{12}C^1H_2{}^{81}Br_2$ (*m/e* 176), appear in the spectrum in the approximate ratio

$$(1 \times 1) : (2 \times 1 \times 1) : (1 \times 1), \quad \text{or}$$
$$1 \quad : \quad 2 \quad : \quad 1$$

Further examples of isotope distributions for various combinations of chlorine and bromine are given in Table 2. It is relatively easy to determine the numbers of chlorine and/or bromine atoms per molecule from the relative abundances of the various molecular ions. In principle, this approach can be extended to the calculation of the elemental composition of any molecule. For example, cholesterol will afford molecular ions

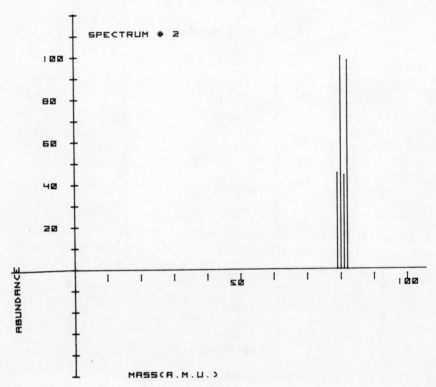

FIG. 3. Mass spectrum of hydrogen bromide.

TABLE 2.   Relative Abundances and Masses of
Combinations of Chlorine and Bromine Isotopes

Combination	Relative abundances	Masses
Cl	3:1	35, 37
Br	1:1	79, 81
Cl_2	9:6:1	70, 72, 74
$ClBr$	3:4:1	114, 116, 118
Br_2	1:2:1	158, 160, 162
Cl_3	27:27:9:1	105, 107, 109, 111
Cl_2Br	9:15:7:1	149, 151, 153, 155
$ClBr_2$	3:7:5:1	193, 195, 197, 199
Br_3	1:3:3:1	237, 239, 241, 243

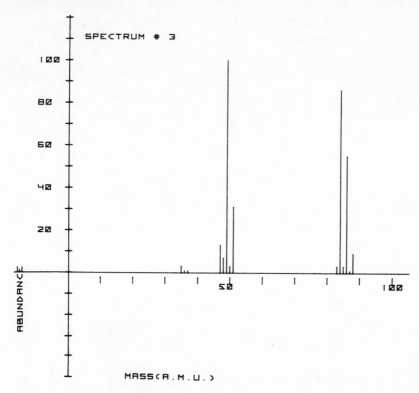

FIG. 4. Mass spectrum of methylene chloride.

ranging from *m/e* 386 ($^{12}C_{27}{}^{1}H_{46}{}^{16}O$) to *m/e* 461 ($^{13}C_{27}{}^{2}H_{46}{}^{18}O$) with a polynomial distribution of relative abundances. Only a few of these ions are of sufficient relative abundance to appear in the spectrum (Fig. 5), but the numbers of carbon and oxygen atoms per molecule can be determined from the relative intensities of the M, M + 1, and M + 2 peaks if they are accurately recorded. This approach to mass spectral interpretation is appealing because of its apparent simplicity, and it provides the basis for a popular textbook by McLafferty (see Section 5, "Suggested Reading"). In practice, however, spectra are rarely recorded with sufficient accuracy to provide a reliable indication of elemental composition. Nevertheless, the presence of certain elements, in addition to chlorine and bromine, can sometimes be inferred. In comparison with the isotope distributions of compounds containing carbon, hydrogen, nitrogen, and oxygen atoms, silicon affords more abundant M + 1 and M + 2 ions, whereas sulfur affords slightly more abundant M + 2 ions. Moreover, compounds containing fluorine, phosphorus, or iodine atoms can sometimes be recognized as such since these elements are monoisotopic.

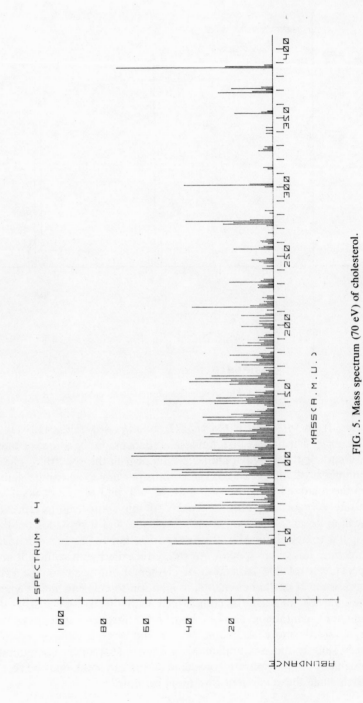

FIG. 5. Mass spectrum (70 eV) of cholesterol.

TABLE 3. Precise Atomic Weights (Based upon $^{12}C = 12$)

Isotope	Atomic weight
^1H	1.007825
^2H	2.014102
^{12}C	12
^{13}C	13.003354
^{14}N	14.003074
^{16}O	15.994915
^{18}O	17.999160
^{19}F	18.998495
^{28}Si	27.976929
^{31}P	30.973765
^{32}S	31.972073
^{35}Cl	34.968851
^{79}Br	78.918329
^{127}I	126.904470

2.1.4. Precise Mass

Nominal atomic weights, rounded to the nearest integer, were used in the examples described above. With the exception of ^{12}C, upon which our atomic weight scale is based, all isotopic species of the various elements have nonintegral atomic weights (Table 3). Also, compounds with the same nominal molecular weight will, if they are of different elemental composition, have different precise molecular weights. For example, nitrogen, carbon monoxide, and ethylene each afford major molecular ions with a nominal m/e value of 28, but with precise m/e values of 28.006148, 27.994915, and 28.031300, respectively. These ions may be resolved by a high-resolution mass spectrometer and their elemental compositions determined.

2.2. Fragment Ions

Energy in excess of that required to ionize a molecule may be dissipated by fragmentation of the molecular ion. The presence of fragment ions in a spectrum provides valuable information concerning the structure of a molecule.

If we reexamine the spectrum of hydrogen chloride (Fig. 2), we find $[M-1]^+$ ions at m/e 35 ($^{35}Cl^+$) and m/e 37 ($^{37}Cl^+$), formed by losses of hydrogen atoms from the respective molecular ions. Similarly, hydrogen bromide (Fig. 3) affords fragment ions of m/e 79 and 81. Note that the ions of m/e 40 and 41 are doubly charged molecular ions. In these simple examples, we have "interpreted" every ion. The spectrum of methylene chloride (Fig. 4) is no more difficult to rationalize.

The spectrum of cholesterol (Fig. 5), however, contains a great number of ions and remains to be fully interpreted. The origin of each ion in the spectrum could be determined if one were to synthesize cholesterol analogs labeled with stable isotopes (^{2}H, ^{13}C, or ^{18}O) in every position. High-resolution spectra would also provide assistance. However, it is neither necessary nor desirable to account for the formation of each and every ion in the spectrum to identify a compound.

There are two basic approaches to identifying compounds by mass spectrometry:

(1) Comparing the spectrum of the "unknown" with those in reference collections to find one with ions of similar *m/e* values and relative abundances.

(2) Partial interpretation of the spectrum of the "unknown" to obtain a possible structure, followed by reexamination of the spectrum in a search for ions which one might expect to be produced by such a compound. A few commonly encountered fragmentation modes are described below.

α-Cleavage. An example is shown in Fig. 6. Acetone is ionized by

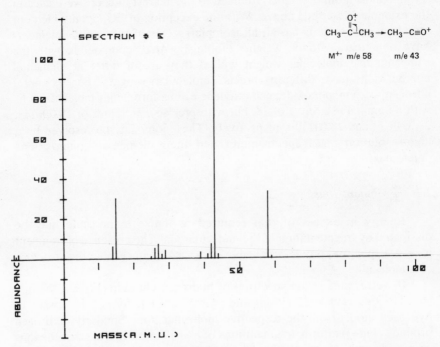

FIG. 6. Mass spectrum of acetone, with scheme (inset) depicting α-cleavage of the molecular ion to afford the ion of *m/e* 43.

the removal of an electron from one of the lone pairs on the oxygen atom. Cleavage of one of the carbon–carbon bonds affords the ion of *m/e* 43 and a methyl radical, which is uncharged and does not appear in the spectrum. [The ion of *m/e* 15 (CH_3^+) is formed by an alternative mode of ionization and fragmentation.]

α-Cleavage is possible with a wide range of ketones, aldehydes, alcohols, acids, esters, and amides, and their sulfur-containing analogs.

β-Cleavage. An example is shown in Fig. 7. Ethylamine is ionized by the removal of an electron from the lone pair on the nitrogen atom. Cleavage of the carbon–carbon bond affords the ion of *m/e* 30. Again, the methyl radical produced as a result of this cleavage does not appear in the spectrum, but an alternative mode of ionization and fragmentation affords the methyl ion observed at *m/e* 15.

β-Cleavage is usually observed for amines, ethers, and thioethers.

γ-Cleavage. This relatively uncommon mode of fragmentation is generally limited to oximes and related compounds. Its mechanism is not fully understood and may proceed via charge localization on the oxygen atom and cleavage accompanied by formation of a five-membered ring, or

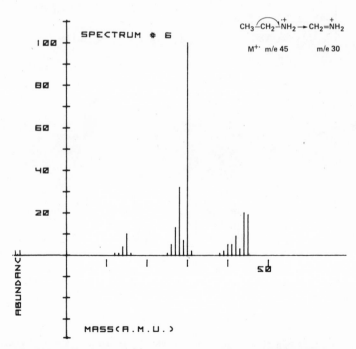

FIG. 7. Mass spectrum of ethylamine, with scheme (inset) depicting β-cleavage of the molecular ion to afford the ion of *m/e* 30.

via charge localization on the nitrogen atom and cleavage accompanied by formation of a four-membered ring.

   Allylic Cleavage.   As mentioned above, molecules can become ionized by the removal of one of a pair of electrons in the π orbital of a double bond. The resulting molecular ion is usually represented as shown for 1-butene in Fig. 8. Allylic cleavage leads to the formation of the ion of *m/e* 41 in this example.

   Note that cleavages adjacent to double bonds are rarely observed.

   Benzylic Cleavage.   This is analogous to the allylic cleavage. Thus ethyl benzene affords an ion of *m/e* 91 as depicted in Fig. 9. The benzyl ion apparently rearranges to a more stable tropylium ion.

   Cleavages adjacent to aromatic nuclei are rarely observed.

   Retro-Diels–Alder Rearrangement.   This rearrangement is the reverse of the Diels–Alder synthesis of cyclohexene analogs from butadienes and olefins. Rearrangement of electrons in the nonaromatic ring of tetralin, for example, affords the fragment ion of *m/e* 104 (Fig. 10). It is not necessary to specify the site of charge localization in either the molecular ion or the fragment ion, so it is conventional to represent the charge as depicted.

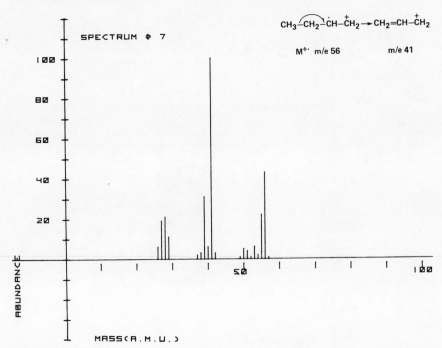

FIG. 8. Mass spectrum of 1-butene, with scheme (inset) depicting allylic cleavage of the molecular ion to afford the ion of *m/e* 41.

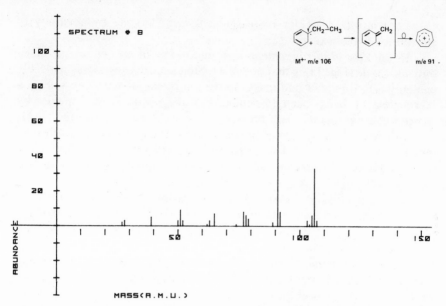

FIG. 9. Mass spectrum of ethyl benzene, with scheme (inset) depicting benzylic cleavage of the molecular ion to afford the ion of *m/e* 91.

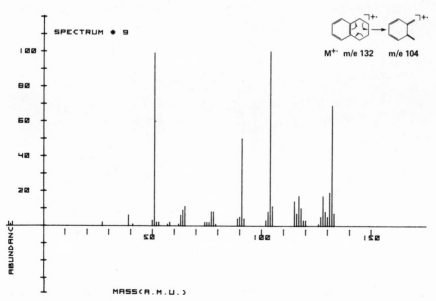

FIG. 10. Mass spectrum of tetralin, with scheme (inset) depicting a retro-Diels–Alder rearrangement of the molecular ion to afford the ion of *m/e* 104.

The retro-Diels–Alder rearrangement is undergone by a variety of substituted cyclohexenes.

McLafferty Rearrangement. A major ion in the spectrum of *n*-butyraldehyde (Fig. 11) is that of *m/e* 44. Simple cleavages of this molecule can only afford ions of odd mass, so the ion of *m/e* 44 must result from a rearrangement. It has been determined that β-cleavage is accompanied by transfer of a hydrogen atom from the γ carbon atom, as shown in Fig. 11.

The McLafferty rearrangement is undergone by many aldehydes, ketones, acids, esters, amides, and related compounds. Aromatic compounds may also undergo a McLafferty rearrangement, as shown in Fig. 12a.

If a compound has two suitable chains attached to an appropriate functional group, either chain may undergo a McLafferty rearrangement, individually or sequentially. The sequential mechanism, referred to as a double McLafferty rearrangement, is illustrated in Fig. 12b.

Complex Rearrangements. Unfortunately, all too many compounds fragment via routes involving more complex rearrangements. One of the

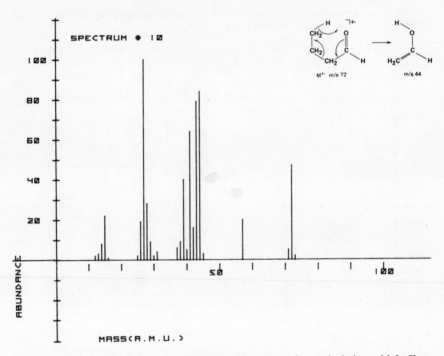

FIG. 11. Mass spectrum of *n*-butyraldehyde, with scheme (inset) depicting a McLafferty rearrangement of the molecular ion to afford the ion of *m/e* 44.

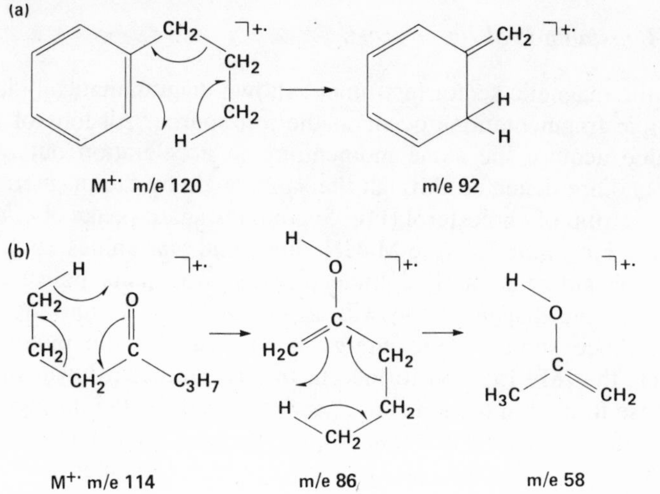

FIG. 12. Schemes depicting (a) an aromatic McLafferty rearrangement, and (b) a double McLafferty rearrangement.

simpler examples is that of cyclohexanol; the formation of the ion of m/e 57 is shown in Fig. 13. An exactly analogous rearrangement leads to the formation of ions of m/e 129, characteristic of trimethylsilyl derivatives of Δ^5-3β-hydroxy steroids such as cholesterol (Fig. 14).

For a more detailed discussion of mass spectral rearrangements, the reader is referred to the excellent books compiled by Budzikiewicz, Djerassi, and Williams (see Section 5).

2.3. Multiply Charged Ions

Multiply charged ions were briefly discussed in Section 2.1.2. They are most often observed in the spectra of polycyclic aromatic hydrocarbons. Ovalene, for example, affords molecular ions with one to four positive charges, at m/e 398, 199, 132.7, and 99.5.

A few compounds have been found to give rise to doubly charged fragment ions with relative abundances greater than those of the corresponding singly charged ions.

Multiply charged ions are easily recognized. Some will appear at nonintegral m/e values, such as those of m/e 132.7 and 99.5 in the spectrum of ovalene. Others, which appear at integral m/e values, will be accompanied by isotope peaks with nonintegral m/e values. Thus, the doubly charged molecular ion of ovalene (m/e 199) will be accompanied by a doubly charged $^{12}C_{31}{}^{13}C\,^1H_{14}$ ion of m/e 199.5.

2.4. Metastable Peaks

In the magnetic sector instrument shown diagrammatically in Fig. 15, most ionic fragmentations occur in the ion source. All ions of the same m/e value acquire the same momentum on acceleration out of the ion source and are deflected through the same angle by the magnet. Thus the mass spectrum of cholesterol (Fig. 5) contains sharp peaks of m/e 386 (the molecular ion), m/e 371 (the M–CH_3 ion), and m/e values corresponding to other fragment ions. If a molecular ion fragments between the ion source and the magnet it too will afford an $[M-15]^+$ ion and a methyl radical. Since some kinetic energy is imparted to the methyl radical, however, the $[M-15]^+$ ion formed in this region has a lower momentum than those formed in the ion source and will be deflected through a greater angle by the magnet, so that it appears in the spectrum at a lower apparent m/e value. If several molecular ions of cholesterol fragment in this portion of the instrument, a range of energies will be imparted to the methyl radicals and these $[M-15]^+$ ions will appear over a range of apparent m/e values. The resulting broad, diffuse peak in the spectrum is termed a metastable peak. Its approximate position may be determined from the formula

$$M^* = M_2^2/M_1$$

where M^* is the apparent m/e value of the metastable peak and M_1 and M_2 are, respectively, the m/e values of the corresponding parent and daughter ions. The metastable peak corresponding to the loss of a methyl radical from cholesterol appears, therefore, at m/e $(371^2/386) = 356.6$.

$[M-15]^+$ ions formed between the magnet and the detector will accompany the molecular ions as they travel toward the detector. Conventional detectors are unable to discriminate between such pairs of ions.

2.5. Ion–Molecule Reactions

The pressure in the ion source of a mass spectrometer is normally maintained sufficiently low that there is little chance of ions colliding with neutral molecules. Under some conditions, however, such interaction may occur, affording ions of higher m/e values than the molecular ion.

The most commonly encountered ions of this type are $[M+1]^+$ ions, formed by the transfer of a hydrogen atom to the molecular ion. Aldehydes, in particular, tend to form $[M+1]^+$ ions, even at low ion source pressures. Trimethylsilyl derivatives occasionally afford $[M+73]^+$ ions by acquisition of trimethylsilyl groups.

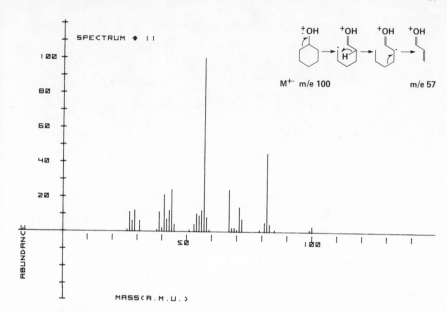

FIG. 13. Mass spectrum of cyclohexanol, with scheme (inset) depicting the formation of the rearrangement ion of *m/e* 57.

Ion–molecule reactions are a potential source of confusion in determining molecular weights by mass spectrometry. This can usually be resolved if something is known of the elemental composition of the sample. Almost all compounds comprised only of H, C, O, F, Si, P, Cl, Br, or I, or which contain an even number of nitrogen atoms per molecule are of even molecular weight. An apparent molecular ion with an odd *m/e* value is either that of a compound with an odd number of nitrogen atoms per molecule, or results from an ion–molecule reaction. Variation of the ion source pressure will change the abundance of such an ion relative to other ions in the spectrum.

Ion–molecule reactions are deliberately induced during chemical ionization mass spectrometry (see Section 3.1).

2.6. *Variation of Electron Energy*

The amount of energy imparted to a molecule during its ionization depends upon the energy of the electron with which it collides. An electron which is accelerated through a greater potential may impart more energy to the ion which it produces. In a conventional mass spectrometer the electrons produced in the ion source are not monoenergetic. Figure 16

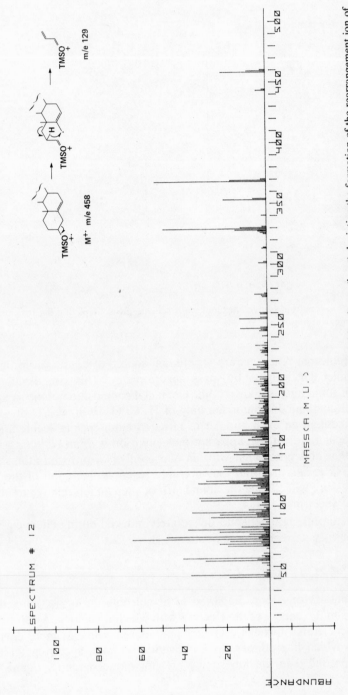

FIG. 14. Mass spectrum of the trimethylsilyl derivative of cholesterol, with scheme (inset) depicting the formation of the rearrangement ion of *m/e* 129.

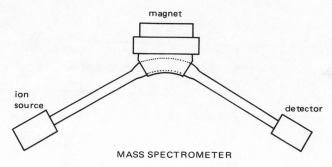

FIG. 15. Schematic diagram of a magnetic sector mass spectrometer.

shows the relationship between electron energy and degree of ionization. It can be seen that electrons with an energy of 70±20 eV have similar effects, so most spectra are obtained with a nominal electron energy of 70 eV.

The use of a lower electron energy affords molecular ions with less energy. The corresponding spectra, therefore, contain molecular ions in greater relative abundance and fewer fragment ions, particularly those with lower m/e values. It should be noted that, while the total ion current is reduced (cf. Fig. 16), the absolute abundance of the molecular ion is not necessarily reduced. The remaining ions are usually those which are of greater diagnostic value: compare the 22.5-eV spectrum of cholesterol (Fig. 17) with the 70-eV spectrum (Fig. 5).

At extremely low electron energies there will be few, if any, fragment ions and the spectra will provide no information on the structure of the

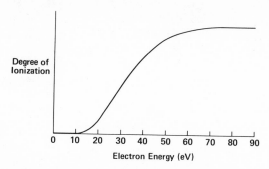

FIG. 16. Relationship between electron energy and degree of ionization for a mass spectrometer with an electron impact ion source.

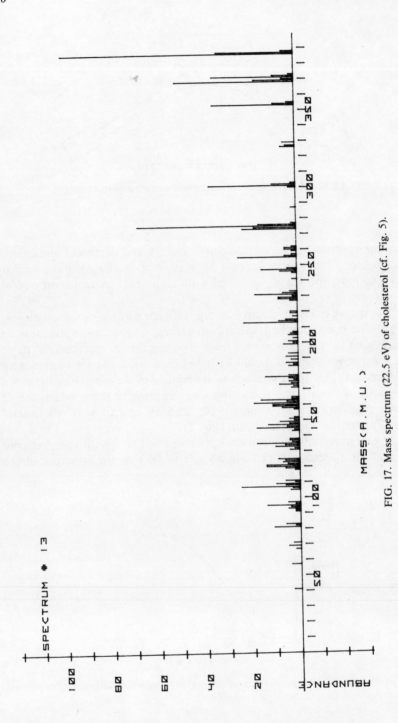

FIG. 17. Mass spectrum (22.5 eV) of cholesterol (cf. Fig. 5).

sample. The use of a combination of low electron energy and high-resolution mass spectrometry, however, affords information on the composition of multicomponent mixtures and is particularly appropriate for petroleum analysis (see Chapter 12).

2.7. Quadrupole Mass Filters

The major factors influencing the relative abundances of the ions produced in the ion source are electron energy, pressure, and temperature (an excessive temperature may induce pyrolysis of the sample prior to ionization). There is little mass discrimination in the magnetic sector instruments so, if these factors are controlled, spectra from different instruments are comparable.

The mass filter of a conventional quadrupole mass spectrometer, however, discriminates against ions of higher m/e value. The degree of discrimination may vary from day to day, depending on how well tuned the instrument is. It is not uncommon, under adverse conditions, for a compound to afford a molecular ion on a magnetic scanning instrument, but not on a quadrupole instrument. (Sales representatives of a company which manufactures quadrupole mass spectrometers have been known to claim that such instruments have enhanced sensitivity for ions of low m/e rather than discrimination against ions of high m/e, but I feel sure that this is said with tongue in cheek.) Thus caution should be exercised in comparing spectra from magnetic scanning instruments with those from quadrupole instruments.

3. Alternative Methods of Ionization

The electron-impact mass spectrum usually informs the analyst of the molecular weight of a compound and provides a substantial amount of structural information. However some samples may not afford a molecular ion, even at reduced electron energy. Compounds with low thermal stability may pyrolyze before ionization, and some thermally stable compounds afford unstable molecular ions, all of which fragment in the ion source. For these samples, alternative methods of ionization are available which give information concerning molecular weights, but which provide little structural information.

3.1. Chemical Ionization

A modified electron-impact ion source is used to obtain ions from a reagent gas at relatively high pressure. The resulting reagent ions react with sample molecules to yield charged products.

When methane is used as the reagent gas, and it is ionized at a pressure of around 1 Torr, the following sequence of reactions takes place:

$$CH_4 + energy \rightarrow CH_4^{+\cdot} + \epsilon^-$$
$$CH_4^{+\cdot} \rightarrow CH_3^+ + H^{\cdot}$$
$$CH_4^{+\cdot} + CH_4 \rightarrow CH_5^+ + CH_3^{\cdot}$$
$$CH_3^+ + CH_4 \rightarrow C_2H_5^+ + H_2$$

The major ionic species in the plasma are CH_5^+ and $C_2H_5^+$. Each of these species may react with a sample molecule. If the sample molecule is a good proton acceptor, an $[M+1]^+$ ion may be produced:

$$R_2C = \ddot{O} + H\text{–}\overset{+}{C}H_4^+ \rightarrow R_2C = \overset{+}{O}H^+ + CH_4$$

Hydrogen atoms may be abstracted from other sample molecules to afford $[M-1]^+$ ions:

$$R\text{–}H + H\text{–}\overset{+}{C}H_4 \rightarrow R^+ + H_2 + CH_4$$

Thus a compound may have a methane chemical ionization spectrum with an $[M+1]^+$ or $[M-1]^+$ ion instead of a molecular ion. Some compounds with both proton-accepting and proton-donating moieties may have a methane chemical ionization spectrum with both an $[M+1]^+$ and an $[M-1]^+$ ion.

Chemical ionization imparts little internal energy to the ion produced, and it undergoes little, if any, fragmentation. Chemical ionization spectra, therefore, usually contain abundant ions in the molecular ion region and a few fragment ions.

Other reagent gases, including isobutane, ammonia, and tetramethylsilane have been used for chemical ionization. If helium is employed as the reagent gas, $M^{+\cdot}$ ions are produced, and they fragment to afford a mass spectrum very similar to that obtained by electron-impact ionization. In this instance, ionization is effected by charge transfer:

$$R_2C = \ddot{O} + \overset{\cdot+}{He} \rightarrow R_2C = \overset{+\cdot}{O} + \ddot{He}$$

3.2. Field Ionization

A strong electrical field is capable of removing an electron from a molecule to afford a molecular ion:

$$M + energy \rightarrow M^{+\cdot} + \epsilon^-$$

Such a field may be established if a high potential difference (typically 10,000 V) exists between a sharp point or edge and a second electrode.

As in the case of chemical ionization, little excess internal energy is imparted during ionization, so field ionization spectra also contain few fragment ions.

3.3. Field Desorption

None of the previously considered ionization methods is of much value for compounds of low thermal stability because they must be introduced to the ion source in the vapor phase. This problem is circumvented in the field desorption ion source. The sample is coated directly onto a removable emitter and is ionized as in the field ionization source, but directly from the solid state.

This procedure has the obvious disadvantage that the ion source has to be partially dismantled between analyses, but molecular weights of unstable oligopeptides, for example, may be determined.

4. Exercises

1. Calculate the isotopic distributions of the molecular ions of:

    (a) vinyl chlorine,

    (b) iodoform, and

    (c) chlorobromomethane.

2. The major molecular ion of a compound has a precise m/e value of 59.9670. Identify the compound and calculate the relative abundance of the ion of nominal m/e 62.

3. How could one distinguish n-butyraldehyde from isobutyraldehyde by mass spectrometry?

4. A compound has a molecular weight of 115 and its mass spectrum contains an abundant molecular ion and fragment ion formed by loss of a methyl radical. If such a fragmentation takes place between the ion source and the magnet of the mass spectrometer, what is the apparent m/e value of the resulting "metastable" peak?

5. Describe how $[M+1]^+$ and $[M-1]^+$ ions may be formed from $C_2H_5^+$ ions during chemical ionization mass spectrometry.

5. Suggested Reading

H. C. HILL, *Introduction to Mass Spectrometry*, Heyden and Son, Ltd., London, 2nd ed., 1972. A good general introduction to the subject.

F. W. McLafferty, *Interpretation of Mass Spectra*, W. A. Benjamin, Inc., Reading, Massachusetts, 2nd ed., 1973. One of the best texts on interpretation of mass spectra, but suffers from an overdependence upon isotope ratio determinations.

J. Roboz, *Introduction to Mass Spectrometry: Instrumentation and Techniques*, John Wiley and Sons, Inc., New York, 1968. Provides a good coverage of instrumentation of historical interest.

H. Budzikiewics, C. Djerassi, and D. H. Williams, *Mass Spectrometry of Organic Compounds*, Holden-Day, Inc., San Francisco, 1967. Although more than 10 years old, this is still the "bible" of organic mass spectrometry, containing a comprehensive review of mass spectral fragmentation processes.

R.A.W. Johnstone, Senior Reporter, *Specialist Periodical Report: Mass Spectrometry*, The Chemical Society, London, Vol. 5, 1979. These biennial reports contain comprehensive reviews of recent research.

Journals devoted to mass spectrometry include:

Biomedical Mass Spectrometry, biomedical applications.

International Journal of Mass Spectrometry and Ion Physics, mostly devoted to instrumentation and physical chemistry.

Mass Spectrometry Bulletin, a monthly bibliography (see Chapter 8).

Mass Spectroscopy (Japan), published by the Mass Spectroscopy Society of Japan.

Organic Mass Spectrometry, mostly devoted to studies of fragmentation processes.

2

Instrumentation for Mass Spectrometry

Brian S. Middleditch

1. Introduction

It is probably safe to say that no two mass spectrometers are alike. Even those of the same model can be modified many times during a production run, each can be equipped with different accessories, and most owners modify their instruments in various ways. Three instruments are described in this chapter, which is not intended as an instruction manual but has been written to acquaint the reader with the basic components of typical instruments.

The basic features of the three instruments described in this chapter are listed in Table 1. The LKB 9000 is referred to in Chapters 3, 10, and 13. The Hewlett-Packard 5992 is similar in many respects to the H.-P. 5930 instrument described in Chapter 4. The Kratos-AEI MS50 is the mass spectrometer employed to obtain the data discussed in Chapter 12.

2. The LKB 9000 Instrument

The first commercially available combined gas chromatograph–mass spectrometer was the LKB 9000 (Figs. 1 and 2) modeled after a prototype constructed by Dr. Ragner Ryhage and his colleagues at the Karolinska Institutet in Stockholm. A modified version, the LKB 9000S, was later manufactured under licence in Japan by Shimadzu.

The gas chromatograph, equipped with a temperature programmer, is coupled to the mass spectrometer via a Ryhage-type jet separator. The

Brian S. Middleditch • Department of Biophysical Sciences, University of Houston, Houston, Texas 77004

TABLE 1. Essential Features of the Instruments Selected for Discussion in This Chapter

Feature	Instrument		
	LKB 9000	H.-P. 5992	AEI MS50
Resolution	Low	Low	Ultrahigh
Ion source	EI	EI	EI, CI, FD
Ion analyzer	Magnetic	Quadrupole	Double focusing
Inlet systems	GC, solid probe, reservoir	GC	GC, solid probe, gas probe, reservoir
GC–MS interface	Steel jet	Membrane, glass jet, direct for capillaries	Membrane, sintered glass

60°-sector, 20-cm-radius magnetic analyzer is equipped with a sweep generator for fast scanning. Spectra are recorded with a 14-stage electron multiplier, an electrometer amplifier, and a wide-band three-channel amplifier connected to a recording oscillograph. Accessories available include a mass marker, multiple ion detector, peak matcher, a heated inlet, and a direct insertion probe inlet (Fig. 3).

2.1. Electron-Impact Ion Source

The function of the electron-impact (EI) ion source is illustrated in Fig. 4. A low pressure (typically, 10^{-6} Torr) is maintained in the ion source by an oil diffusion pump backed by a mechanical pump. Passage of a

FIG. 1. The LKB 9000 mass spectrometer.

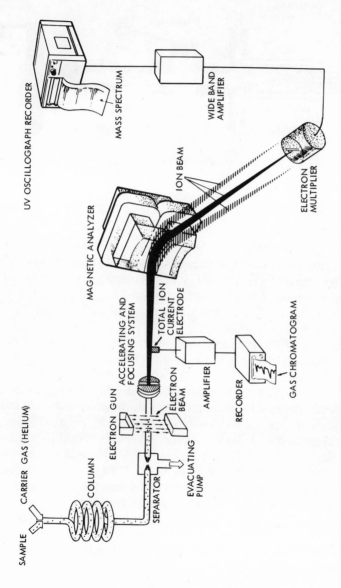

SAMPLE

CARRIER GAS (HELIUM)

COLUMN

ELECTRON GUN

SEPARATOR

EVACUATING PUMP

ELECTRON BEAM

AMPLIFIER

RECORDER

GAS CHROMATOGRAM

ACCELERATING AND FOCUSING SYSTEM

TOTAL ION CURRENT ELECTRODE

MAGNETIC ANALYZER

ION BEAM

ELECTRON MULTIPLIER

WIDE BAND AMPLIFIER

MASS SPECTRUM

UV OSCILLOGRAPH RECORDER

FIG. 2. The LKB 9000 mass spectrometer (schematic).

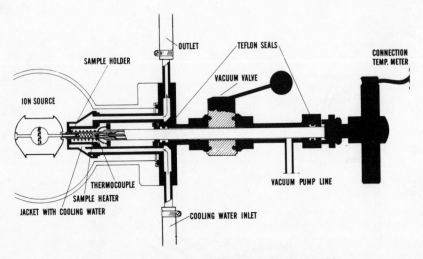

FIG. 3. Direct-insertion-probe inlet of the LKB 9000.

current of about 4 A through the rhenium filament elevates it to red heat and promotes the emission of the electrons. The shield focuses the electron beam onto the trap. The mean energy of the electrons is determined by the potential difference between the filament and trap. To obtain a mean electron energy of 70 eV, for example, a potential difference of 70 V is selected.

The presence of the permanent magnet in the ion source causes the electrons to take a spiralling path to the trap. This increase in path length results in an increase in ionization efficiency over a direct path from the filament to the trap. Sample molecules drift into the electron beam and become ionized. Many of the molecular ions decompose in the ion source to afford fragment ions, as previously described. The positively charged ions are accelerated out of the ion source and into the flight tube through a series of slits normally held at a potential of -3500 V with respect to the ion source. In some instruments (but not the LKB 9000), there is a repeller electrode in the ion source, at a higher potential than the other electrodes, which assists in the expulsion of ions into the flight tube.

The electron energy can be adjusted between 5–100 eV with a stability greater than 1%. As discussed in Chapter 1, it is conventional to use an electron energy of 70 eV for most applications, although 20 or 22.5 eV are now frequently employed.

Heat from the filament usually affords a temperature of about 200°C, but there is an auxiliary heater under thermostat control which is capable of raising the ion source temperature to 310°C. A temperature of 250°C is usually employed.

2.2. Flight Tube

After leaving the ion source, the ion beam passes through a variable entrance slit. The voltages on the individual extraction, focusing, and deflection electrodes may be varied to focus the ion beam onto a second variable slit, the exit slit, in front of the detector.

The entrance and exit slits have maximum widths of 0.4 and 0.8 mm, respectively, and are adjusted by micrometer controls from outside the flight tube. Wide slits allow many ions to reach the detector and, therefore, afford high sensitivity. If the slits are closed down, greater resolution will be obtained at the expense of sensitivity. The actual slit widths to be used for any particular application reflect a compromise between the resolution and sensitivity required or attainable. There is some dispersion of the ion beam as it travels along the flight tube, due mainly to repulsion between the ions, so the exit slit is usually set wider than the entrance slit. Typical slit widths are 0.1 mm (entrance) and 0.3 mm (exit). An additional micrometer control outside the flight tube allows rotation of the exit slit ±5° to ensure that both slits are in the same plane.

2.3. Total Ion Current Detector

An additional electrode is positioned between the ion source and the magnet. Approximately 90% of the ion beam passes through a hole in this

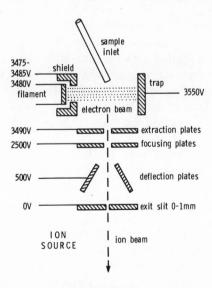

FIG. 4. Electron-impact ion source (schematic).

electrode and continues along the flight tube, but the remainder is collected
and amplified. The resulting signal is recorded on a potentiometric recorder. When the gas chromatographic inlet to the instrument is employed,
this recorder provides chromatograms, so no auxiliary detector is required.
This total ion current (TIC) detector also provides an indication of the
amount of sample entering the ion source via other inlet systems.

2.4. Electromagnet

A portion of the flight tube, at the bend, is flattened and placed
between the poles of an electromagnet. This magnet deflects ions in the
ion beam according to their m/e values.

The relationship between m/e value of an ion being focused on the
detector, the accelerating voltage V, the magnetic field strength B, and the
radius R of the ion beam as it passes through the bend in the flight tube
is

$$m/e = R^2B^2/2V \tag{1}$$

R cannot be varied, and V is usually held constant, so the m/e value of
ions focused onto the detector is proportional to B^2. For singly charged
ions, those of lower mass are deflected more than those of higher mass.
By increasing the current through the magnet coils (which increases B)
ions of increasingly high m/e value are swept past the detector. The
magnet current can be increased at a rate sufficient so that the entire mass
range of the instrument is covered in as little as 1 sec (or as much as 4
min). With the usual accelerating voltage of 3.5 kV, the mass range is
m/e 2–200. Equation (1) indicates that the mass range can be extended by
decreasing the accelerating voltage. On the LKB 9000 two alternative
values of the accelerating voltage may be selected, extending the mass
range to an upper limit of m/e 2400 (Table 2). One drawback of reducing
accelerating voltage to extend the mass range is that the sensitivity is also
decreased.

One does not normally scan the full mass range of the instrument
when recording a spectrum: suitable upper and lower limits of m/e are
selected. An m/e range of 10–500 would be adequate for the analysis of
sterols, for example (cf. Fig. 14, Chapter 1). Spectra are scanned from low
to high mass with this instrument and, upon completion of each mass
spectral scan, the magnet current drops to its lower preset limit very
rapidly so that a further spectrum can be recorded almost immediately.

An approximate indication of the m/e value of an ion focused on the
detector at any time is provided by a mass meter. The operation of this
device depends upon the Hall effect; if a constant current is maintained
through a cube of germanium, a varying magnetic field at right angles to

TABLE 2.   Effect of Accelerating Voltage V on
Mass Range for the LKB 9000

V	Maximum m/e
3500	1200
2330	1800
1750	2400

the direction of the current will induce a varying potential difference across the other two faces of the cube. This varying potential difference is displayed on a multirange meter calibrated with an m/e scale.

A more sophisticated mass marker accessory, also based upon a Hall probe, provides an m/e calibration directly on the mass spectrum and on a visual display unit (Fig. 5). This accessory can be calibrated to an accuracy of about 0.3 amu throughout the m/e range of the instrument (with an accelerating voltage of 3.5 kV).

2.5.   Spectrum Detection and Recording

During a mass spectral scan, ion beams for ions of individual m/e values arrive sequentially at the detector. The first dynode of the electron multiplier produces electrons when ions strike it. These electrons are attracted toward the second dynode. When they impinge upon it, they produce a larger number of electrons which, in turn, are attracted toward the third dynode. The number of electrons produced at the second dynode is dependent upon the potential difference between the first two dynodes; a higher gain is obtained by using a higher voltage difference. This amplification is repeated at each of the 14 dynodes so that a weak ion beam can be converted to a measurable electric current.

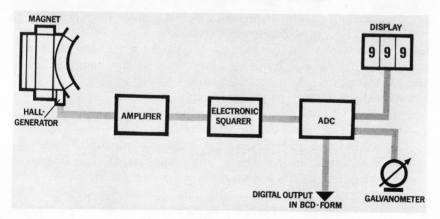

FIG. 5. Mass marker of the LKB 9000 (schematic).

An electrometer preamplifier connected to the electron multiplier provides additional magnification of the signal before it is carried to the galvanometer amplifier in the control unit of the instrument.

The galvanometer amplifier has three channels with gains set at X1, X10, and X100. Each of these channels is recorded simultaneously so that a wide dynamic range is obtained. This feature is useful for two reasons:

(i) If the molecular ion (or other important ion) is much less abundant than the base peak and cannot be seen on the X1 channel, it is often visible on the X10 or X100 channel. It is therefore possible to measure ions with relative abundance of less than 0.01%.

(ii) If a mixture which is being analyzed contains components with a wide range of concentrations, the spectrum of each component can be recorded without changing the electron multiplier gain. Thus 1 μg of one component would give a spectrum on the X1 trace with an intensity comparable to that of the spectrum of 10 ng of a second component recorded on the X100 trace.

The three signals are recorded using an oscillographic recorder. If the instrument is equipped with a mass marker, the mass scale appears on a fourth channel. This high-speed recorder produces spectra on light-sensitive paper. The paper must be exposed to light to develop the image, but overexposure will cause the image to fade. It is possible to use a chemical fixative to preserve the image, but this is rarely done. The recordings can be stored in the dark for many years without significant deterioration. They can be rolled up and stored in the boxes in which the paper was supplied. It is easier to read the spectra later if they are stored flat, so some laboratories keep them face-down in drawers designed for holding blueprints.

2.6. Gas Chromatographic Inlet

The standard gas chromatograph supplied with the instrument can be used at temperatures ranging from 25°C above ambient to 400°C, although temperatures above 300°C are rarely employed for GC–MS. It can be operated isothermally or by linear temperature programming at rates as high as 20°C min^{-1}.

As with most other gas chromatographs, glass or metal columns with external diameters of 0.25 or 0.125 in. can be used. The flow rate of carrier gas through the column is usually 30 ml min^{-1}.

The inlet system incorporates a two-stage stainless steel Ryhage-type molecular separator. In this device, more fully described in Chapter 3, approximately 99% of the carrier gas is removed from the effluent from the gas chromatograph column, while only 25% of the sample is lost. There is a cutoff valve between the separator and the ion source (Fig. 6).

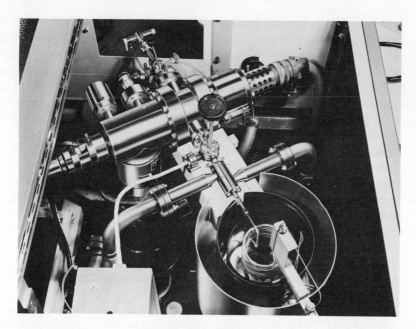

FIG. 6. GC–MS interface of the LKB 9000.

This valve is kept closed while the solvent is eluting from the column (so that the vacuum can be maintained in the ion source) and is then opened to permit the compounds of interest to enter the ion source.

The optimum flow rate of gas entering the separator is about 30 ml min⁻¹, so it cannot be used directly with capillary columns, which require flow rates lower than 5 ml min⁻¹. One method of coupling a capillary column to a Ryhage-type separator is to add "make-up" gas through a tee-piece (fitted between the column and separator) to bring the total flow rate up to 30 ml min⁻¹. An alternative is to remove the first stage of the separator, since the second stage alone can handle a flow rate of 5 ml min⁻¹.

Different operating procedures are employed depending on whether the spectral scans are initiated manually or under the control of an automated data acquisition system.

2.6.1. Manual Operation

The description which follows will outline the steps involved in obtaining a mass spectrum of a pure compound via the gas chromatographic inlet.

A solution of the sample with a concentration of about 1 mg ml^{-1} is made up. The gas chromatograph oven is set at a temperature which will afford a retention time of about 5 min, and the injection port temperature is set about 20°C higher. An aliquot of the sample (1–5 μl) is injected into the gas chromatograph. In a little under a minute (for a 2 m × 2-mm-i.d. column) the solvent will elute from the column. This event can be observed as there is a temporary increase in the pressure of the manifold connected to the first stage of the molecular separator. After the solvent has been pumped away, the isolation valve can be opened to allow the gas chromatograph effluent to enter the ion source. One then watches the TIC recorder as the chromatogram is produced. When the peak of interest appears, the operator waits until the apex of the peak is reached before pressing the "scan" button to obtain a spectrum. A second mass spectrum is recorded after the TIC signal returns to baseline so that the contribution of column bleed and instrument background to the sample spectrum can be assessed. If 1–5 μg of sample is injected into a column with a low loading of a modern silicone stationary phase, maintained at a temperature of less than 250°C, there is usually little contribution from column bleed.

The mass spectrum should be recorded at the apex of the gas chromatographic peak, and the scan time should be short in relation to the width of the peak. If these precautions are not observed, the concentration of sample in the ion source may not be constant throughout the scan and a "biased" spectrum could be produced. With a little practice, it is relatively easy to record a spectrum at the apex of every peak in a chromatogram and even at partially resolved shoulders on those peaks.

In Section 2.3 we saw that the total ion current chromatogram was produced by sampling about 10% of the ion beam leaving the ion source. Even though only about 1% of the carrier gas enters the ion source, it can contribute significantly to the total ion current. This may lead to excessive noise and drift of the signal. These problems are circumvented by setting the electron energy of the ion source below the ionization potential of helium (24.5 eV) while the chromatogram is monitored. Values frequently employed are 20 or 22.5 eV. During actual mass spectral scans, the electron energy can be temporarily increased to 70 eV (or any other value). To protect the total ion current recording from distortion during this time, the recorder pen is frozen at the same level for the duration of the scan.

This process produces many yards of recordings on light-sensitive paper, and these recordings have to be measured manually. This task can be exceedingly tedious. Although not as cumbersome as it might appear it is nevertheless more efficient to automate at least a portion of this procedure.

2.6.2. Automated Operation

One could program a computer to mimic the steps involved in manual operation so that a mass spectrum would be scanned at the apex of each chromatographic peak. The LKB 9070 data acquisition system, however, exploits the full capacity of the PDP-8/M computer around which it is built in extending the range of tasks to be performed. The total ion current signal is not used by the data acquisition system. Instead, spectra are continually scanned throughout the analysis and stored. A reconstructed total ion chromatogram can be obtained from these data as the analysis proceeds or at a later time. This chromatogram differs from the conventional total ion current recording in one important respect: it comprises only the signals produced by ions within the *m/e* range scanned. Thus if an *m/e* range is chosen so that it excludes that of helium (*m/e* 4), the carrier gas does not contribute to the reconstructed total ion chromatogram and any electron energy can be used throughout the analysis.

Any of the acquired spectra can be recalled at will on a display screen, and can be output as a line diagram or in numerical form. Background spectra may be subtracted from sample spectra, compounds can be identified by searching libraries of standard spectra, and single ion chromatograms can be reconstructed. The range of tasks which can be performed using this system is continually being extended, and many users have developed additional capabilities. The ingenuity of these routines is limited only by the imagination of the programmer.

2.7. Other Inlet Systems

Two useful alternatives to the gas chromatographic inlet are available as optional accessories.

2.7.1. Direct Inlet

The gas chromatograph, of course, is used as an inlet system if spectra of individual components of mixtures are to be recorded. There are also advantages to using this inlet for single substances: The solvent or reagent (if a derivative has been made) is efficiently removed, minor impurities may be separated, and dust is excluded from the ion source. For samples which are not sufficiently volatile or stable for gas chromatography, the direct inlet (Fig. 3) can be used. They are introduced to the ion source through a water-cooled vacuum lock. A small amount of the sample, in the solid state or in solution, is placed in a quartz vial at the tip of the direct-insertion probe. Air and solvent vapor (if applicable) are removed by evacuating the air lock, and the tip of the probe is then inserted into the ion source. The sample may be vaporized by heating to temperatures as high as 300°C.

2.7.2. Heated Inlet

This is basically a small reservoir from which a sample in the vapor state can be bled continuously into the ion source through a fine leak. It is commonly used for introducing perfluorokerosene or other reference compounds to the ion source to facilitate calibration of the instrument. The reservoir itself is equipped with two inlet systems. The most frequently used inlet is through a pair of silicone septa. The space between the septa is evacuated to minimize the influx of air to the reservoir during sample introduction. Solid samples can be lowered through a pool of gallium on a sintered glass thimble at the top of the reservoir. The sample evaporates and passes through the frit, while air is excluded by the molten gallium. (The melting point of gallium is 29.8°C.) The membrane inlet can be used at temperatures up to 220°C, and the gallium inlet up to 350°C. The evaporation chamber has a volume of 160 ml. A gold capillary connects this reservoir to the ion source.

2.8. Multiple Ion Detector

As will be seen in Chapter 4, focusing ions of a particular *m/e* value on the detector throughout an analysis produces single ion chromatograms. The multiple ion detector is a device which enables one to continually monitor up to four *m/e* values (or eight with an additional module) throughout a chromatogram.

Ions are normally accelerated through a potential of 3.5 kV as they leave the ion source. The magnet current can be adjusted so that ions of a single *m/e* value are focused on the detector. If the magnet current is held constant and the accelerating voltage is reduced, ions of a higher *m/e* value are now focused on the detector [cf. equation (1)]. Rapid alternation of the accelerating voltage will cause ions of both *m/e* values to be continually focused onto the detector so that single ion chromatograms can be produced for each *m/e* value.

This technique is used for selective detection of components of mixtures which are unresolved by gas chromatography (where each affords a spectrum containing an ion with a characteristic *m/e* value absent from the spectrum of the other), or for quantitative analysis using an internal standard which may be monitored at an *m/e* value different from that of the sample.

The *m/e* range of the latest version of the multiple ion detector (Fig. 7) is 30–100% of the highest *m/e* value selected. Measurement times of 6, 20, 60, or 200 msec can be selected individually for each channel. Amplification factors of 1–500 and background subtraction can also be applied to each channel. A range of filters with time constants of 0.5–64 sec is available; the value used should be sufficient to remove noise without "clipping" the signal.

2.9. Similarity to Other Instruments

The LKB 9000 is an example of a magnetic-deflection single-focusing mass spectrometer—a type often regarded as a "conventional" mass spectrometer. Such instruments are manufactured by many other companies, including duPont, JEOL, Kratos-AEI, Varian, and VG-Micromass. Comparable performance and a similar range of accessories are available on many of the other instruments, although each (including the LKB 9000) has its complement of unique features.

It is not easy to compare the performance of the different instruments because different definitions of resolving power, sensitivity, and stability are employed.

FIG. 7. Multiple ion detector of the LKB 9000.

3. The Hewlett-Packard 5992 Instrument

The Hewlett-Packard Company manufactures a range of mass spectrometers (see p. 47), all with quadrupole mass filters. The 5992 was chosen for illustration because of its revolutionary design. This instrument has essentially the same capabilities as many other quadrupole mass spectrometers, yet it is compact enough to sit on a laboratory bench (Fig. 8). Moreover, the data system, based on a programmable calculator no larger than a typewriter, is almost as powerful as those built around minicomputers.

3.1. General Description

The most notable external feature of the instrument (apart from its size) is the lack of dials, knobs, and switches. There is only an on–off switch and a knob for the isolation valve between the gas chromatograph and mass spectrometer. All of the operating commands are entered by using the calculator keyboard, and the calculator is also used to report on the status of the various components of the instrument. This lack of controls can be disconcerting to experienced mass spectroscopists since they may feel a lack of control over the instrument, but this fear soon dissipates as they learn to work with the calculator to become truly in control.

The compactness of the instrument is due mainly to the fact that the entire mass spectrometer is located inside the diffusion pump (Fig. 9). A welcome design feature of the vacuum system is that the top sealing flange is held in place by atmospheric pressure rather than the usual set of bolts.

The gas chromatograph is rather conventional. The standard version is designed for on-column injection through a silicone rubber septum into a 0.25-in.-o.d. column. A purge-and-trap device (based on the principle

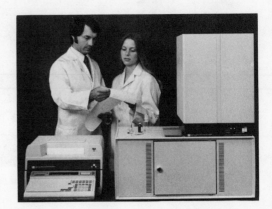

FIG. 8. The Hewlett-Packard 5992 mass spectrometer.

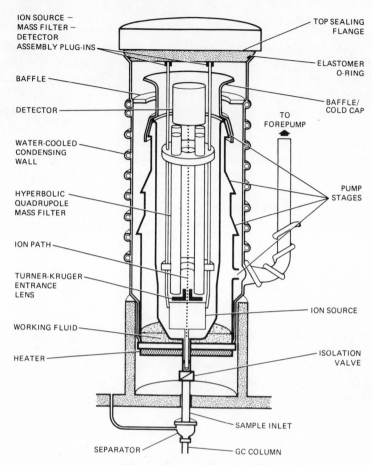

FIG. 9. Vacuum and analyzer assemblies of the Hewlett-Packard 5992.

described in Chapter 5) is available as an alternative inlet system. An additional accessory allows capillary columns to be employed. Two molecular separators are available: a silicone membrane separator and single-stage glass-lined jet separator (see pp. 78–80). There is a minimum of dead volume between the GC column and the isolation valve and between this valve and the ion source.

The GC oven can be used at temperatures of 90–280°C with the membrane separator and 90–350°C with the jet separator. The lower limit can be extended to −50°C with a subambient option by using liquid nitrogen or carbon dioxide. For temperature-programmed operation, linear program rates of 0.128–16°C min^{-1} can be employed. The most recent version of this instrument (5992B) automatically opens the isolation valve at the appropriate time.

3.2. The Mass Analyzer

The quadrupole mass analyzer is very different from the magnetic deflection analyzer. The quadrupole unit, as its name suggests, consists of four parallel rods (Fig. 9). The rods opposite one another are connected electrically. Between the two pairs of rods, there is a constant voltage upon which is superimposed a varying (radiofrequency) voltage. These parameters can be varied to permit an ion of any selected m/e value to travel from the ion source to the detector. The other ions are deflected from this path and are not detected. The quadrupole unit can be used to constantly monitor ions of a single m/e value or to scan an entire spectrum. When scanning complete spectra, the mass filter is stepped in increments of 0.1 amu, and 1–8 measurements can be made of ion beam intensity before stepping to the next value. The relationship between samples per 0.1 amu and scan rate is shown in Table 3. Greater precision is obtained by using slower scan rates, but higher scan rates may be required to obtain unbiased spectra, particularly if capillary columns are employed for GC.

The internal profile of the quadrupole rods is hyperbolic, so that the problems with high mass sensitivity (p. 21) encountered in instruments with round rods are minimized. These problems are further reduced by the use of a Turner–Kruger entrance lens (Fig. 9). Without this device, some ions would be deflected in the fringe fields between the ion source and mass filter, this loss being greater for ions of higher m/e value. In the Turner–Kruger lens, an additional electrode is raised to an increasingly high voltage throughout a mass spectral scan to minimize this loss.

3.3. Use of the Instrument

The capabilities of the Hewlett-Packard 5992 are best illustrated by describing analyses which may be performed using the instrument.

TABLE 3. Relationship between Samples per 0.1 amu and Scan Rate (amu per sec) for the Hewlett–Packard 5992 Instrument

Samples	Scan rate
1	620
2	330
4	190
8	100

3.3.1. AUTOTUNE

One of the features of this instrument which makes it so easy to use is the AUTOTUNE program, which automatically adjusts the ion source and analyzer parameters for optimum performance. This is done using perfluorotri-*n*-butylamine (PFTBA) for calibration. A typical AUTOTUNE report is shown in Fig. 10. Instrument parameters before and after tuning are listed,

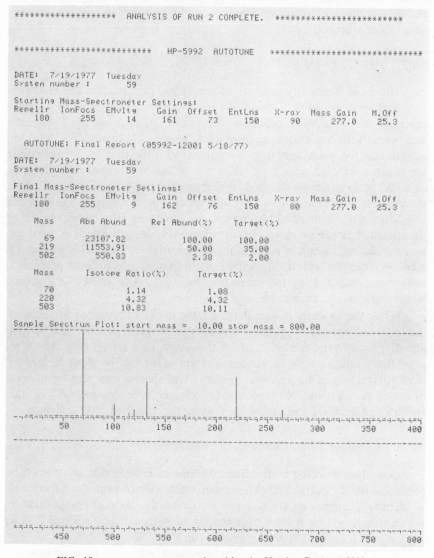

FIG. 10. AUTOTUNE report produced by the Hewlett-Packard 5992.

and the relative abundances of the ions of *m/e* 69, 219, and 502 are reported. These should be greater than the "target" values of 100, 35, and 2, respectively. The heights of the first isotope peaks (*m/e* 70, 220, and 503) should be close to the target values. A line diagram of the spectrum is also given.

If any of the AUTOTUNE criteria are not met, this is indicated on the report. The DIAGNOSTICS software can then be used to determine the reason for the poor performance. If the problem is merely one of the burned-out filament, a second filament can be switched in without dismantling the instrument.

3.3.2. PEAKFINDER

The most frequently used set of programs is included in the PEAKFINDER software. All of the conditions for a conventional GC–MS analysis are entered through the calculator keyboard. After a sample is injected and the solvent has eluted, spectra are scanned continually. The total ion abundance for each spectrum is compared with that of the previous spectrum after each scan and, when the instrument senses that the apex of a GC peak has been reached, the appropriate spectrum is recorded on magnetic tape. There is room for 100 spectra on the tape if 330 ions per spectrum are recorded. The capacity of the tape can be increased or decreased if fewer or more ions, respectively, are recorded. A total ion current chromatogram is printed at two sensitivities (X1, X10) during the analysis, together with a similar pair of chromatograms for ions of a single *m/e* value selected by the operator (Fig. 11). Any of the instrument parameters can be changed during an analysis.

When the analysis is complete, a table of retention times for each of the recorded spectra is printed. A PLOT/TAB program can be used to plot line diagrams in the format used for the AUTOTUNE report (Fig. 10) or to print the tabulated data in numerical form. This can be done before or after subtraction of background spectra. The spectra can also be plotted by using an optional X–Y plotter for clearer reproduction. All of the spectra in Chapter 1 were plotted by using the standard 5992 software.

3.3.3. Compound Identification

Operators working with a limited range of compounds soon learn to recognize their spectra. The experienced mass spectroscopist can frequently identify a compound (or at least determine the class of compounds to which it belongs) by a brief examination of its spectrum. Several aids to compound identification are available with the 5992.

3.3.3.1. Library Search. The standard software tape contains a

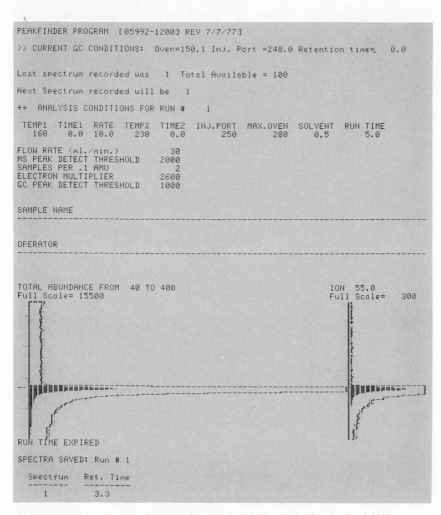

FIG. 11. PEAKFINDER recording produced by the Hewlett-Packard 5992.

library of 375 spectra of drugs (see p. 184), a library of 517 pollutants, and space for two user-generated libraries. Each entry in this library consists of only the ten most significant ions in each spectrum. When interpreting spectra, we give most attention to the more abundant ions and those of greatest *m/e* value, so the factor used in ranking significance of the ions in a spectrum is the product of *m/e* value and relative abundance. An example is shown in Fig. 12 for the spectrum recorded in Fig. 11. The index of significance has been normalized so that the most significant ion (*m/e* 194) has a value of 100. This abbreviated spectrum is then compared with those in any one of the four libraries, and the ten most similar spectra are

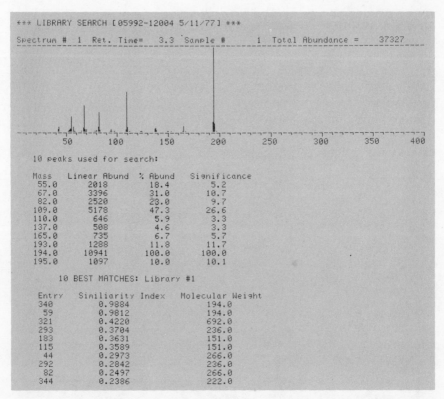

FIG. 12. LIBRARY SEARCH report for the spectrum recorded in Fig. 11, produced by the Hewlett-Packard 5992.

located. A similarity index (S. I.) is then calculated for each of the selected library spectra by using the following formula:

$$\text{S.I.} = \sum_{m=1}^{1000} A_m\, a_m \Big/ (\sum_{m=1}^{1000} A_m^2 \cdot \sum_{m=1}^{1000} a_m^2)^{1/2}$$

where A_m is the abundance of the ion at the mass m in the "unknown" spectrum and a_m is the abundance of the ion at mass m in the library spectrum. The similarity index can range from 0 to 1, with a high number indicating close similarity. In the example given in Fig. 12, library spectra 340 and 59 are both very similar to the "unknown" spectra. These library spectra are both duplicate entries for caffeine, and the identification was made correctly.

   3.3.3.2. Communications Interface.   An optional device allows the operator to use the data system of the 5992 as a terminal for communication with remote time-sharing computers by telephone connection. Thus one can make use of the Mass Spectral Search System (Chapter 9), the Cornell University facility (p. 182), or the NIH service (p. 181). Once a compound has been identified, the terminal can also be used for bibliographic searches. *The Mass Spectrometry Bulletin* (p. 191) can be searched by using the Mass Spectral Search System, and many other data bases are now accessible.

3.3.3.3. Other Aids. If the spectrum of the compound of interest has not been recorded in any of these libraries, the 5992 can provide some additional assistance in interpreting the data. An empirical formula calculator will list possible elemental compositions of any ion entered through the keyboard, and a molecular weight calculator will determine the isotope abundances of any combination of elements selected. If these features are used in conjunction with compilations of mass spectral data (Chapter 7), the "Molecular Weight Index to the Merck Index," or even the molecular formula index of a catalog from a supplier of organic chemicals, there is a reasonable chance that the compound can be identified.

3.3.4. Selective Ion Monitoring

This topic is covered more fully in Chapter 4. The SIM software for the 5992 can be used to monitor up to six ions per analysis, each with different sampling times. Unlike the magnetic deflection instruments, there is no limit on the mass range which can be covered during a single analysis. After an analysis has been performed, the GC peak areas are calculated and can be normalized to facilitate the interpretation of data.

3.3.5. On-Line Library Search

This is the ultimate in automation of GC–MS. For each mass spectral scan, a small library of abbreviated spectra is searched and the identity of the best match is printed directly on the chromatogram. If this procedure is used for determining volatiles from the EPA priority pollutant list in drinking water, for example, the appropriate library would be loaded into the data system. A tube containing the water sample would be attached to the purge-and-trap device (p. 39), and the whole analysis would be performed without operator intervention. Helium would be bubbled through the sample and displaced volatiles would be trapped in a tube containing Tenax. When purging is complete, the trap would be heated to desorb the volatiles, which would be flushed onto the GC column. The GC–MS analysis would then be initiated. Each time a compound is eluted from the column, its spectrum would be recorded, background subtracted, the library searched, and the identity and relative concentration printed on the chromatogram. An example of an on-line search is shown in Fig. 13.

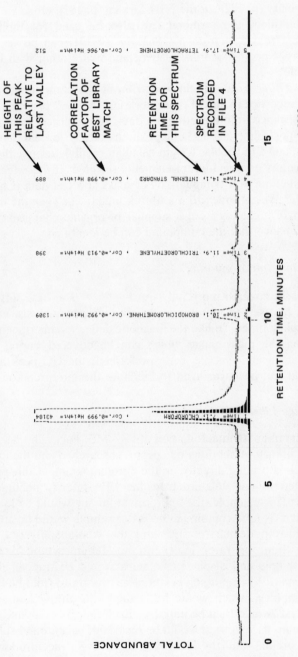

HEIGHT OF
THIS PEAK
RELATIVE TO
LAST VALLEY

CORRELATION
FACTOR OF
BEST LIBRARY
MATCH

RETENTION
TIME FOR
THIS SPECTRUM

SPECTRUM
RECORDED
IN FILE 4

5 Time= 17.9, TETRACHLOROETHENE   , Cor.=0.966 Height= 512

4 Time= 14.1, INTERNAL STANDARD   , Cor.=0.998 Height= 889

3 Time= 11.9, TRICHLOROETHYLENE   , Cor.=0.913 Height= 398

2 Time= 10.1, BROMODICHLOROMETHANE, Cor.=0.992 Height= 1309

1 Time= 7.1, CHLOROFORM   , Cor.=0.999 Height= 43184

TOTAL ABUNDANCE

RETENTION TIME, MINUTES

0   5   10   15

FIG. 13. Drinking water analysis using purge-and-trap device with Hewlett-Packard 5992.

3.4. Other Accessories

The capabilities of the data system can be extended by installing a flexible disk drive. There is more storage capacity than on the standard magnetic tapes, and data can be accessed more rapidly. This added capacity permits the storage of background spectra from the valleys between GC peaks. This is particularly useful for temperature-programmed GC where the amount of column bleed changes during the analysis.

In the Hewlett-Packard 5993 instrument, the programmable calculator is replaced by a full-sized computer, so that an even greater variety of data manipulations can be performed.

3.5. Similarity to Other Instruments

There is no other bench-top automated mass spectrometer on the market. Hewlett-Packard manufactures other quadrupole mass spectrometers with a broader range of inlet systems, ion sources, and data systems. Other quadrupole mass spectrometers are available from Extranuclear, Finnigan, Nuclide, and Varian.

4. The Kratos-AEI MS50 Instrument

The MS50 (Fig. 14) is the most powerful of a long line of mass spectrometers manufactured in Manchester, England during the past

FIG. 14. The Kratos-AEI MS50 instrument.

quarter century. The company originally known as Metropolitan-Vickers merged with Associated Electrical Industries (AEI), which, as a result of successive corporate mergers then came under the control of Marconi and GEC. Recently, the AEI Scientific Apparatus group was spun off from this corporate giant, and is now owned by Kratos of San Diego, California. The research and manufacturing facilities remain in England.

The instruments discussed so far are examples of low-resolution mass spectrometers, capable of measuring m/e values only to the nearest integer. The advantages of more precise mass measurement were referred to in Chapter 1, and many examples of the application of this technique will be presented in Chapter 12. AEI was the first company to manufacture a high-resolution mass spectrometer (the MS9), and its MS50 is the only ultrahigh-resolution instrument available.

4.1. Resolution

There are several definitions of resolution employed by mass spectroscopists. The most widely used is the "10% valley" definition. According to this definition, the resolution of an instrument is equal to the highest m/e value for which an ion can be separated from another ion of adjacent m/e value and identical abundance with a valley (between the ions) no higher than 10% of the height of the ions on the recorded spectrum. In our example, the resolution is 500. The typical high-resolution instrument can be tuned to give a resolution of about 30,000. This does not mean that an ion of m/e 30,000 can be distinguished from an ion of m/e 30,001 since the mass range of the instrument would not extend this far and a compound with this molecular weight would be insufficiently volatile to enter the ion source. Rather, this is an indication of the equivalent resolution in the usable mass range. At such a resolution, for example, ions $C_{22}H_{36}{}^{+\cdot}$ (m/e 300.2817) and $C_{21}H_{32}O^{+\cdot}$ (m/e 300.2463) would be easily resolved. The conventional high-resolution mass spectrometer is incapable of resolving some important doublets (Table 4), but this can be achieved by the MS50 when tuned to its maximum resolution of 150,000. A very good example of the performance of this instrument is given in Fig. 15, which is a portion of the spectrum of a petroleum fraction. Note that all of these ions have a nominal mass of 232.

As with all mass spectrometers, resolution is decreased as scan rate is increased. This relationship is shown for the five scan rates available on the MS50 (Table 5). Spectra are usually scanned exponentially, so that the range m/e 800–80 would be scanned in the same time as the range m/e 80–8, for example. Linear scans can be performed over narrow mass ranges, as in Fig. 15.

TABLE 4. Resolution Required to Resolve
Doublets with Similar m/e Values (at m/e 200)

Doublet	m/e Difference	Resolution
$^{13}C-^{12}CH$	0.004471	45,000
C_3-SH_4	0.003273	70,000
H_2-D	0.001548	130,000

4.2. Inlet Systems

The MS50 has a full range of inlet systems for the introduction of solid, liquid, and gaseous samples, and for GC–MS.

An all-glass heated inlet system contains two reservoirs, with capacities of 1 liter and 20 ml. They are situated in an oven which may be heated at 350°C. Valves between the reservoirs, pumps, and the ion source are magnetically operated. The glass line to the ion source is in a separately heated oven.

A direct-insertion lock can be used with a heated probe (50–350°C) for solids or an alternative probe for the introduction of gases and volatile liquids. Two of these vacuum locks can be installed on the same instrument so that, for example, a reference compound and a solid sample can be introduced to the ion source simultaneously.

Two GC–MS interfaces can be used. One uses a silicone membrane, and the other a sintered glass frit.

4.3. Ion Sources

Three ion sources are available: EI, combined EI/CI, and combined EI/FD.

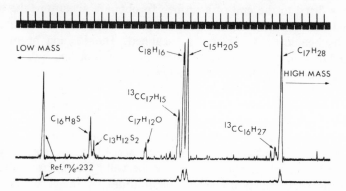

FIG. 15. Partial spectrum of a petroleum fraction, recorded using a Kratos-AEI MS50.

TABLE 5. Relationship between Scan Rate (Seconds per Decade) and
Resolution (10% Valley) for the Kratos-AEI MS50

Scan rate	Resolution
2	1,000
3	3,000
10	10,000
30	20,000
100	40,000
Static	150,000

4.3.1. Electron-Impact Ion Source

The standard ion source (Fig. 16) is relatively conventional. It can be operated at electron energies of 50–100 eV and with electron beam currents of 10–500 μA. It can be heated to 300°C or water cooled. The four inlet ports allow versatility of inlet systems.

4.3.2. Combined EI/CI Ion Sources

A 5-in. diffusion pump with a 600-liter sec^{-1} capacity is used in place of the 3-in. (150-liter sec^{-1}) pump normally installed, so that an ion source pressure of 1 Torr can be maintained. Electron energies of 5–400 eV can be used.

4.3.3. Combined EI/FD Source

The emitter for field desorption is a fine wire treated so that it is covered by "whiskers." This produces a large surface area and many sharp points and edges for the production of strong electrical fields. The emitter is fitted to the end of a probe. To obtain a spectrum, the sample is coated onto the emitter, the probe is inserted into the ion source, the emitter is heated, and spectra are scanned. The FD ion beam passes through holes in the EI source block and ion repeller, before passing through the entrance slit into the analyzer tube.

4.4. Ion Analyzer

The low-resolution single-focusing magnet deflection instrument is only partially successful in separating ions according to their m/e values. Not all of the ions with the same m/e value will have precisely the same velocity as they travel down the flight tube, so there will be some dispersion of the deflected ion beam, which contributes to the overlap of adjacent ion beams.

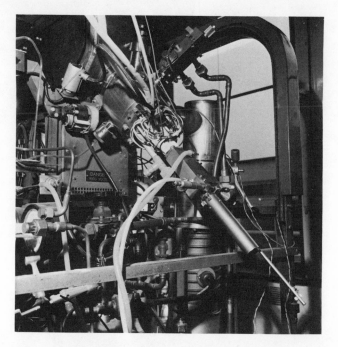

FIG. 16. Ion source region of the Kratos-AEI MS50.

A solution to this problem is to use the electrostatic analyzer for velocity focusing in addition to the magnet analyzer. In the electrostatic analyzer, the centifugal force (mv^2/r) is counterbalanced by the centripetal force (Ee) on an ion, where v is its velocity, r is the radius of the electrostatic sector, and E is the field strength, so that

$$mv^2/r = Ee$$

Thus for ions of a single m/e value, only those of a single velocity will traverse the electrostatic sector if r and E are constant.

In the MS50 (Nier–Johnson geometry), the electrostatic sector precedes the magnetic sector (Fig. 17), ions are focused on a point, and spectra can be scanned in the normal way. [In some instruments (Mattauch–Herzog geometry) these sectors are reversed, ions are focused in a plane, and photoplate detection can be employed.]

The electrostatic analyzer has a radius of 15 in. and a sector angle of 90°. The magnetic analyzer has a similar sector angle, but a radius of 12 in.

One further impediment to obtaining good separation of ions is the presence of distorting fringe magnetic fields at the entrance to and the exit from the magnetic sector. The ion beam, in passing through the entrance

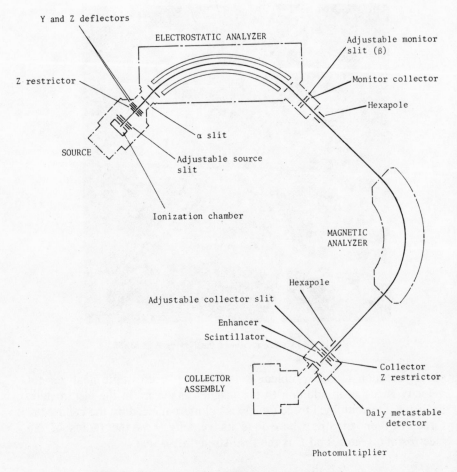

FIG. 17. The Kratos-AEI MS50 instrument (schematic).

slit into the flight tube, acquires a rectangular cross section. These fringe fields impart curvature to the cross section so that it is crescent-shaped by the time that it reaches the detector. Overlap of adjacent ions can be reduced to some extent by reducing the length of the slits as well as their width, but this reduces sensitivity. In the MS50, this problem is overcome by placing hexapole units before and after the magnetic sector to restore the rectangular cross section of the ion beam. These hexapole units can also be used to rotate the ion beam so that rotation of the collector slit is unneccessary.

A further refinement in the MS50 is the presence of an X–Y lens which electronically adjusts the focal length of the ion path and effectively simulates horizontal movement of the magnet.

Thus the number of mechanical adjustments required during tuning is kept at a minimum, and this contributes greatly to stability. Further stability is provided by using a frame for the instrument constructed from 12 × 8-in. box-section girders, and by water cooling the magnet power unit.

4.5. Detector

A Daly fluorescence detector is employed in the MS50 (Fig. 18). The ion beam passes through the collector slit and it is further accelerated as it passes through a slit in the enhancer. The positively charged scintillator decelerates the ion beam and it is drawn back toward the enhancer. Secondary electrons are produced when the ions strike the enhancer. These electrons are attracted toward the scintillator, in which they produce photons. These, in turn, are detected by a photomultiplier tube outside the vacuum chamber (the glass plate forms part of the vacuum wall).

With the photomultiplier outside the vacuum chamber, it does not become contaminated with pump oil, samples, or (with GC–MS) column bleed.

The main advantage of using the Daly detector is that metastable ions are readily detected. This greatly assists the mass spectroscopist in determining fragmentation pathways. In the simplest mode, the voltage on the scintillator is reduced to a level of about 50 V below that of the accelerating voltage in the ion source. Normal ions are not repelled, but the "metastable" ions are. They will strike the enhancer and be detected. By manipulating the accelerating voltage and the field strength of the electrostatic analyzer it is possible to determine the *m/e* values of all of the daughter ions or the *m/e* values of all the precursors of any ion in a spectrum.

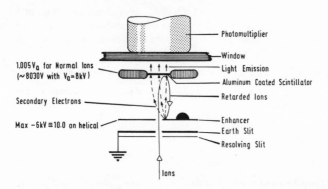

FIG. 18. The Daly fluorescence detector of the Kratos-AEI MS50

4.6. Matrix Control System

With such a large number of components and variable parameters, it could take some time to make the appropriate settings for an analysis. The matrix control system allows the MS50 to be preprogrammed for as many as 10 different operations. The required set of operating parameters can then be effected rapidly.

4.7. Data Handling

With the vast amount of data generated by an ultrahigh-resolution mass spectrometer, automated data acquisition and analysis are mandatory. The Kratos-AEI DS50 data system (which can also be used with low-resolution mass spectrometers) is designed around the Data General Nova 2/10 computer. The operations which it is capable of performing are similar to those of other data systems, but elemental compositions of each ion in every spectrum can also be determined.

4.8. Similarity to Other Instruments

The MS50 is the only ultrahigh-resolution mass spectrometer available. High-resolution instruments are marketed by duPont, JEOL, Kratos-AEI, and Varian.

5. Exercises

1. In the LKB 9000, the usual carrier gas flow rate is 30 ml min^{-1}. In the jet separator, 99% of the carrier gas is removed and up to 75% of the sample is retained. Determine the enrichment factor of the separator.

2. For a fixed magnet current in the LKB 9000, an accelerating voltage of 3.5 kV focuses ions of m/e 386 onto the detector slit. In a SIM analysis, ions of m/e 386 and 388 are to be monitored sequentially. Determine the difference in accelerating voltage required to focus the ions of m/e 388 onto the detector slit.

3. Compare the performance of a magnetic analyzer with that of a quadrupole analyzer.

4. Discuss the factors which limit the m/e range of a magnetic scanning mass spectrometer such as the LKB 9000.

5. Determine the resolution required to distinguish between molecular ions of:

   (a) CO and C_2H_4,

   (b) $C_{10}H_{21}CHO$ and $C_{12}H_{26}$, and

   (c) $C_{20}H_{41}CHO$ and $C_{22}H_{46}$.

6. Suggested Reading

J. H. BEYNON, *Mass Spectrometry and Its Applications to Organic Chemistry*, Elsevier, Amsterdam (1960).

K. BIEMANN, *Mass Spectrometry: Organic Chemistry Applications*, McGraw-Hill, New York (1962).

R. W. KISER, *Introduction to Mass Spectrometry and Its Applications*, Prentice-Hall, Englewood Cliffs, New Jersey (1965).

J. ROBOZ, *Introduction to Mass Spectrometry*, Wiley-Interscience, New York (1968).

H. D. BECKEY, *Field Ionization Mass Spectrometry*, Pergamon Press, Oxford (1971).

Combined Gas Chromatography–Mass Spectrometry

Charles J. W. Brooks and Charles G. Edmonds

1. Introduction

1.1. Gas-Phase Techniques in Analytical Chemistry

Before the advent of gas chromatography (GC) in 1952, fractional distillation was the most widely used analytical method based on volatilization of samples. Extremely refined distillation techniques had been developed, especially in the petroleum industry, but these had several important limitations. The power of separating compounds of closely similar boiling points was inadequate; the methods usually required large samples and were quite inappropriate for even the milligram range; minor components of mixtures were not readily detected, and thermal decomposition of samples was a frequent problem.

The efficiency of liquid chromatographic techniques for the analysis of complex mixtures of organic natural products became apparent from the extensive applications, in 1940–1955, of adsorption chromatography on alumina, especially in the steroid field. Characterization of the separated components was generally based on classical methods together with absorption spectrometry. The concurrent development of liquid–liquid partition chromatography, and of the allied technique of paper chromato-

Charles J. W. Brooks and Charles G. Edmonds • Department of Chemistry, University of Glasgow, Glasgow G12 8QQ, Scotland. *Charles G. Edmonds' present address:* Divisaõ Química, Instituto Nacional da Pesquisas de Amazônia, Manaus, Amazonas, Brasil

graphy, led to the recognition of the value of chromatographic properties *per se* for characterizing organic compounds. The applicability of paper chromatography, in particular, to very small samples quickly led to the acceptance of chromatographic mobilities as structurally indicative parameters. Destructive spray reagents could be used for detection, where the analyzed sample amounted only to an expendable aliquot of the total material.

The gap in techniques that then remained was between the uses of liquid chromatography (primarily for samples of fairly low volatility) and of gas–solid chromatography (primarily for gaseous materials). This was filled by gas chromatography, introduced by A. J. P. Martin (Nobel Laureate) in collaboration with A. T. James in 1952. Although GC is capable of preparative use with recovery of most of the material, it has been far more extensively applied purely to provide information about the composition of samples, with no attempt being made to collect the separated components. The extremely high sensitivity of detectors permits the use of correspondingly small samples, and thus facilitates GC of compounds of relatively high molecular weight and low vapor pressure. Where the general structural features of samples are already known (in classes of natural products, for example) GC can provide highly definitive information through correlations established with reference compounds, using a range of stationary phases and/or derivatives. Except for very simple compounds, gas chromatographic data cannot yield reliable evidence of structure in the absence of prior information.

In mass spectrometry (MS), direct measurement is made of the *m/e* values of ions arising from the impact of electrons (12–70 eV) on a molecule, and generally including the molecular ion, formed by loss of one electron: thus the molecular weight can be inferred. Further structural evidence can be obtained from fragment ions—many of which result from interpretable modes of breakdown—and from characteristic isotopic ratios. At high resolution, the accuracy of mass measurement allows determination of the elemental composition of ions, thus greatly strengthening the assignment of structures.

1.2.   Combined Gas Chromatography–Mass Spectrometry

Since GC and MS are both applicable to sample amounts on the order of micrograms, the combination of the two techniques was developed as a powerful method for the analysis and characterization of small amounts of compounds present in complex mixtures.

The most obvious problem associated with the coupling of the two types of instruments is the difference in operating pressures. The gas chromatographic effluent is at about 1 atm, but the analyzer tube of the mass spectrometer is normally at or below 10^{-5} Torr. Interface systems must accordingly reduce the pressure of carrier gas without causing excessive loss of sample. Effective devices for sample enrichment became commercially available in 1965, and combined gas chromatograph–mass spectrometers have since undergone rapid development. The diagnostic power of the direct dynamic coupling of the two instruments lies in the consistency which must be obtained between the chromatographic retention time and mass spectrometric data for each component examined. Previously used techniques involved trapping of samples from GC prior to MS. This process was cumbersome, unsuitable for trace and unstable components, wasteful owing to losses in trapping, and uncertain in regard to the identical nature of the samples examined at the two stages. Furthermore, it entailed loss of the essential correlation between the retention time and the time of recording of the mass spectrum.

The inherent advantages of GC–MS can be more expeditiously exploited by the use of computers for the continuous acquisition and evaluation of the data.

1.3. Role of GC–MS in Relation to Other Techniques

Equipment for GC–MS is expensive and requires careful usage in order to minimize the practical problems arising from both constituent techniques. In general, it is most economic to carry out substantial preliminary work on samples by simpler techniques, gaining as much information as possible before proceeding to GC–MS.

The removal of nonvolatile components is particularly important, while studies of a range of derivatives by TLC (thin-layer chromatography) and GC can yield provisional evidence of probable structural types. For biological samples of complex composition, fractionation by chromatographic and classical procedures into groups of related functional type is almost invariably necessary; each group can then be converted to the most suitable derivatives, and analyzed under the most appropriate GC–MS conditions.

2. Gas Chromatography

2.1. Introduction

Gas chromatography is a technique for the separation and characterization of compounds according to their distribution between a mobile gas phase ("carrier gas") and a stationary phase. If the latter is a solid, the distribution usually depends on adsorption effects. In gas–liquid chromatography (GLC), the stationary phase is a liquid or gum, coated in a thin film either on the inner wall of the chromatography column (open-tubular GLC), or on a finely divided inert granular support (packed-column GLC). In general, conditions are designed to minimize adsorption, and to achieve partition chromatography, in which the mobility of a compound is determined by its relative solubility in the mobile and stationary phases. Since the sample concentration in the gas phase is primarily dependent upon its vapor pressure at the operating temperature, the gas-chromatographic behavior of a compound is broadly related firstly to its molecular weight and secondly to its interaction with the stationary phase. Gas–liquid chromatography is thus complementary to the techniques of liquid–liquid, liquid–solid, and liquid–gel chromatography in which separations are generally less sensitive to molecular weight.

The introduction of GC in 1952 led to an immense area of applications in which GC has transcended the scope of earlier techniques. Its advantages can be briefly summarized as follows:

Efficiency and Versatility of Separation. The resolving power of GC exceeds that of all other chromatographic techniques, with the partial exception of high-pressure liquid–solid chromatography. Many separations not demanding a high degree of resolution can be achieved with simple columns and very short analysis times.

Sensitivity of Detection. Most gas chromatographs are equipped with hydrogen-flame ionization detectors, allowing the convenient study of samples of $0.1–1\mu g$ ($10^{-7}–10^{-6}$ g). Analysis at the 1-pg (10^{-12}-g) level is possible with selective detectors and particular compounds.

Power of Characterization. Compounds are characterized by their retention volumes under given conditions: these are highly reproducible, and for very simple compounds the retention volumes on several stationary phases may afford definitive identification. In general, characterization is based on careful comparative studies of various derivatives and on correlations with reference compounds.

Quantitative Applications. The validity of each estimation can be monitored by the simultaneous qualitative analysis of the sample components. This affords a major advantage over many spectrophotometric or spectrofluorimetric analyses and over radioimmunoassay techniques.

Range of Application. The high sensitivity of detection methods, mentioned above, is such that only about 1 μg of sample need be volatilized. Accordingly, GC is applicable to compounds of high molecular weight (1000–2000) provided that functional groups are modified (where necessary) to improve volatility or stability. Even short-lived compounds can often be analyzed satisfactorily, by virtue of the inertness of the carrier gas, the short time of analysis, and the possibility of dispensing with extensive preliminary purification of samples.

The chief limitations of GC are its inapplicability to nonvolatile samples (except via pyrolysis) and the fact that it yields no direct information on the molecular weight or structure of a substrate.

2.2. Basic Principles

The gas chromatograph is an analytical instrument which provides a simple, rapid, sensitive, and accurate method for the separation, identification, and determination of volatile compounds. Its construction is shown in Fig. 1. An inert carrier gas of high purity is passed, via a flow controller, to the injector. Sample, as solid, liquid, or solution, is vaporized in the heated injector and swept as a narrow concentrated band onto the column without serious disruption of carrier gas flow. The separation process occurs under controlled temperature conditions, most often isothermal or in a linear temperature program, and the column effluent is passed on to the thermostated detector. The signal from the detector is converted to an analog voltage by an amplifier or other electronic device, and this is recorded as a function of time by a strip chart potentiometric recorder. Figure 2 shows the typical features of such an analog chromatogram. The elapsed time between the injection of the sample and the emergence of the peak maximum is the uncorrected retention time. Subtraction of the time required for elution of a totally unretained compound yields the corrected retention time. This parameter, proportional to retention volume at constant flow rate, is a characteristic property of each compound under specified chromatographic conditions. The area of the peak, the integral of the analog signal, is proportional to the quantity of the component. Gas chromatography thus produces concurrent information on the amount and on the identity of materials analyzed.

The separation of two compounds in GC depends on the difference in their distribution coefficients with respect to the mobile and stationary phases at a given temperature. In practical terms this requires the selection of the most suitable stationary phase. Separating power, usually termed "resolution," is further determined by the column efficiency; this is reflected by the degree of broadening of the initially compact band of vapor as it passes through the column. The rate of peak broadening is a function of column design and operating conditions, and is quantitatively related to the "height equivalent to a theoretical plate" (HETP), a concept derived from distillation theory. The number of theoretical plates, n, may conveniently be calculated by the expression

$$n = 5.54 \, (w_{1/2}/t_{\mathrm{dr}})^2$$

where $w_{1/2}$ is the peak width at one-half height and t_{dr} is the uncorrected

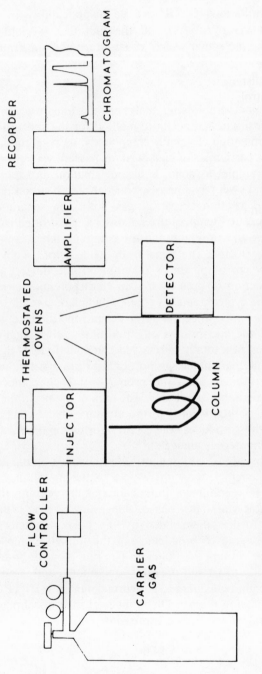

FIG. 1. Schematic diagram of the components of a conventional gas chromatograph.

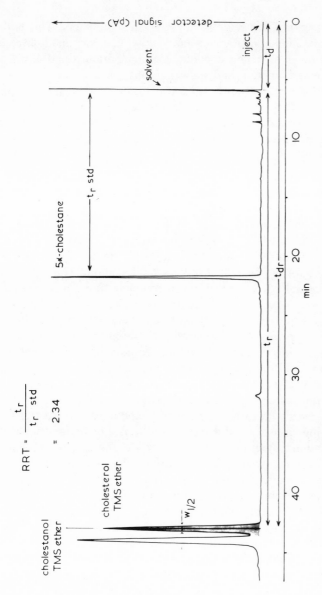

FIG. 2. Chromatogram of the separation of cholesterol and cholestanol as trimethylsilyl (TMS) ether derivatives. 5α-Cholestane was included as an internal standard. The trace is labeled illustrating the typical features of a gas chromatogram. Chromatographic conditions: 50 m × 0.5-mm-i.d. Silanox-type OV-1 glass open-tubular column installed in a Pye Model 104 gas chromatograph equipped with flame ionization detector, oven temperature, 265° C; helium carrier gas, 5 ml min⁻¹.

retention time (Fig. 2). The HETP is then obtained by

$$\text{HETP} = L/n$$

where L is the length of the column, normally in centimeters. This is the preferred measure of column efficiency, and allows comparison of different column lengths under specified conditions of temperature and gas flow rate, sample type and amount, and stationary phase.

In gas–liquid chromatography, several theories have been developed to explain column effects leading to peak broadening. The rate theory of van Deemter *et al.* is perhaps the most useful. The theory relates column efficiency measured as average HETP to the linear gas velocity $\bar{u}$ as

$$\text{HETP} = A + B/\bar{u} + C\bar{u}$$

The terms on the right-hand side of this equation represent the three principal causes of peak broadening.

The *multiple-path* term A arises from the tortuous narrow passages in the bed of the particles which constitute the packed column. These paths are devious and of different lengths, and result in an approximately Gaussian spreading of the peak as it traverses the column. The *molecular diffusion* term B results from the diffusion of solute molecules longitudinally. The *resistance to mass transfer* term C describes the ease of passage of solute molecules between the stationary-liquid and moving-gas phases.

Since the carrier gas is compressible, the linear gas velocity is not constant throughout the column. Thus the average height equivalent to a theoretical plate (HETP) is determined for an average linear gas velocity $\bar{u}$. The contribution of molecular diffusion, B, increases with decreasing flow rate and with decreasing density of carrier gas. Resistance to mass transfer, C, increases with increasing flow rate and density of carrier gas, and with increasing viscosity and thickness of the film of stationary phase. Therefore an optimum flow rate is a compromise between the B and C terms. The choice of carrier gases of various densities is nearly self-canceling and is normally dictated by the requirements of the detector. The A term is constant for a given column and is independent of flow rate. In packed-column GLC, optimum performance is obtained with small uniform inert support particles, coated with a uniformly thin film of phase, and evenly packed in columns of small diameter.

In open-tubular columns the multipath term A disappears, offering a potential advantage of these over packed columns. In practice, however, the plate heights are of the same order for columns of either type. The comparatively unrestricted gas flow in open-tubular columns permits the use of a wide range of column lengths. High numbers of theoretical plates can be obtained by the use of long columns; alternatively, shorter columns are effective for rapid yet efficient analyses. Thus it is the "open" nature of these columns compared with packed columns that is most significant, rather than the generally narrow diameter which the popular term "capillary column" might imply. Columns in which the phase forms a film on the wall of the column are briefly designated as "wall-coated, open-tubular" (WCOT). The surface area available for stationary phase may be increased by modifying the inner wall of the column to produce a porous-layer, open-tubular column (PLOT) or by depositing a thin layer of support material to produce a support-coated, open-tubular column (SCOT); these modifications provide increased sample capacity.

2.3. Features of the Technique

2.3.1. Sampling

Gas samples may be injected via a gas-tight syringe or by means of a mechanical inlet manifold. Liquids or solids in solution may be conveniently injected by means of a microsyringe through a self-sealing septum port. With injections in a volatile solvent, the early part of the chromatogram is obscured by the "solvent peak." This occupies part of the chromatogram; moreover, the solvent may also have deleterious effects on the column or the detector. For samples that are not too volatile, these problems can be avoided by the use of a dry sampling technique. In a common method, sample solution is quantitatively transferred by evaporation to inert sample carriers which may be handled, stored, and sequentially introduced from a cold zone to the vaporization zone of the gas chromatograph. Several examples which appear below demonstrate dry injection achieved by transferring sample solution to a "falling needle" suspended magnetically in a cold region within a closed sampling device. A stream of carrier gas enters the injector. A portion, vented to atmosphere, passes the suspended needle and removes the solvent. The dry needle is then lowered into the flash heater zone and the sample is immediately vaporized and swept onto the column by the remainder of the gas stream.

2.3.2. Thermostated Areas

The operating temperature to be used depends on the dimensions of the column, the particular stationary phase, and its concentration. These are selected to suit the samples being analyzed. In the case of the injection device the temperature must be sufficient to allow rapid vaporization of solid or liquid samples but not so high as to cause decomposition. The column temperature must be controlled within very narrow limits. Isothermal and temperature-programmed modes may be selected according to the requirements of the analysis. The temperature of the detector is determined by its normal operating limitations and by the temperatures of the rest of the system. Close attention to the control of the temperature of all parts of the system is necessary, especially when easily condensable samples are being studied.

2.3.3. Stationary Phases

The column is the heart of the chromatographic system. In addition to the various column types mentioned above, great flexibility in analysis is obtained by the choice of the appropriate stationary phase from among more than 300 that are available. Where column efficiency is insufficient to obtain a satisfactory separation, phase selectivity may often succeed. As a general rule, stationary phases should not be widely different in polarity from the samples being chromatographed. Where compounds of different classes but similar boiling points are to be separated, a phase of appropriate selectivity must be used. The chromatography of a given sample on several distinctive phases is often of great importance in characterization and in confirming homogeneity.

2.3.4. Detectors

Most analytical work is carried out on samples ranging from 1 ng to a few micrograms and sensitive nonspecific detectors are required that are linear in response and insensitive to changes in flow rate and temperature. While differing in operating principle, the argon ionization detector and the flame ionization detector have been found to be very suitable in these respects. In the argon detector the sample remains practically unchanged, while in the flame ionization detector it is largely destroyed by combustion in a hydrogen flame. The nondestructive thermal conductivity detector is applicable wherever the highest sensitivity is not needed. There is also a useful place for detectors that are designed to respond to specific classes of compounds. The electron capture detector, which is responsive to halogenated compounds and conjugated carbonyl compounds, and alkali flame ionization detectors, which are largely specific for phosphoros and nitrogen compounds, are good examples of such devices. Radioactivity detectors are useful for studies involving isotopically labeled materials. Multielement detectors or several specific detectors in parallel may be valuable for special analytical applications.

2.3.5. Quantitative Analysis

In general, the recorded peak areas are a function of the amounts of sample eluted. In programmed temperature gas chromatography, conveniently measured relative peak heights are as satisfactory as relative peak areas for the evaluation of relative concentrations. However, in isothermal gas chromatography it is advisable to determine peak areas. These areas may be evaluated in a number of ways: manual methods such as triangulation or excision and weighing are effective, but mechanical or electronic integrators are much faster and are essential for routine quantitative analysis, especially of complex mixtures.

Analytical calibration may be undertaken directly or by using a known concentration of a reference compound (of different but comparable retention time) as an internal standard. The latter procedure improves precision in analyses based on the injection of solutions by microsyringe by compensating for changes in the volumes injected. The practically instantaneous vaporization that occurs in a properly designed injection system preserves the original relative concentration of solutes.

Detectors must be checked for their linear response over the concentration ranges in question. Calibration should be carried out for each particular compound since there may be marked variation in "response factors." The flame ionization detector is particularly suitable as a general detector by virtue of its linear response over a very wide concentration range.

2.4. Standardization of Retention Behavior

Largely because of the long-term stability of gas chromatographic columns (column temperature and flow rate being controllable within narrow limits), reproducible conditions of separation are easily attainable in gas chromatography. While separation is the outstanding feature of the technique, retention data provide powerful ancillary information. In isothermal chromatography, retention data may be recorded as dimensionless numbers derived from relative retentions with respect to a conveniently selected internal standard. These ratios, which are sensibly constant for each compound (for a given stationary phase at a particular temperature), are characteristic physical properties, and may be applied effectively as indicators of structural features.

Figure 2 depicts a chromatogram of the separation of cholesterol (5-cholesten-3β-ol) and 5α-cholestan-3β-ol as their trimethylsilyl (TMS) ether derivatives. 5α-Cholestane is included as internal standard and the main features are indicated. As mentioned above, the sample is characterized by its retention volume corrected for the "dead" volume of the column. For constant gas flow rate it is not necessary to compute volumes, retention times being proportional thereto. The retention time t_r is obtained by subtracting dead time t_d from the uncorrected retention time t_{dr}. The dead time, the time required for an unretained component to traverse the column, is here approximated by the time required for hexane (solvent) to emerge. This approximation can be made only because the operating temperature is relatively high. The calculation of relative retention time (RRT) is illustrated in Fig. 2.

2.4.1. Behavior of Homologous Series

At constant temperature it has been shown that for compounds forming a homologous series, log (retention volume) is a linear function of the number of carbon atoms per molecule. This relationship is very precise for all but the few early members of simple carbon-chain series. Useful correlations hold also for quasihomologous compounds such as steroids. The predictive value of linear plots is very good, and graphs of log (retention time) against carbon number have been extensively applied in structural analysis. Figure 3 shows the behavior of homologous *n*-alkanes, *n*-alkanols, and *n*-alkanol TMS ethers.

As illustrated in Fig. 3, the slopes of the plots are approximately equal over a relatively wide span, and the introduction of a particular functional group in a particular location is associated with a characteristic logarithmic increase in retention. Similar retention increments may be observed in series of analogous compounds similar in structure in the region of the site of substitution. The actual magnitude and direction of such changes are, of course, dependent on the interaction of substituent groups with the stationary phase. Qualitative assessment of functional groups is greatly facilitated by comparisons of retention behavior on columns of differing polarity. In linear-temperature-programmed gas chromatography at moderate rates of temperature rise ($2-10°C$ min^{-1}) it is found that consecutive members of a homologous series are eluted at approximately equal intervals. The retention time is thus a nearly linear function of carbon number under these conditions.

2.4.2. Group Retention Factors

Clayton examined the retention of a large number of steroids and defined group retention factors as the ratios of retention times of monosubstituted and corresponding unsubstituted steroids. He showed that

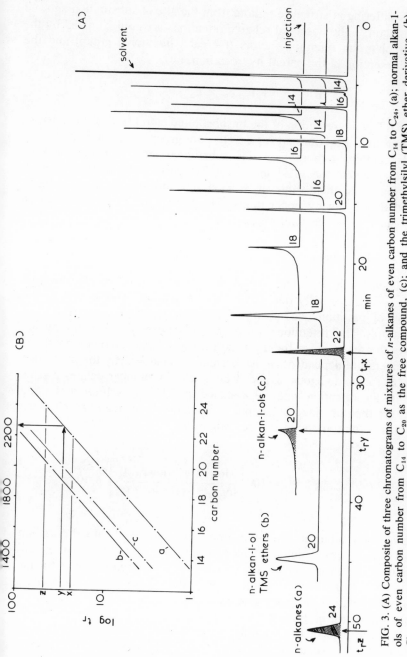

FIG. 3. (A) Composite of three chromatograms of mixtures of *n*-alkanes of even carbon number from C_{14} to C_{24}, (a); normal alkan-1-ols of even carbon number from C_{14} to C_{20} as the free compound, (c); and the trimethylsilyl (TMS) ether derivative, (b). Chromatographic conditions: 50 m × 0.5-mm-i.d. Silanox-type OV-1 glass open-tubular column installed in a Pye Series 104 gas chromatograph equipped with a flame ionization detector, oven temperature 200° C; helium carrier gas, 10 ml min⁻¹. (B) Semilogarithmic graph demonstrating the linear relationship between the log of retention time (t_r) and carbon number in homologous series of *n*-alkanes (a), *n*-alkan-1-ols (c), and *n*-alkan-1-ol TMS ether derivatives (b). Arrows demonstrate the Kovats retention index of *n*-eicosan-1-ol (y) determined by the bracketing even-numbered *n*-alkanes, docosane (x), and tetracosane (z).

steroids differing in structure remote from the site of substitution showed similar factors. The observed relative retentions are approximated by the product of group retention factors (for each functional group) and the relative retention of the parent hydrocarbon, i.e.,

$$\mathrm{RRT} = \mathrm{RRT}_{\mathrm{nucleus}} \times k_a \times k_b \times \cdots$$

where RRT is the retention of the unsubstituted steroid, $\mathrm{RRT}_{\mathrm{nucleus}}$ that of the parent hydrocarbon, and k_a, k_b, etc., are group retention factors. For steroids containing more than one substituent the correlation is most satisfactory where the substituents are far apart. The regularity of such correlations, and their value in detailed structure elucidation, has been substantiated by many applications. The logarithms of group retention factors may be used as additive parameters.

2.4.3. Retention Indices

Relative retention values expressed with respect to a single reference compound are markedly dependent on temperature. This is due to the variation of partition coefficients of different types of compound with temperature. The retention index system devised by Kovats, which employs the members of the n-alkane series as fixed reference points, is less temperature dependent. In isothermal chromatography the linear plot of $\log (t_r)$ for homologous alkanes against carbon number defines a scale such that the retention index of n-alkane, n-C_nH_{2n+1}, is $100n$. Retention indices are then obtained by simple interpolation. This may be expressed algebraically for compound x of retention time $t_r(x)$ as

$$I^{\%\ stationary\ phase}_{x,\ temperature} = 100 \left[\frac{\log t_r(x) - \log t_r(z)}{\log t_r(z+1) - \log t_r(z)} \right] + 100z$$

where t_r is the adjusted retention time and z and $(z+1)$ are the carbon numbers of the bracketing n-alkanes [i.e., $t_r(z) < t_r(x) < t_r(z+1)$]. Retention indices are easily obtained graphically as illustrated in Fig. 3 for the data in Table 1. In linear-temperature-programmed gas chromatography, a similar interpolation may be made employing the linear plot of t_r against carbon number (or carbon number $\times$ 100).

The advantage of the retention index system is its consolidation of retention data in a standard numberical form, relatively invariable over a moderate range of temperature (typically 1 index unit per °C). Qualitative assessment of structure and functional groups may speedily be accom-

TABLE 1. Retention Index Increments (ΔI) of *n*-Alkanes for the Introduction of 1-ol and 1-yl TMS Ether Functions (Based on Chromatogram Shown in Fig. 3)

Carbon number	*n*-Alkane		*n*-Alkan-1-ol			*n*-Alkan-1-yl TMS ethers		
	t_r (min)	I	t_r (min)	I^a	ΔI	t_r (min)	I^a	ΔI
14	1.25	1400	3.35	1665	265	4.80	1760	360
16	2.65	1600	6.95	1865	265	9.85	1960	360
18	5.50	1800	14.40	2070	270	20.10	2160	360
20	11.28	2000	29.65	2270	270	40.50	2360	360
22	23.12	2200	—	—	—	—	—	—
24	46.45	2400	—	—	—	—	—	—

[a] Calculated Kováts retention indices rounded off to nearest 5 units.

plished by comparison of absolute values of retention indices and of their increments.

2.5. Derivatives and the Discriminative Power of Gas Chromatography

With the exception of pyrolysis gas chromatography, volatility and stability are prerequisites of samples for GC. Table 2 shows a selection of compounds for which the preparation of derivatives is necessary to make them amenable to gas chromatography. Compounds that are practically involatile because of strong intermolecular associations, such as acids, zwitterionic amino acids, and hydrogen-bonded polyhydroxy compounds, must be derivatized to increase volatility. Esterification of acidic groups and acetylation or trimethylsilylation of hydroxyl or amino functions are commonly used transformations. The presence of sensitive or interactive functional groups may give rise to thermal degradation. Protective derivatives may mitigate or avoid this problem.

Compounds of high polarity may show inconveniently long retention times or adsorption effects which cause "tailing" of peaks (see Fig. 3c, and others). For ionizable samples these effects may be reduced by arranging for the chemical congruence of the samples and stationary phase, e.g., basic phases for amines, acidic phases for acids. However, convenience and the risk of adsorption of highly polar samples make it generally useful to prepare less polar derivatives that are not as prone to adsorption, and are more suitable for analysis on generally used stationary phases.

A particularly important class of derivatives is the group (generally halogen-containing) designed for use with the electron capture detector. Other derivatives are designed to take advantage of detectors that respond

TABLE 2.   Examples of Compound Types for Which Derivative Formation Is
Essential for Satisfactory Gas-Chromatographic Analysis

Type	Compound	Formula	Example of derivative
Polyhydroxy acid	Tartaric acid	(1)	Dimethyl ester di-TMS ether
Amino acid	Alanine	(2)	N-Acetyl methyl ester
Polyol	Inositols	(3)	Hexa-TMS ether
Sugar	Glucose	(4a)	Penta-TMS ether
		(4b)	Oxime penta-TMS ether
Sphingosine	Sphinganine	(5)	N-Acetyl di-TMS ether
Catecholamine	Noradrenaline	(6)	N,O,O,O-Tetraacetyl Tri-O-TMS ether Schiff base
Prostaglandin	Prostaglandin D$_2$	(7)	Methyl ester O-methyloxime di-TMS ether
Corticosteroid	Cortisone	(8)	3,20-Di-O-methyloxime di-TMS ether

selectively to particular elements. Radioactivity detection is useful where
labeling can be achieved by the formation of derivatives with radioactive
reagents, which provide the high specific activities generally required.

The judicious choice of derivatives offers much scope for the control
of chromatographic retention. The prolongation of retention time through
formation of derivatives is often useful in moving a desired peak out of an
area of high interfering peaks. Similarly, fortuitous overlap of peaks of
differing functionality may be eliminated by preparing a derivative. Analytical group separations may also be obtained, as, for example, those of
ketosteroid O-alkyl oximes from nonketonic steroids in complex mixtures.

Separations of closely related compounds may be enhanced by
derivative formation. A classic example is the enhanced resolution of
epimeric steroid alcohols as their TMS ethers. Separation of enantiomers
may be effected through derivatization with a chiral reagent. This is
illustrated for mixtures of chiral terpenoid secondary alcohols, derivatized
as their esters of (+)-*trans*-chrysanthemic acid, an optically pure terpenoid acid. The resulting diastereomeric derivatives are easily distinguished

by gas chromatography on a packed column of moderate efficiency (Table 3).

Retention changes accompanying formation of derivatives provide indications of the number and nature of functional groups present. This is illustrated in Fig. 4. Characterization of functional groups in the compounds present in the mixture is achieved by the preparation of derivatives specific to each type. In Fig. 4 this is illustrated by the distinction of hydroxyl and carbonyl functions in different compounds through the preparation of TMS ether and *O*-methyloxime derivatives. Retention indices for these comparisons are recorded in Table 4. The retention

TABLE 3. Kováts Retention Indices for Four Terpenoid Secondary Alcohols as (+)-*Trans-*Chrysanthemate Esters[a]

Alcohol	Configuration	I^b	ΔI
(+)-Menthol	(S)	1880	
(−)-Menthol	(R)	1895	15
(+)-Isomenthol	(S)	1880	
(−)-Isomenthol	(R)	1895	15
(+)-Neomenthol	(S)	1835	
(−)-Neomenthol	(R)	1850	15
(+)-Fenchol	(R)	1825	
(−)-Fenchol	(S)	1810	15

menthol (+) S　　isomenthol (−) R　　neomenthol (−) R　　fenchol (+) R

menthyl chrysanthemate

[a] From M. T. Gilbert, Ph.D. Thesis, University of Glasgow, 1975, with permission.
[b] Chromatographic conditions: 5-m 1% SE-30 installed in a Pye Series 104 gas chromatograph equipped with flame ionization detector, 143°C, nitrogen carrier gas 40 ml min⁻¹.

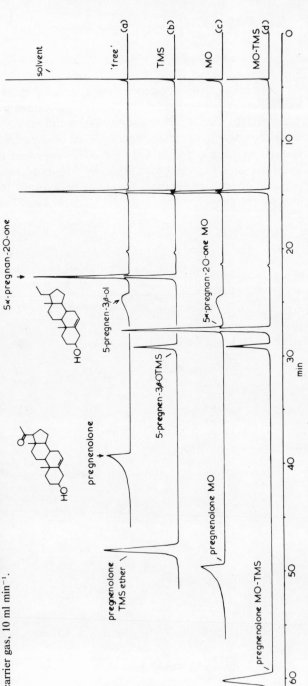

FIG. 4. Composite of four chromatograms of the separation of a mixture of 5α-pregnane, 5α-pregnan-20-one, 5-pregnen-3β-ol, and pregnenolone as the free, untreated mixture (a), treated to form the trimethylsilyl (TMS) ether derivative of 3β-hydroxy functions (b), the O-methyloxime (MO) derivative of the 20-keto functions (c), and both derivatives (MO-TMS) where possible (d). Chromatographic conditions: 50 m × 0.5-mm-i.d. Silanox-type OV-1 open-tubular column installed in a Pye Series 104 gas chromatograph equipped with flame ionization detector, oven temperature, 225° C; helium carrier gas, 10 ml min⁻¹.

TABLE 4. Kováts Retention Indices[a] and Retention Index Increments (ΔI) for the Compounds and Various Derivatives Chromatographed in Figs. 4 and 5

Column	Compound/derivative	5α-Pregnane	5α-Pregnan-20-one	5-Pregnen-3β-ol	Pregnenolone
50-m-Silanox OV-1, 225°C	Free	2240	2425	2470	2650
	TMS (ΔI)	—	—	2530 (+60)	2715 (+65)
	MO (ΔI)	—	2505 (+80)	—	2730 (+80)
	MO-TMS (ΔI)	—	—	—	2790 (+140)
6-ft 1% OV-1, 200°C	Free	2250	2430	2460	2645
	TMS (ΔI)	—	—	2525 (+65)	2715 (+70)
	MO (ΔI)	—	2495 (+65)	—	2715 (+70)
	MO-TMS (ΔI)	—	—	—	2790 (+145)
6-ft 1% QF-1, 175°C	Free	2315	2815	2730	3220
	TMS (ΔI)	—	—	2625 (−105)	3120 (−100)
	MO (ΔI)	—	2630 (−180)	—	3035 (−185)
	MO-TMS (ΔI)	—	—	—	2930 (−290)

[a] Measured retention indices rounded off to nearest 5 units.

increments for each derivatized functional group are constant for the same position within the molecule. Where two functional groups are present the total increment is the sum of the two.

Detailed interpretations of this type should be based on retention indices and retention index increments using at least two stationary phases. The value of selective phases lies in the variety of characteristic separation modes. Compounds which do not separate on a nonselective phase may be readily resolved, or the order of elution may be reversed on a more polar phase. This is demonstrated in Fig. 5 for the steroid mixture discussed above. On the relatively nonselective phase OV-1, the ketosteroid is eluted before the hydroxysteroid, while on the "ketone-selective" phase QF-1 this order is reversed. Derivatization of the hydroxyl and keto functions, reducing their polarity, causes a reduction in the retention index on QF-1 (cf. Table 4) in contrast to their behavior on nonpolar phases. This decrement is likewise additive where more than one functional group is present.

3. Gas Chromatography–Mass Spectrometry

The direct combination of a gas chromatograph and a mass spectrometer constitutes an analytical system of unparalleled capability. Mass spectra obtained rapidly on components emerging from a gas chromatographic column afford informative correlations of mass spectrometric and chromatographic data. Full use is made of the separating power of GC together with the structural information derivable from MS. This gives exceptional power of discrimination between closely similar structures. Two themes have been of particular significance in the application of the technique: first, the extremely high sensitivity that can be achieved in detection and quantitative estimation, by using the mass spectrometer as a mass selective detector; second, the application of derivatization to control both the chromatographic and mass spectrometric behavior of compounds with the aim of enhancing the informative elements in the data.

3.1. The Interface

In the synergy of the two instruments the connecting device is of critical importance. The gas chromatograph has, under normal circumstances, an outlet pressure of 760 Torr, whereas the analyzer pressure of the mass spectrometer is normally, at most, 10^{-5} Torr. There is also the unavoidable dilution of the sample vapor by carrier gas. At a carrier gas flow rate through the gas chromatograph of 30 ml min^{-1}, a 1-μg sample

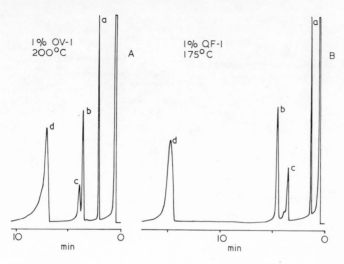

FIG. 5. Chromatograms on packed columns of the apolar stationary phase OV-1 and the ketone-selective stationary phase QF-1 of a mixture of 5α-pregnane (a), 5α-pregnan-20-one (b), 5-pregnen-3β-ol (c), and pregnenolone (d). Chromatographic conditions : 1% phase as noted, coated on 100–120 mesh Gas-Chrom Q, 6 ft × 3-mm-i.d. glass column installed in a Perkin-Elmer Model F11 gas chromatograph equipped with flame ionization detector, temperature as noted, nitrogen carrier gas 40 ml min⁻¹.

(MW = 300) emerging from the column over a period of 10 sec would be present in the gas stream to the extent of ≈0.001% (v/v). A number of molecular separators have been devised, and four of these in wide use are shown in Fig. 6. These devices depend on the difference between physical properties of the carrier gas and sample to obtain the necessary reduction of pressure together with sample enrichment.

The operational parameters used to evaluate the performance of separators are the separation factor and the yield, or efficiency. The separation factor N, also called the enrichment factor, is defined as the ratio of the sample concentration in the carrier gas entering the mass spectrometer to the sample concentration in the carrier gas emerging from the chromatographic column. Thus

$$N = \frac{c_{\text{MS}}}{c_{\text{GC}}} \quad \text{or} \quad N = \frac{(p_s/p_{\text{cg}})_{\text{MS}}}{(p_s/p_{\text{cg}})_{\text{GC}}}$$

where c_{MS} and c_{GC} are the respective sample concentrations and p_s and p_{cg} are the partial pressures of the sample and carrier gas. The efficiency or yield, Y, is the most important factor in the evaluation of the performance

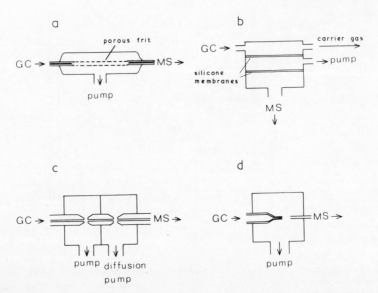

FIG. 6. Schematic diagrams of the four most common devices for the interface of a gas chromatograph to a mass spectrometer: the fritted tube effusion separator of Watson and Biemann (a), a two-stage permeable membrane separator of Llewellyn and Littlejohn (b), a two-stage jet-orifice separator of Becker and Ryhage (c), and the single-stage glass-jet separator of Story and co-workers (d).

of the system. Yield is defined as the percentage of the total sample that enters the mass spectrometer. Thus

$$Y = (Q_{MS}/Q_{GC}) \times 100\%$$

where Q_{GC} and Q_{MS} are the respective quantities of sample leaving the chromatograph and entering the mass spectrometer.

Although the definition of yield is independent of any separation process, it is algebraically related to the separation factor by the equation

$$N = \frac{(V_{GC}/V_{MS})Y}{100\%}$$

where V_{GC} and V_{MS} are the carrier gas volumes measured at 760 Torr delivered from the gas chromatograph and to the mass spectrometer. From the last equation above it is clear that N is inversely proportional to the fraction of total carrier gas volume that enters the mass spectrometer. A yield approaching 100% is possible only in the absence of a separator, or with a device such as the hydrogen/silver-palladium separator in which the carrier gas is physicochemically extracted. In these cases N may be very high. For all other separator designs some sample is lost in the separation process. The separator designs in Fig. 6 are described more fully below.

3.1.1. The Effusion Separator

The preferential effusion of carrier gas through narrow passages under molecular flow conditions is the operating principle of a number of molecular separators. The device of Watson and Biemann (Fig. 6a) employs an ultrafine (10^{-4} cm) porous glass frit. The carrier gas (normally helium) and sample enter via an inlet restrictor causing a pressure drop to about 1 Torr. The rate at which gas effuses through the glass frit to the external vacuum is inversely proportional to the square root of the molecular weight and directly proportional to the partial pressure of each component. Thus the carrier gas is stripped from the sample. The enriched stream is then passed to the mass spectrometer through a second restrictor. Inexpensive construction, together with the chemical inertness of the glass surfaces, have made this design very popular. The same principle has been applied using metal or ceramic frits, and a device with an adjustable effusion path between narrow slits has produced good results.

3.1.2. The Semipermeable Membrane Separator

Sample enrichment obtained by the preferential passage of organic vapors through a thin polymer barrier is the operating principle of the separator of Llewellyn and Littlejohn (Fig. 6b). Effluent from the gas chromatograph is brought into contact with a thin silicone rubber membrane supported on a glass frit. The inorganic carrier gas is insoluble in the polymer and passes out of the device. Organic material of considerable solubility in the polymer diffuses through the film. In a single-stage design this vapor is passed directly to the mass spectrometer. An alternative is a two-stage design shown, where a second chamber and membrane are evacuated by a rotary pump. A major disadvantage of the membrane separator is its severe temperature dependence. An optimum temperature exists for each compound, depending on boiling point, molecular weight, and functionality. The device is also handicapped by a significant time lag between the entry of compounds and their transfer to the mass spectrometer. This causes peak broadening and tailing, making the separator unsatisfactory for critical chromatographic work. An operating temperature limit of 250°C, as well as the risk of membrane rupture, also causes difficulties. A device which operates in the opposite sense, preferentially removing helium carrier gas through the wall of a Teflon tube, has found some application, but its range of operating temperatures is extremely restricted.

3.1.3. The Jet Separators

The differential diffusion of gases in the expanding stream emerging from a jet orifice may be used to obtain sample enrichment in a GC–MS interface. The construction may be of one (Fig. 6d) or two (Fig. 6c) stages. Effluent from the gas chromatograph passes through a jet orifice and expands rapidly into the evacuated peripheral volume. Enrichment is obtained by the diffusive removal of carrier gas from the heavier components in the jet stream. The enriched stream enters the collecting orifice opposite. Similar processes occur in a second stage if it is present. Adjustment of jet dimensions and positions in single or two-stage units provides separators suited to a variety of applications with packed and open-tubular columns. Jet separators are designed to function efficiently at specific flow rates, and below these rates performance rapidly deteriorates. Construction may be in glass or stainless steel. The metal versions are characterized by low catalytic activities since the stream traverses the device at supersonic speed.

3.1.4. Direct Coupling

The simplest way of introducing the gas chromatographic effluent is to split the gas stream at the exit of the column and introduce that portion which the mass spectrometer can accept directly. The generally low capacity of pumping systems limits the proportion of column effluent diverted to the mass spectrometer, and thus reduces sensitivity. The aforementioned devices were designed to enhance sensitivity by improving the transfer of sample. As pumping technology has advanced, direct introduction of the column effluent into the mass spectrometer without a molecular separator has become more practicable. This has proved particularly useful with open-tubular columns where the comparatively low gas flow gives opportunity for high sample utilization. By omitting the separator, significant advantage may be obtained in simplicity of construction and reduced cost.

3.2. Additional Requirements

From the foregoing it may be seen that the chromatographic requirements, selection of interface, and mass spectrometer pumping capacity are closely interrelated considerations. Interfaces may be unsuitable for certain applications (e.g., a membrane separator for high-resolution open-tubular columns) or incompatible with particular pumping systems. In general, pumping efficiency of the highest practical level is desirable. The criterion for suitability of a mass spectrometer for GC–MS is the volume of a gas which may be admitted before the pressure rises to a level at which instrument resolution and sensitivity are impaired. The suitability and adaptability of the interface for the types of chromatography antici-pated are more important factors in the choice of a commercial instrument combination.

GC–MS operation places high demands on instrument sensitivity. In a full scan of subnanogram quantities of compounds, minor peaks may represent fewer than ten ions, and ion detection systems of high inherent sensitivity are required. The electron multiplier is now almost universally employed in this role. This device produces current amplification of 10^3–10^8 without significant noise and with a negligible time constant. A useful alternative to the use of the electron multiplier is photoplate recording. A more detailed account of this and other ion detection techniques is given in Chapter 2.

A mass spectrum characterizes a molecule not only by the masses of the molecular and fragment ions but also by their relative abundances. In the MS sampling systems, such as the standard batch inlet or the direct insertion probe, the pressure of sample should be constant throughout what may be a relatively slow scan. However, the gas chromatographic peak is a profile of changing sample concentration with time. If the time required for scanning is comparable to the duration of the emerging peak, significant distortion of the true relative peak heights may occur. To minimize this, the mass spectrum must be recorded very rapidly—typically from m/e 50 to m/e 500 in 5 seconds or less—making change in sample concentration during the scan insignificant. This also requires a fast response time of the recording system. Resolution and sensitivity are determined by the slowest component of the detector–amplifier–recorder systems. The preferred rate of scanning is selected for each analysis by optimizing those variables that are controllable.

During the course of a GC–MS analysis it is essential to obtain a simultaneous chromatogram. From this record the instrument operator may assess the separation obtained and choose the appropriate points to record mass spectra. A fraction of the effluent may be split to an auxiliary detector: a hydrogen-flame ionization detector is most commonly used. A short interval between the detector signal and the entry of the sample into the mass spectrometer may facilitate inspection of peak shape, and so allow spectra of unresolved peaks or "shoulders" to be properly recorded. Alternatively, part of the mass-spectral ion current may be used to obtain a GC–MS record. A collector located at the exit of the ion source intercepts a fraction of the ion beam. With this device, the energy of the electron beam must be below the ionization potential of the carrier gas to prevent the carrier gas ions from swamping sample ionization. The electron energy may be increased to the desired value for the mass spectrum at the moment of scanning.

Certain limitations are imposed on the operation of the gas chromatograph in GC–MS. Choices of stationary phase and operating temperature are governed by the stability of the stationary phase. Column bleed results in "background" ions which may hinder interpretation of spectra and limit the sensitivity in trace analysis. Column bleed may be minimized by careful conditioning at or above the analysis temperature prior to use. The nature and flow rate of carrier gas may be determined by the interface. For example, the fixed configuration of jets in the two-stage jet-orifice separator of Becker and Ryhage permits optimal operation only at a helium flow rate of 30 ml min^{-1}. This flow rate is appropriate for most packed-column applications. However, with the generally lower flow rates (1–10 ml min^{-1}) employed with open-tubular columns, the separator does not perform satisfactorily. This difficulty may be overcome by modification of the interface or by adding gas to column effluent in a laminar fashion by means of a device such as one shown in Fig. 7. In this way the gas flow

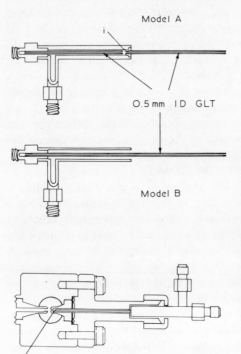

FIG. 7. Diagrams of two variants of a device used in the authors' laboratory for open-tubular column GC–MS with the two-stage jet-orifice separator of the LKB 9000. The construction employs glass-lined metal tubing (GLT) and in Model A (top) column effluent enters and is combined in laminar flow with makeup gas to a total flow of 30 ml min^{-1} within the device at point i. In Model B (middle) the principle is identical, but the combination occurs in the column connector assembly of the mass spectrometer (bottom) at j.

to the separator may be adjusted to the required 30 ml min⁻¹ without affecting gas chromatographic resolution.

3.3. Derivatives for GC–MS

There are two general ways in which derivatives may be employed to affect mass spectra. The derivative may, without altering the fragmentation mode, introduce mass changes that facilitate interpretation. This may be accomplished by isotopic or other minor structural substitution. Alternatively, a particular mode of fragmentation may be induced either to assist structural elucidation or to produce specific ions for analytical purposes.

3.3.1. Production of Mass Shifts

The simplest mass shifts are in the molecular ion, for which any effect of a change in molecular mass on the fragmentation is least significant. Specific adjustment of the isotopic composition to alter the patterns of molecular or fragment ions may be informative. For example, the replacement of exchangeable hydrogens with deuterium is useful (see p. 94). A novel technique for the recognition of metabolites in pharmacological and biochemical investigations involves the mass spectrometric analysis of partially labeled drugs or other precursors. Typically one or more isotopes are present in such proportions that the labeled center produces conspicuous "twin ions" or other well-defined patterns of ions. Metabolites unaltered at the site of labeling are readily detected by GC–MS inasmuch as their mass spectra preserve the characteristic patterns.

Another important application of isotope substitution in GC–MS is to provide an internal standard in the quantitative analysis of the corresponding unlabeled parent compound (or vice versa). The distinction between the two compounds, easily assessed from the mass spectra, is usually obtained with several deuterium atoms at sites in the molecule not seriously affected by fragmentation. The difference in retention time is generally small, and the chromatographic properties of "unknown" and reference compounds, being very similar, simplify extraction and work-up procedures and increase the reliability of analysis. Large amounts of labeled material may be added to the sample containing the unlabeled parent prior to analysis. Adsorption effects in the gas chromatography at low sample levels are thus minimized by the large excess of "carrier" co-chromatographing with the sample.

Functional group derivatives produce generally larger mass shifts in the molecular ion and may substantially alter fragmentation. Table 5 shows a selection of derivatizing reagents for various functional groups and the

TABLE 5. Examples of Derivatives Used for GC-MS

Typical substrates	Group Z	Derivative	Abbreviation	Δm
I. Acyl derivatives : general reaction RXH→RXCO·Z				
Alcohols (ROH)	HCO	Formyl		28
Phenols (ArOH)	CH_3CO	Acetyl	Ac	42
Enols ($>C=C<OH$)	CD_3CO	Trideuteroacetyl		45
	CF_3CO	Trifluoroacetyl	TFA	96
	C_2F_5CO	Pentafluoropropionyl	PFP	146
Thiols (RSH)	C_2F_5CO			
Amines (RNH$_2$)	C_3F_7CO	Heptafluorobutyryl	HFB	196
Oximes (RR'NH) (RR'NOH)	C_6F_5CO	Pentafluorobenzoyl		194
II. Alkyl and silyl derivatives : general reaction RXH→RXZ				
Substrates as listed under *I*— yielding ethers	CH_3	Methyl	Me	14
	CD_3	Trideuteromethyl		17
	C_2H_5	Ethyl	Et	28
	$CH_2C_6H_5$	Benzyl		90
Carboxylic acids— yielding esters	C_6F_5	Pentafluorophenyl		166
	$Si(CH_3)_3$	Trimethylsilyl	TMS	72
	$Si(CD_3)_3$	(d_9)-Trimethylsilyl		81
	$Si(CH_3)_2tBu$	t-Butyldimethylsilyl	TBDMS	114
	$Si(CH_3)_2CH_2Cl$	Chloromethyldimethylsilyl	CMDMS	106/8
	$Si(CH_3)_2CH_2Br$	Bromomethyldimethylsilyl	BMDMS	150/2
	$Si(CH_3)_2CH_2I$	Iodomethyldimethylsilyl	IMDMS	198
	$Si(CH_3)_2C_6F_5$	"Flophemesyl"		224

III. *Derivatives of carbonyl compounds : general reactions*

Type (i): RR'CO→RR'CZ ;   Type (ii): RR'CO→RR'C$\underset{O}{\overset{O}{\diagdown}}$Z

		Compound	Abbr.	Value
	^{18}O	[^{18}O]-labeled compound		2
	(i) N.OCH$_3$	Methyloxime	MO	29
	N.OC$_2$H$_5$	Ethyloxime	EO	43
	N.O sec-Bu	sec-Butyloxime	sec-BuO	71
	N.O i-Pentyl	iso-Pentyloxime	i-PO	85
Aldehydes (RCHO)	N.OCH$_2$C$_6$H$_5$	Benzyloxime	BO	105
	N.OCH$_2$C$_6$F$_5$	Pentafluorobenzyloxime		195
	N.O.Si(CH$_3$)$_3$	Trimethylsilyloxime		87
Ketones (RR'CO)	N.N(CH$_3$)$_2$	Dimethylhydrazone	DMH	42
	(ii) —CH$_2$ / —CH$_2$	Ethyleneacetal		44

IV. *Derivatives of primary amines : general reaction*

RNH$_2$→RN=Z

		Compound	Value
Primary amines (RNH$_2$)	C(CH$_3$)$_2$	Acetone Schiff base	40
	C(CH$_2$)$_3$	Cyclobutanone Schiff base	52
	C(CH$_2$)$_4$	Cyclopentanone Schiff base	66
	C(CH$_2$)$_5$	Cyclohexanone Schiff base	80
	CH.N(CH$_3$)$_2$	Dimethylaminomethylidene	55
	C:S	Isothiocyanate	42

continued overleaf

TABLE 5 (continued)

V. *Derivatives of bifunctional substrates : general reaction*

Typical substrates	Group Z	Derivative	Abbreviation	Δm
Vicinal or 1:3-diols, hydroxy amines, hydroxy acids, etc.	BCH_3	Methaneboronate		24
	B-n-Bu	n-Butaneboronate		66
	B-t-Bu	t-Butaneboronate		66
	B-C_6H_{11}	Cyclohexaneboronate		92
	B-C_6H_5	Benzeneboronate		86
Vicinal diols and hydroxy amines	CH_2	Methylidene		12
	$CH(CH_3)$	Ethylidene		26
	$C(CH_3)_2$	Isopropylidene (Acetonide or oxazolidine)		40
Vicinal diols	$CH(C_6H_5)$	Benzylidene		88
	$Si(CH_3)_2$	Dimethylsilylidene (Dimethylsiliconide)		56
	CO	Carbonate		26
	SO	Sulfite		46

VI. Derivatives of bifunctional substrates : miscellaneous

γ-Hydroxy acids		Lactone	−18
γ-Amino acids		Lactam	−18
Diols	(X = O or NH)	Oxide	−18
α-Amino acids		Methylthiohydantoin	55
		Phenylthiohydantoin	117
	(R′ = CH$_3$ or C$_6$H$_5$)		

associated mass increments (ΔM) for the molecular ion. Implicit in the application of functional group derivatives for GC–MS analysis is the association of characteristic retention changes with derivative formation. These changes depend on the chromatographic conditions employed and are interpreted in structural terms in conjunction with concomitant changes in mass spectra.

3.3.2. Modification of Fragmentation

Derivatives are frequently effective in yielding molecular ions where such ions appear in low abundance (or not at all) in the parent compounds. Ions formed by simple losses of radicals may also be indicative of the molecular weight. For example, the TMS ether derivatives of alcohols sometime produce more prominent molecular ions than their parent compounds. The usually prominent $[M-90]^{+\cdot}$ ion, resulting from the elimination of trimethylsilanol, gives an additional indirect indication of the molecular weight.

Derivatives are also of great importance for their directive effects on mass spectrometric fragmentation. A useful example is the enhancement of α-cleavage in methyl and TMS ether derivatives of secondary alcohols:

$$\underset{\underset{\displaystyle R}{|}}{R-CH-\overset{+\cdot}{O}R''} \longrightarrow \underset{\underset{\displaystyle R}{|}}{HC \overset{+}{=} OR''}$$

This type of α-cleavage is also involved in the highly characteristic fragmentation of TMS ethers of Δ^5-3-hydroxysteroids which yield complementary ions of m/e 129 and $[M-129]^+$ (see p. 96).

Figure 8 illustrates the retention increments associated with a series of O-alkyloxime derivatives (from methyl to benzyl) and also the alteration of a fragmentation of these derivatives associated with the 20-keto group in a steroid. The 3β-hydroxy group of pregnenolone has been derivatized as the TMS ether, and the retention index increments (ΔI values) for the oximation of the 20-keto function [O-methyl ($+80$), O-ethyl ($+140$), O-sec-butyl ($+270$), O-iso-pentyl ($+315$) and O-benzyl ($+785$)] are illustrated. The ring D fragment, together with the molecular ion, shows corresponding mass increments as R is altered from methyl to benzyl.

3.3.3. Reagents for Derivative Formation

Table 5 lists a variety of functional group derivatives suitable for GC–MS analysis, together with their mass increments. It should be pointed out that while certain derivatives may be used with a wide range

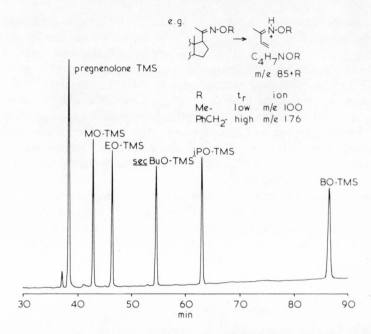

FIG. 8. Chromatogram of a mixture of the trimethylsilyl (TMS) ether and *O*-alkyloxime-trimethylsilyl ether derivatives of pregnenolone demonstrating retention index increments for the series of alkyl substituents: *O*-methyl (+ 80 units), *O*-ethyl (+ 140 units), *O*-*sec*-butyl (+ 270 units), *O*-*iso*-pentyl (+ 315 units), and *O*-benzyl (+ 785 units). Chromatographic conditions : 40 m × 0.5-mm-i.d. Silanox-type OV-1 glass open-tubular column, installed with makeup gas adaptor in the LKB 9000, linear temperature program 1° min⁻¹ from initial temperature 220°C, helium carrier gas 5 ml min⁻¹. Inset scheme shows prominent feature of 20-*O*-alkyloxime electron impact (EI) spectra. Mass spectrometer conditions: source temperature 270°C, electron energy 70 eV.

of individual functional groups (Classes I and II) some are capable of selective application. For example, a number of trimethylsilylating agents exhibit reactivities different enough to discriminate between particular types of hydroxylic groups in steroids. However, under forcing conditions, enol ethers may be obtained, even of the highly hindered 11-oxosteroids. While diazomethane is potentially reactive with many compounds, in practice it is largely selective for methylation of carboxylic acids, since its numerous other reactions are generally much slower. In contrast, there are reagents which normally react with only a small range of functional groups (Classes III and IV).

A valuable group of reagents comprises those that are suitable for forming cyclic derivatives of compounds possessing bi- or multifunctional reactivity (Classes V and VI). This kind of process is of great importance because its selectivity depends on the particular relative disposition and stereochemistry of the reactive groups in the substrate. The formation of a ring limits the range of possible conformations, and gives rise to distinctive chemical, chromatographic, and mass spectrometric properties.

For multifunctional compounds, the use of "mixed" derivatives prepared by sequential treatment with reagents of distinctive selectivity may be of value. Such derivatives offer the possibility of controlling both gas chromatographic and mass spectrometric properties according to the information sought. The use of isotope-labeled or homologous reagents can further assist in the recognition of structurally significant ions. While the classical connotation of the term "derivative formation" implies a simple, potentially reversible change, it is not inappropriate to consider other transformations which facilitate analysis by causing distinctive changes in chromatographic behavior and fragmentation. The use of cholesterol oxidase to convert 3β-hydroxysteroid-5-enes into 4-en-3-ones (p. 121) is an example of this type.

4. Selected Applications

4.1. Samples Not Requiring Derivatization

As has been indicated earlier, there are comparatively few sample types for which derivative formation is generally inappropriate. Saturated and aromatic hydrocarbons are among these. However, there are also samples which may be conventionally presented for GC–MS as derivatives: typical of these would be the methyl esters of fatty acids. In this section, examples are given of GC–MS analyses of various mixtures examined without conversion to derivatives.

n-Alkanes and the corresponding 1-alkenes are well distinguished by MS, as exemplified for the C_{18} hydrocarbons in Fig. 9. Alkenyl ions $[C_nH_{2n-1}]^+$, are preponderant over alkyl ions, $[C_nH_{2n+1}]^+$, in the alkene mass spectrum, but are relatively minor in that of the alkane. The alkene is also distinguished by prominent radical ions, $[C_4H_6]^{+\cdot}$ and $[C_4H_8]^{+\cdot}$ at m/e 43 and 56, respectively. For the gas chromatographic separation of alkanes from 1-alkenes, high column efficiency is necessary, as indicated in Fig. 9, because of the small difference in retention ($\Delta I = 5$ on OV-1 phase).

The steroids 5α-androstane and 5α-16-androstene are easily distinguished by GC (Fig. 10) since the introduction of the C-16 double bond causes a decrement of about 40 index units on OV-1. Again the mass spectrum of the olefin shows distinctive fragmentation affording several even-mass ions (m/e 56, 80, 94, 108, 148) in high abundance. Possible

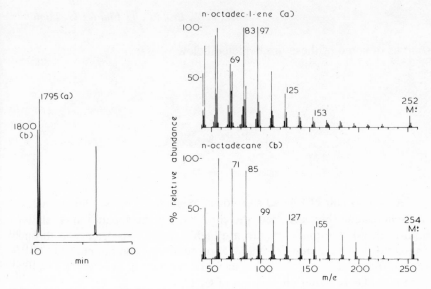

FIG. 9. Total ion current (TIC) chromatogram (left) and electron-impact (EI) mass spectra (right) of the GC–MS analysis of a mixture of *n*-1-octadecene (a) and *n*-octadecane (b). Retention indices appear on the chromatogram. Chromatographic conditions: 50 m × 0.5-mm-i.d. Silanox-type OV-1 glass open-tubular column, installed with makeup gas adaptor in the LKB 9000, temperature 200° C, helium carrier gas 4 ml min^{-1}. Mass spectrometer conditions: source temperature 270° C, electron energy 70 eV.

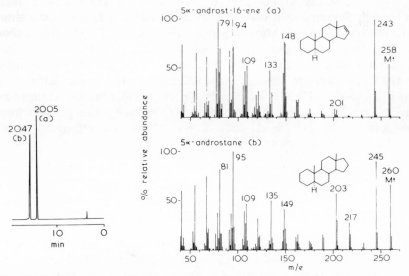

FIG. 10. Total ion current (TIC) chromatogram (left) and electron-impact (EI) mass spectra (right) of the GC–MS analysis of a mixture of 5α-16-androstene (a) and 5α-androstane (b). Retention indices appear on the chromatogram. Chromatographic and mass spectrometer conditions: same as Fig. 9 except column temperature (230°C).

sources of some of these are depicted below:

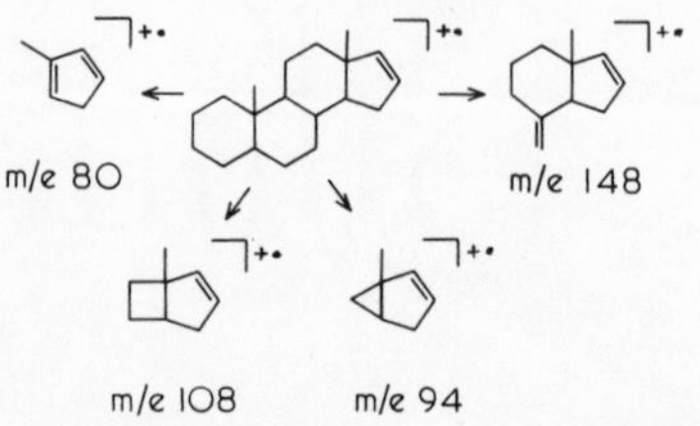

m/e 80   m/e 148

m/e 108   m/e 94

A second pair of related steroids is exemplified by 5α-cholestane and 5α-2-cholestene (Fig. 11). Here the chromatographic separation is relatively difficult but the mass spectra are highly distinctive. Cleavage through ring D yields ions at *m/e* 217 and 215, respectively, and the nuclear double bond position at C-2 is well characterized by the ion at *m/e* 316 resulting from retro-Diels–Alder elimination of C_4H_6:

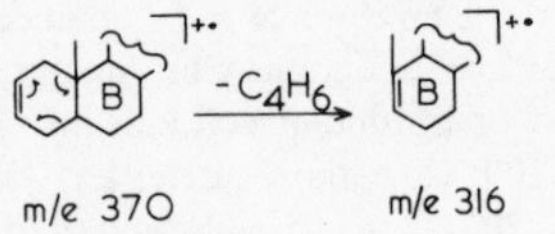

m/e 370   m/e 316

In the analysis of essential oils and perfumery materials, it is often necessary to submit samples directly to GC–MS. A representative group of monocyclic terpenoids (five ketones and a phenol) afforded, on a short packed column, the gas chromatogram shown in Fig. 12. The separations improve in the order : (a,b) (stereoisomers); (c,d) (structural isomers); (e,f) (structural isomers with different functionality); the gas chromatographic distinction between menthone (a) and isomenthone (b) is essential because the mass spectra (Fig. 13) are closely similar. In other instances, fragmentations are informative; thus piperitone (d) is easily distinguished from pulegone (c) by the radical ions of *m/e* 82 and 110 arising by retro-Diels–Alder-type reactions:

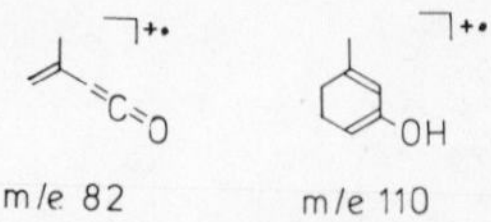

m/e 82   m/e 110

Piperitenone (f) yields an abundant molecular ion, since simple fragmentations are inhibited by the combined effects of the nuclear and exocyclic double bonds. In thymol (e) the aromatic nucleus confers similar stability but promotes the benzylic cleavage of one methyl group affording a base peak at *m/e* 135.

The complementary value of GC and MS is further illustrated by the

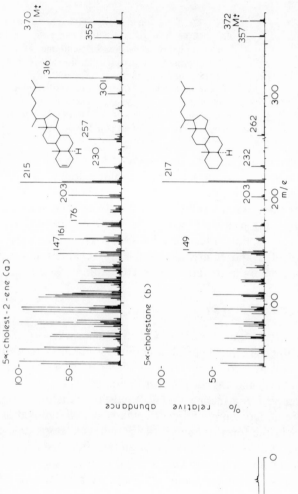

FIG. 11. Total ion current (TIC) chromatogram (left) and electron-impact (EI) mass spectra (right) of the GC–MS analysis of a mixture of 5α-2-cholestene (a) and 5α-cholestane (b). Retention indices appear on the chromatogram. Chromatographic and mass spectrometer conditions: same as Fig. 9 except column temperature (240°C).

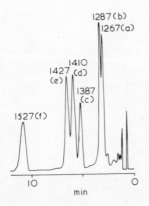

FIG. 12. Total ion current (TIC) chromatogram of the GC–MS analysis of a mixture of monoterpenoid ketones and one phenol, consisting of menthone (a), isomenthone (b), pulegone (c), piperitone (d), thymol (e), and piperitenone (f) obtained on the LKB 9000. Retention indices appear on the chromatogram. Chromatographic conditions: 3 m × 3.5-mm-i.d. glass column packed with 1% OV-17 coated on 100–120 mesh Gas-Chrom Q, temperature 130°C, helium carrier gas 30 ml min⁻¹.

examples in Figs. 14 and 15. The five isomeric substituted salicylates are adequately separated by GC on a Carbowax 20M column. The ethyl ester (a) is fully distinguished by the loss of C_2H_6O, yielding the base peak at m/e 136. The methoxy compounds (b,c) differ from their nuclear methyl isomers (d,e) in giving prominent ions at m/e 110, presumably equivalent to resorcinol and catechol molecular ions, respectively. The mass spectra of b and c, and of d and e, are too similar for reliable identification of these positional isomers in the absence of GC data.

4.2. "On-Column" Transformations

In the majority of applications of GC–MS, alterations of the sample during GC are undesirable. Typical examples of changes resulting from inadequately controlled conditions are the elimination of the elements of water from alcohols, and the Beckmann fission of *O*-alkyloximes. There is, however, scope for effecting certain kinds of transformation during GC–MS. One relatively simple process is the selective removal of certain functional types in the form of nonvolatile derivatives. Boric acid can, for example, absorb hydroxylic compounds. Catalytic hydrogenation has also been developed for application in GC, and has proved especially useful in petroleum analysis and in the study of insect pheromones of alkene or alkadiene types.

A reaction of particular value in the GC–MS of ketones is base-catalyzed deuterium–hydrogen exchange; this takes place fast enough to reach a high degree of completeness during the 10–20 minutes of a convenient retention time. Figure 16 illustrates an application of *in transitu* deuteriation to a sesquiterpenoid ketone. (It should be noted that protection of the allylic alcohol group by trimethylsilylation is essential, and that silicone stationary phases cannot be used as they are unstable toward alkali.) The mass increment of 3 in the molecular ion upon deuteriation

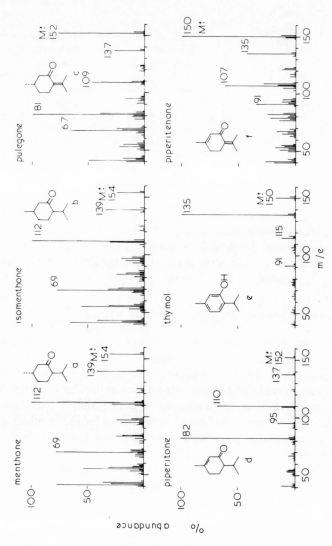

FIG. 13. Electron-impact (EI) mass spectra obtained in the GC–MS analysis shown in Fig. 12. Mass spectrometer conditions: source temperature 270°C, electron energy 70 eV.

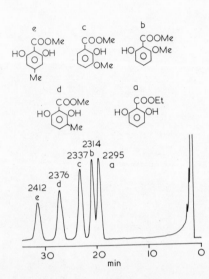

FIG. 14. Chromatogram of the separation of a mixture of isomeric hydroxybenzoates consisting of ethyl γ-resorcylate (a), methyl 6-methoxysalicylate (b), methyl *o*-vanillate (c), methyl 3-methyl-γ-resorcylate (d), and methyl *p*-orsellinate (e). Chromatographic conditions: 3 m × 3.5-mm-i.d. glass column packed with 2% Carbowax 20M coated on 100–120 mesh Gas-Chrom Q installed in a Perkin-Elmer Model F11 gas chromatograph equipped with flame ionization detector, temperature 135°C, nitrogen carrier gas 40 ml min⁻¹.

clearly indicates the number of active hydrogen atoms in the sample, and corresponding increments in the ions $[M-15]^+$ and $[M-43]^+$ are observed. The application of this method is most pertinent where preliminary deuterium–hydrogen exchange is made difficult by the inadequate quantity, high volatility, or instability of the sample.

4.3. Derivatives Promoting Informative Fragmentations

The structural information derivable from a mass spectrum varies greatly; many compounds fail to undergo well-defined fragmentations, while others give rise to easily interpreted spectra. In all cases, the study of derivatives can provide additional evidence of molecular features. The selection of derivatives designed to enhance or to confirm particular modes of fragmentation is an extremely important art in GC–MS. In addition to the two simple illustrations given in this section, examples will be found on pages 103 to 125.

Figure 17 shows the mass spectra of a simple sterol, 5-pregnen-3β-ol, and its TMS ether. Salient features of the free sterol spectrum are the abundant molecular ion (*m/e* 302) and ions from loss of methyl (*m/e* 287), H_2O (*m/e* 284), and both methyl and H_2O (*m/e* 269). Probable assignments for other major ions are indicated at the top of page 97. While the spectrum is indicative of a nuclear-unsaturated sterol, it affords no evidence for the 5-ene structure. The TMS ether, however, yields two very prominent ions, *m/e* 129 and *m/e* 245 $[M-129]^+$, which are highly characteristic of the Δ^5-3-hydroxy TMS ether system; only the 3α-epimer would be ex-

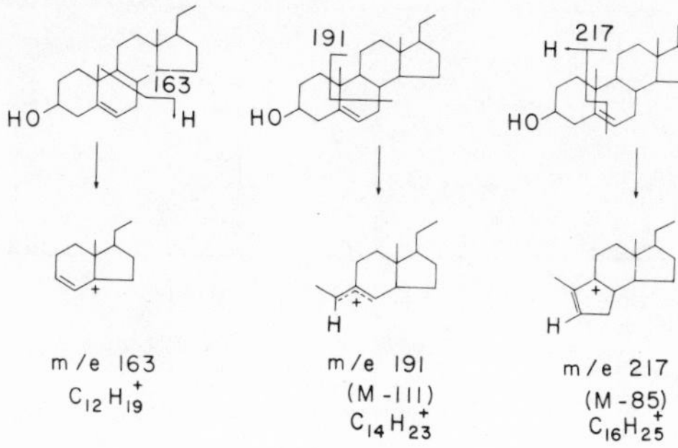

m/e 163	m/e 191	m/e 217
$C_{12}H_{19}^{+}$	(M-111)	(M-85)
	$C_{14}H_{23}^{+}$	$C_{16}H_{25}^{+}$

pected to give similar results, and this can be easily distinguished by its much shorter retention time in GC. The formation of the ion of *m/e* 129 is depicted below:

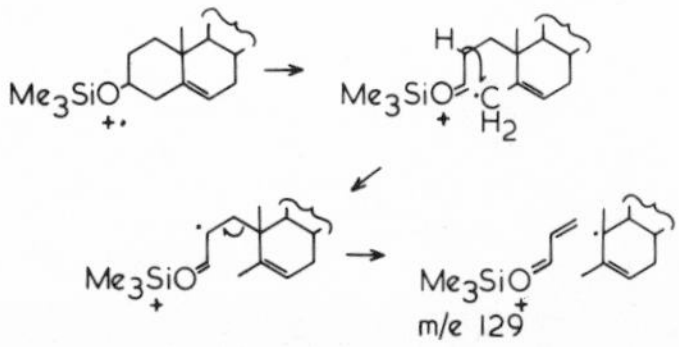

The genesis of the [M−129]$^{+}$ ion is by an analogous process with charge retention on the nuclear fragment.

A further illustration of the effects of trimethylsilylation is given in Fig. 18. In the uppermost spectrum, the characteristic ions at *m/e* 129 and [M−129]$^{+}$ are observed for 5-androsten-3β-ol TMS ether. The introduction of the 17β-hydroxyl-17α-methyl group leads to the mass spectrum shown in the center diagram: the ion at *m/e* 129 remains predominant, but [M−129]$^{+}$ is now largely replaced by [M−129−18]$^{+}$ (*m/e* 229) resulting from elimination of H_2O from the tertiary alcohol group. The ions at *m/e* 253 [M−90−18−15]$^{+}$ and *m/e* 213 can be represented as shown below:

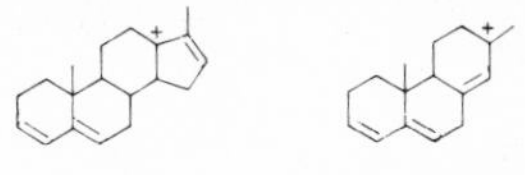

m/e 253   m/e 213

There is no ion in the spectrum that would be sufficiently characteristic of the tertiary alcohol group, though the ion of *m/e* 71 probably constitutes $CH_2=CH—C(Me)=^{+}OH$ derived from ring D. Trimethylsilylation of the tertiary alcohol transforms this situation, as shown in the lower spectrum of Fig. 18: the ion at *m/e* 143, corresponding to *m/e* 71 just mentioned,

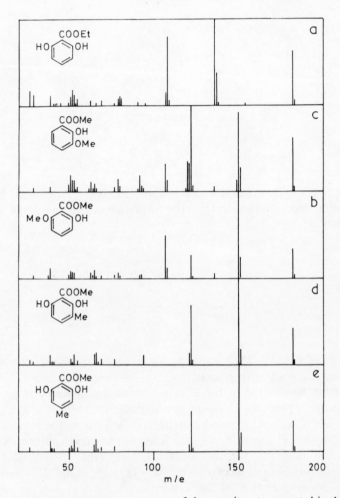

FIG. 15. Electron-impact (EI) mass spectra of the constituents separated in the chromatogram shown in Fig. 14. Mass spectrometer conditions: LKB 9000, GC inlet, source temperature 270°C, electron energy 70 eV.

strikingly indicates the presence of the 17-methyl-17-trimethylsilyloxy grouping. Loss of trimethylsilanol, accompanied by cleavage across ring D, removing C-16 and C-17 with its substituents (with charge retention on the nuclear residue) gives rise to m/e 227, while cleavage at C-15/16 affords m/e 213. Other prominent ions in the spectrum are associated with successive losses of trimethylsilanol moieties from the *bis*-TMS derivative: m/e 358 $[M-90]^{+\cdot}$, m/e 343 $[M-90-15]^{+}$, m/e 268 $[M-90-90]^{+\cdot}$, and m/e 253 $[M-90-90-15]^{+}$.

FIG. 16. Partial mass spectra of the unchanged (upper) and on-column deuteriated (lower) sesquiterpenoid ketol, brachylaenolone, as the trimethylsilyl ether derivative in GC–MS analysis on the LKB 9000. Chromatographic conditions: normal GC–MS, 2 m × 3.5 mm-i.d. glass column packed with 1% SE-30 coated on 100–120 mesh Gas-Chrom Q, temperature 130°C; *in transitu* deuteriation GC–MS, 2 m × 3.5-mm-i.d. glass column packed with 1% Apiezon-L and 1% Ba(OH)₂ coated on 100–120 mesh Gas-Chrom Q, temperature 145°C. Prior to analysis, the column was saturated with successive injections of D₂O. Helium carrier gas 30 ml min⁻¹. Mass spectrometer conditions: source temperature 270°C, electron energy 70 eV.

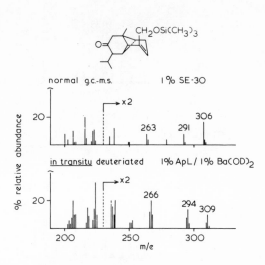

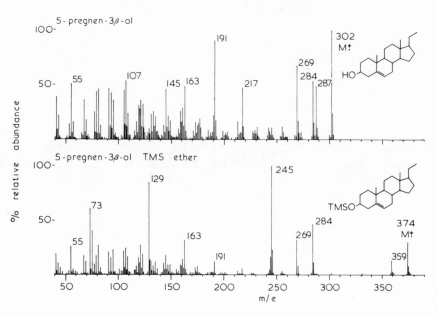

FIG. 17. Electron-impact (EI) mass spectra of 5-pregnen-3β-ol as the free compound (upper) and the trimethylsilyl (TMS) ether derivative. Mass spectrometer conditions: LKB 9000, GC inlet, source temperature 270°C, electron energy 70 eV.

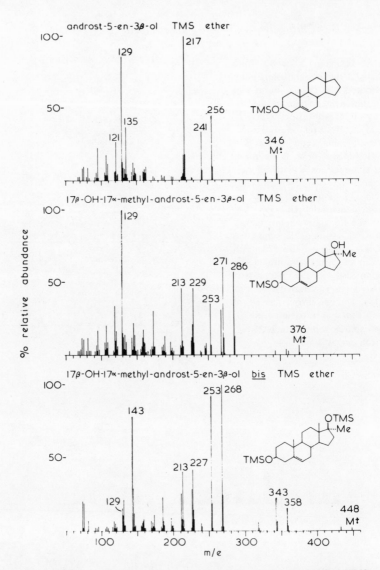

FIG. 18. Electron-impact (EI) mass spectra of 5-androsten-3β-ol trimethylsilyl (TMS) ether (upper), 17β-hydroxy-17α-methyl-5-androsten-3β-ol as the 3β-TMS ether (center), and the 3β,17β-*bis* TMS ether derivative (lower). Mass spectrometer conditions: LKB 9000, GC inlet, source temperature 270°C, electron energy 70 eV.

4.4. Analysis of Closely Related Unsaturated Steroids

A combined gas chromatograph–mass spectrometer equipped with packed columns is a uniquely powerful tool for the simultaneous separation and identification of the components of mixtures. For many demanding applications, high efficiency of separation is essential in order to achieve reliable results. This is particularly important for complex mixtures of compounds of biological origin. Open-tubular columns offer high resolving power and, after some earlier difficulties, a number of investigators have reported procedures for the preparation of reproducible, thermostable glass open-tubular columns. These have an advantage over the long-known metal columns in their low catalytic activity. Their suitability for compounds sensitive to adsorption or decomposition has led to an increasingly wide application to biochemical problems.

The location of unsaturation within the steroid nucleus is often not easily deduced from mass spectral data. In these circumstances ambiguities may often be settled by chromatographic resolution of compounds. An example of this problem and of the utility of open-tubular columns is shown in the GC–MS analysis of a mixture of Δ^5-, Δ^4-, and $\Delta^{5(10)}$-estren-17-ones (Fig. 19). The compounds are only poorly separated on a packed column (see inset chromatogram). However, the total ion current chromatogram with an open-tubular column shows "baseline" resolution of the components of the mixture. Mass spectra may be obtained without

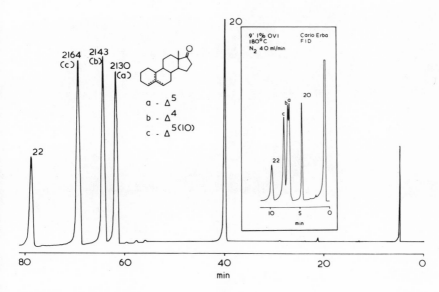

FIG. 19. Total ion current (TIC) chromatogram of the separation of a mixture of 5-estren-17-one (a), 4-estren-17-one (b), and 5(10)-estren-17-one (c). C_{20} and C_{22} n-alkanes are included as co-injected standards, and retention indices appear on the chromatogram. Chromatographic conditions: 50 m × 0.5-mm-i.d. Silanox-type OV-1 open-tubular column installed with makeup gas adaptor in the LKB 9000, temperature 200°C, helium carrier gas 5 ml min⁻¹. Inset: packed-column chromatogram of the same mixture demonstrates the advantage of the more efficient open-tubular column.

cross-contamination. Notable features in the mass spectra of the resolved isomeric steroids as shown in Fig. 20 are the prominence of the molecular ion in each case, and their great similarity. The common fragment of m/e 230 $[M-28]^{+ \cdot}$ may be the result of the elimination of carbon monoxide from ring D or from the loss of C_2H_4:

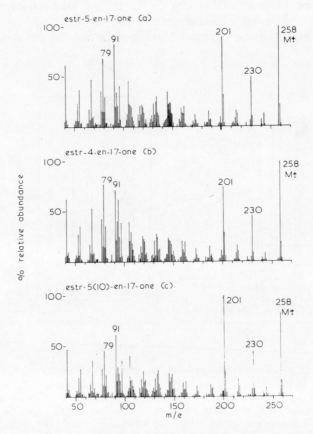

The former is likely as analogous ions appear in the mass spectra of other 17-keto steroids.

FIG. 20. Electron-impact (EI) mass spectra of the components of the mixture separated in the chromatogram shown in Fig. 19. Mass spectrometer conditions: LKB 9000, source temperature 270°C, electron energy 70 eV.

A metastable ion at m/e 276 in all three spectra (not shown in the line diagrams) indicates that the fragment ion of m/e 210, which is the base peak from the $\Delta^{5(10)}$ isomer, results at least in part from elimination of C_2H_5 from the ion at m/e 230. This might arise by the process depicted below:

$$m/e\ 201$$

The mass spectra contain no ion which could be said to be diagnostic of the structure, possibly because of isomerization of the double bond sites in the ion source. It is obvious that the specificity of the retention data is critical to the identification of these steroids by GC–MS.

The separation of 5α- and Δ^5-3-hydroxysteroids is relevant to the study af many natural mixtures, notably in mixtures of sterols of marine origin and in the urinary steroid metabolites of the newborn human and other mammals. Under normal packed-column conditions employing non-selective stationary phases, compounds differing in this respect are not satisfactorily distinguished. However, the higher efficiency offered by open-tubular columns can bring about a complete resolution (see Fig. 2). Two such separations are illustrated in Fig. 21, which refers to a six-component mixture consisting of 5α-16-androsten-3α-ol (a), 5β-16-androsten-3α-ol (b), 5,16-androstadien-3β-ol (c), 5α-16-androsten-3β-ol (d), 5-androsten-3β-ol (e), and 5α-androstan-3β-ol (f) as their TMS ethers. As expected, the $3\alpha/3\beta$ hydroxy epimers (a,d) and the $5\alpha/5\beta$ epimers (a,b), separable by packed-column chromatography, are well resolved. The $\Delta^{5,16}/5\alpha$-Δ^{16} pair is well differentiated from the ring D saturated compounds.

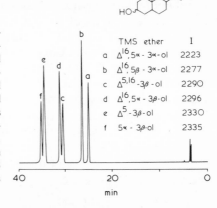

FIG. 21. Total ion current (TIC) chromatogram of the separation of a mixture of 5α-16-androsten-3α--ol (a), 5β-16-androsten-3α-ol (b), 5,16-androstadien-3β-ol (c), 5α-16-androsten-3β-ol (d), 5-androsten-3β-ol (e), and 5α-androstan-3β-ol (f) as the trimethylsilyl (TMS) ether derivatives. Retention indices are tabulated on the chromatogram. Chromatographic conditions: 50 m × 0.5-mm-i.d. Silanox-type OV-1 glass open-tubular column installed with makeup gas adaptor in the LKB 9000, temperature 230°C, helium carrier gas 5 ml min^{-1}.

	TMS ether	I
a	Δ^{16},5α - 3α-ol	2223
b	Δ^{16},5β - 3α-ol	2277
c	$\Delta^{5,16}$ -3β -ol	2290
d	Δ^{16},5α - 3β-ol	2296
e	Δ^5 -3β-ol	2330
f	5α - 3β-ol	2335

The 3β-trimethylsilyloxy Δ^5/5α pairs are resolved by virtue of the improvement in chromatographic resolution afforded by the glass open-tubular column. Mass spectra obtained at the apex of each component peak corresponded closely with those of the individual derivatives. Examples of the mass spectra of a separated 5α/Δ^5 pair are shown in Fig. 22 for 5,16-androstadien-3β-ol and 5α-16-androsten-3β-ol as TMS ethers. The mass spectra show the expected similarities, e.g., ions corresponding to $[M-15]^+$, $[M-90]^{+\cdot}$, and $[M-90-15]^+$. The ions of m/e 129 and $[M-129]^+$ are prominent in the $\Delta^{5,16}$ compound, indicating the Δ^5-3-hydroxy TMS ether structure (origin as on p. 96). Corresponding ions are of low abundance in the 5α compound. In the absence of the dominant influence of the Δ^5 double bond, the Δ^{16} double bond enhances ions associated with the loss of the C-18 methyl group, viz., m/e 331 $[M-15]^+$ and m/e 241 $[M-90-15]^+$. Ions of m/e 107 and m/e 148 are analogous to fragments derived from the Δ^{16} hydrocarbon (p. 91).

4.5. Studies of Drug Metabolism

GC–MS is of great importance in the investigation of the metabolism of drugs *in vivo* and *in vitro*. General pathways of biotransformation are increasingly well known, and analytical procedures can thus be aimed at the expected metabolites, while allowing also for the exploration of unexpected products. The use of radioactive isotope labeling remains important for revealing the distribution of metabolites in experiments carried out in animals or by *in vitro* techniques, but is rarely justifiable in human subjects. Labeling with stable isotopes usually has—or is assumed to have—little effect on toxicity, and is now widely used in human metabolic studies. A mass increment of about 4 amu affords adequate distinction by MS from isotopic forms present in natural abundance, without causing any major alteration in retention time. This feature can be applied in quantitative analysis, one of the isotopic forms—usually the unnatural one—being used as an internal standard for determination of a much smaller amount of the other, and acting as a "carrier" to protect it from adsorption during GC. Further details of these techniques and their applications are given in Chapters 4 and 11.

One aspect of metabolism in which GC–MS has unique power concerns the stereoselectivity of chemical transformations. It is well known that the physiological and pharmacological effects of enantiomeric compounds are different; moreover, drugs that contain no asymmetric center may well undergo metabolic changes such as hydroxylation with the production of optically active derivatives. Gas chromatography is the most effective technique for discriminating between enantiomeric pairs, or diastereomeric sets, of compounds (cf. Table 4). For the direct analysis of

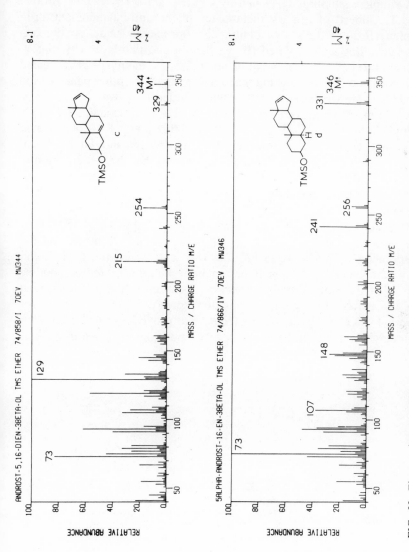

FIG. 22. Electron-impact (EI) mass spectra of components (c) and (d) of the mixture separated in the chromatogram shown in Fig. 21. Mass spectrometer conditions: LKB 9000, source temperature 270°C, electron energy 70 eV.

enantiomers, chiral stationary phases are required. Alternatively, treatment with a chiral reagent can afford diastereomers separable on conventional columns.

A simple example is furnished by a recent study (by C.J.W. Brooks and M.T. Gilbert) of urinary metabolites of the anti-inflammatory drug ibuprofen (Brufen) [($\pm$)-2-(4-isobutylphenyl)propionic acid]. Acidic urinary metabolites were isolated (by extraction and thin-layer chromatography) after a single dose of the drug, and GC of the methyl esters revealed a peak corresponding to starting material together with other peaks eluted later. It was shown, however, that the excreted drug was no longer in racemic form. After conversion of the isolated acid to its amide with (+)-β-phenylethylamine, GC (Fig. 23) indicated that about 80% of (+)-acid was present, as a result of stereoselective metabolism. In a later publication, a research group at the Upjohn Co. showed that the ($-$)-enantiomer of Brufen underwent a remarkable enzyme-catalyzed inversion at the chiral center during metabolism, whereas the (+)-enantiomer remained unaltered.

Several metabolites of Brufen were characterized by GC–MS. Retention indices of derivatives of two of these are given in Table 6, and the corresponding mass spectra in Fig. 24. The center spectrum (b) is dominated by the ion of *m/e* 265 resulting from benzylic cleavage [corresponding to *m/e* 177 from Brufen methyl ester; cf. spectrum (a)] greatly enhanced by the trimethylsilyloxy group. The bottom spectrum (c) was recorded for the methyl ester of a hydroxy metabolite that proved somewhat resistant to trimethylsilylation; the implied tertiary position of substitution was

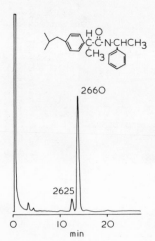

FIG. 23. Chromatogram of the amides of (+)-α-phenylethylamine with ($\pm$)-2-(4-isobutylphenyl)propionic acid [2625 ($-$)-acid, 2660 (+)-acid] isolated from urine after ingestion of the antirheumatic drug Brufen. Retention indices appear on the chromatogram. Chromatographic conditions: 5 m × 3.5 mm i.d. glass column packed with 1% OV-17 coated on 100–120 mesh Gas-Chrom Q, installed in a Pye Series 104 gas chromatograph equipped with flame ionization detector, temperature 245°C, nitrogen carrier gas 40 ml min^{-1}.

TABLE 6. Retention Data for Derivatives of 2-(4-Isobutylphenyl)propionic Acid and Two of Its Metabolites

Fig. 24 reference[a]	Compound	Kováts retention indices[a]		
		$I^{1\% \text{ OV-1}}_{150°C}$	$I^{1\% \text{ OV-17}}_{170°C}$	$I^{1\% \text{ QF-1}}_{150°C}$
(a)	2-(4-Isobutylphenyl)propionic acid (*Brufen*) methyl ester	1515	1725	1800
(b)	2,4'-(1-Hydroxy-2-methylpropyl)-phenylpropionic acid methyl ester trimethylsilyl ether	1700	1860	2135
(c)	2,4'-(2-Hydroxy-2-methylpropyl)-phenylpropionic acid methyl ester	1660	1935	2090

[a]Chromatographic conditions: 1.5 m × 3.5-mm-i.d. silanized glass columns packed with the phase and percentage indicated, coated on 100–120 mesh Gas-Chrom Q; installed in a Pye Series 104 gas chromatograph equipped with flame ionization detector, temperature as indicated, nitrogen carrier gas 40 ml min^{-1}.

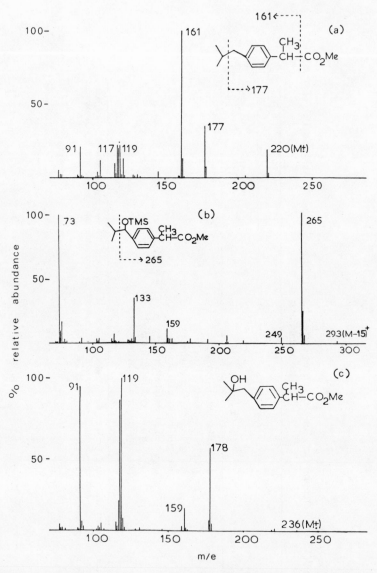

FIG. 24. Electron-impact (EI) mass spectra of the parent drug Brufen (a) and its secondary hydroxy (b) and tertiary hydroxy (c) metabolites as the methyl ester TMS ether and methyl ester derivatives, respectively. Mass spectrometer conditions: LKB 9000, source temperature 270°C, electron energy 70 eV.

indicated by the ion of *m/e* 178 attributable to loss of acetone by a McLafferty-type rearrangement:

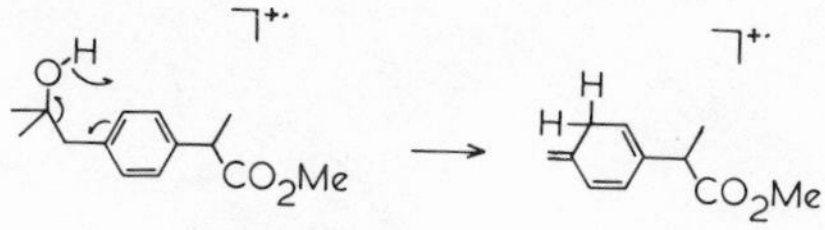

m/e 178

When the TMS ether was made, the base peak of the spectrum at *m/e* 131 ($Me_2C=^+OSiMe_3$) confirmed the structural assignment. It should be noted that for full elucidation of structures, determination of the stereochemical composition at the original center would be required for each metabolite, together with similar information for new chiral centers such as that in the secondary hydroxy metabolite mentioned above. GC–MS is an essential technique for exploring these details, and for monitoring the application of other stereochemically diagnostic methods.

4.6. *Protection of Thermally Labile Samples by Derivative Formation*

Some examples have already been given of the use of derivatives in the protection of functional groups (cf. pp. 73, 83). This theme is illustrated here with respect to the analysis of urinary steroids and of biogenic amines.

Many of the steroid metabolites present in urine are stable enough for direct GC, but it is generally more satisfactory to convert them to TMS ethers, or other derivatives, that are less liable to undergo adsorption or to be adversely affected by imperfections in the gas-chromatographic conditions. There are, moreover, steroids for which the formation of derivatives is essential; the major examples are in the corticosteroid group. Thus the dihydroxyacetone side chain typical of cortisone is largely degraded during GC, affording several products, notably the corresponding 17-ketone:

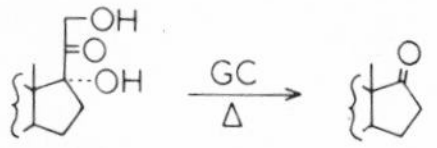

The 20,17-ketols are also unstable, undergoing rearrangement to D-homo-steroids such as:

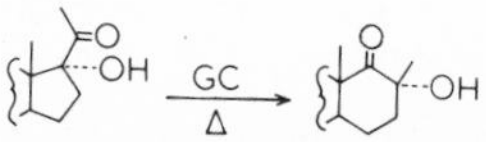

Protection of the dihydroxyacetone side chain is not satisfactorily achieved by trimethylsilylation of the two hydroxyl groups, but a number of effective derivatives have been developed, such as

$$
\begin{array}{ccc}
\begin{array}{l}\text{—OSiMe}_3\\ \text{=NOMe}\\ \cdots\text{OH}\end{array} &
\begin{array}{l}\text{—OSiMe}_3\\ \text{=NOMe}\\ \cdots\text{OSiMe}_3\end{array} &
\begin{array}{l}\text{—OSiMe}_3\\ \text{=NOR}\\ \cdots\text{OSiMe}_3\end{array} \\[2em]
\begin{array}{l}\text{=O}\quad\overset{\text{O}}{\underset{\text{O}}{}}\text{SiMe}\end{array} &
\begin{array}{l}\text{=O}\quad\overset{\text{O}}{\underset{\text{O}}{}}\text{BR}\end{array} &
\begin{array}{l}\text{—OSiMe}_3\\ \text{—OSiMe}_3\\ \cdots\text{OSiMe}_3\end{array}
\end{array}
$$

Among these, the O-methyloxime TMS ethers introduced by Gardiner and Horning in 1966, and homologous oxime TMS ethers remain the most convenient for application to urinary steroid samples. Oximation is carried out first and is followed by trimethylsilylation, preferably under conditions suitable for reaction of the sterically hindered 11β- and 17α-hydroxyl groups. The resulting mixture of TMS ethers (of nonketonic steroids and steroids with the unreactive 11-keto group) and oxime TMS ethers is most suitably analyzed by programmed-temperature GC because of the wide range of retention index values represented. This is illustrated by the chromatogram of TMS and O-methyloxime TMS derivatives of reference steroids shown in Fig. 25. The components for which O-methyloxime formation is essential are the dihydroxyacetone-type steroids (peaks h and i).

Characterization of the steroids depends on the accurate determination of retention parameters—particularly important for distinguishing stereo-isomers—and on the correlation of the GC data with mass spectra recorded at defined retention times. The establishment of relationships between structural features, retention data, and mass spectra of many reference steroids is a prerequisite for the identification of steroids in biological samples. Salient features of data for the compounds represented in Fig. 25 are summarized in Table 7. The following notes briefly refer to some diagnostic points.

(i) The O-methyloximes (of monoketones) are distinguished by their odd molecular weights; where the molecular ion is weak the ions $[M-31]^+$, due to loss of methoxy radical, and $[M-121]^+$, due to further loss of trimethylsilanol, are indirectly helpful.

(ii) The isomeric steroids androsterone (b) and etiocholanolone (c) give rise to similar mass spectra, and their gas chromatographic properties provide the essential distinction.

(iii) The characteristic ions produced by 20-ketosteroid O-methylox-imes at m/e 100 and 87 (cf. Fig. 8) are prominent for 5α-pregnan-20-one (a) and pregnenolone (b) derivatives; in the latter case, the ions at m/e 129 and 288, $[M-129]^+$, are also abundant, indicating the Δ^5-3-OTMS grouping.

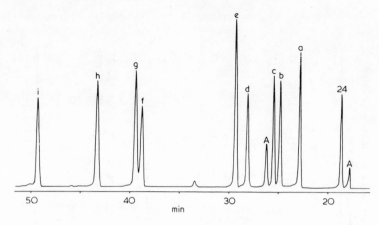

FIG. 25. Chromatogram of the separation of a synthetic mixture of hydroxy and keto steroid standards derivatized as the trimethylsilyl (TMS) ethers and *O*-methyloxime trimethylsilyl ethers (MO-TMS). *n*-Tetracosane (24) is included as internal standard and two prominent peaks (A) result from artifacts entering in the derivatization process. a, 5α-Pregnan-20-one; b, androsterone; c, etiocholanolone; d, dehydroepiandrosterone; e, 5α-androstane-3β,17β-diol; f, pregnenolone; g, 3β-hydroxy-5α-pregnan-20-one; h, THS (tetrahydrocortexolone); i, THE (tetrahydrocortisone). Chromatographic conditions: 50 m × 0.5-mm-i.d. Silanox-type OV-1 glass open-tubular column installed in a Pye Series 104 gas chromatograph equipped with flame ionization detector, linear temperature program 1°C min⁻¹ from initial temperature 210°C, helium carrier gas 5 ml min⁻¹.

In the spectrum of 3β-hydroxy-5α-pregnan-20-one O-methyloxime TMS ether, the ion [M−31]⁺ is predominant, but that of *m/e* 100 is also of high abundance.

(iv) The base peak of *m/e* 129 in (e) does not arise from a Δ⁵-3-OTMS group, but largely from the 17β-OTMS group. The molecular ion indicates a saturated steroid, and the absence of a peak at [M−129]⁺ is in accord with this.

(v) Mass spectra of the corticosteroid derivatives (h,i) show prominent molecular ions and a preponderance of ions [M−31]⁺, [M−121]⁺, and [M−211]⁺ resulting from losses of methoxyl and trimethylsilanol moieties. Ions of *m/e* 103 and, in the case of THE (i), [M−103]⁺ arise from the primary OTMS group at C-21.

To summarize, it may be said that the identification of the molecular ion is of initial importance. Recognition of the probable origin of the principal fragment ions must then be based on careful judgment of the spectrum as a whole, assessment of the compatibility of the retention data, and comparison with data for reference compounds of as similar structure as possible.

Application of *O-iso*-pentyloxime TMS derivatives to the GC–MS of

TABLE 7. Retention Indices and Salient Mass Spectral Features of the O-Methyloxime-Trimethylsilyl Ether (MO-TMS) Derivatives of the Standard Compounds Separated in the Chromatogram Shown in Fig. 25

Fig. 25 reference	Compound	I^a	$M^{+\cdot}$ (%)	Base peak	Prominent fragment ions[b]						
a	5α-Pregnan-20-one	2490	331 (1)	100	300 (33)	243 (6)	232 (4)	217 (6)	173 (5)	119 (5)	87 (52)
b	Androsterone (3β-hydroxy-5α-androstan-17-one)	2535	391 (1)	270	360 (50)	300 (7)	213 (12)	107 (18)	75 (55)		
c	Etiocholanolone (3α-hydroxy-5β-androstan-17-one)	2550	391 (1)	270	360 (45)	300 (4)	213 (13)	107 (16)	75 (47)		
d	Dehydroepiandrosterone (3β-hydroxy-5-androsten-17-one)	2605	389 (10)	129	358 (38)	299 (13)	284 (16)	268 (45)	260 (39)	105 (12)	73 (31)
e	5α-Androstane-3β,17β-diol	2630	436 (11)	129	346 (18)	331 (11)	256 (16)	241 (26)	215 (22)	73 (73)	75 (73)
f	Pregnenolone (3β-hydroxy-5-pregnen-20-one)	2795	417 (20)	100	402 (46)	386 (55)	312 (31)	288 (40)	239 (25)	129 (61)	87 (67)
g	3β-Hydroxy-5α-pregnan-20-one	2810	419 (10)	388	404 (15)	298 (8)	243 (11)	107 (11)	100 (71)	87 (35)	75 (37)
h	THS (3α,17α,21-Trihydroxy-5β-pregnan-20-one)	2875	595 (23)	564	474 (22)	384 (15)	255 (23)	244 (17)	147 (38)	103 (38)	73 (375)[c]
i	THE (3α,17α,21-Trihydroxy-5β-pregnane-11,20-dione)	2975	609 (65)	488	578 (30)	506 (40)	398 (38)	147 (35)	103 (34)	73 (360)[c]	

[a] Chromatographic conditions as noted in Fig. 25.
[b] Mass spectrometer conditions: LKB 9000, source temperature 270°C, electron energy 70 eV.
[c] Abundance determined for base peak of m/e 100 or above.

urinary steroids is illustrated in Fig. 26. The higher mass of the *iso*-pentyl group leads to increases in the retention times of the ketonic derivatives, and thus separates them to a greater extent from nonketonic steroids. The chromatogram represents a "steroid profile" characterizing the sample. The identities of the steroids are elucidated as far as possible by GC–MS correlations as outlined above. Relevant data for the five annotated peaks in Fig. 26 are cited in Table 8.

The materials eluted within the first 20 min are endogenous, nonsteroidal materials and contaminants arising in sample preparation and derivatization. Notable is the large peak eluting at about 18 min ($I = 2514$), which is a contaminant from the silylating reagent. The next portion of the chromatogram, up to the point of elution of the first prominent ketosteroid (16α-hydroxy-DHA), is the region containing the bulk of hydroxy steroids. This first ketosteroid appears as two peaks separated by about 15 index units. These arise from the *syn*- and *anti*-isomers of the *O*-alkyloxime derivative. The presence of this isomerism for steroid derivatives, observable at packed-column resolving efficiency for 3-ones, 4-en-3-ones, and 16-ones unsubstituted at C-17 and for certain 17- and 20-ones at the higher efficiency provided by open-tubular columns, is a complicating factor in the use of *O*-alkyloxime derivatives to obtain group separations in steroid profile analysis. The odd mass number molecular ion (indicating a nitrogen atom present in the molecule) shows prominently with a base peak of m/e 129, implying a Δ^5-3-OTMS structure. Cleavage of the N–O bond of the *O*-*iso*-pentyloxime results in an ion at m/e 446 [M−87]$^+$ and this, combined with m/e 356 [M−87−90]$^+$ and m/e 266 [M−87−90−90]$^+$, confirms the derivative as an *O*-*iso*-pentyloxime *bis*-trimethylsilyl ether. A fragmentation across ring D produces an ion, containing the oxime oxygen, nitrogen, C_{17}, and C_{16} with its OTMS, of m/e 145. Comparison with GC and MS data of the authentic compound confirm the compound as 16α-hydroxydehydroepiandrosterone.

The peak at $I = 3076$ has some mass spectral features similar to those previously discussed: m/e 446 [M−87]$^+$, m/e 356 [M−87−90]$^+$, m/e 266 [M−87−90−90]$^+$, and m/e 129. A prominent ion at m/e 230 provides evidence for a 16-keto-17-hydroxy arrangement in ring D:

m/e 230

Similar ions retaining the keto or alkyloximino group appear in the spectra of other *O*-alkyloxime TMS ether derivatives of this compound (16-one *bis*-TMS ether, m/e 145; 16-methyloxime *bis*-TMS ether, m/e 174; 16-*sec*-

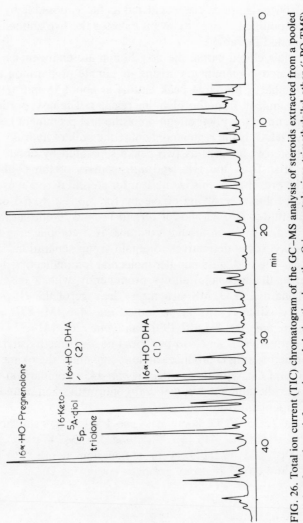

FIG. 26. Total ion current (TIC) chromatogram of the GC–MS analysis of steroids extracted from a pooled sample of newborn infant urine and derivatized as the *O-iso*-pentyloxime trimethylsilyl ether (*i*-PO-TMS) derivatives. Chromatographic conditions: 50 m × 0.5-mm-i.d. Silanox-type OV-1 glass open-tubular column installed with makeup gas adaptor in the LKB 9000, linear temperature program 1°C min⁻¹ from initial temperature 210°C, helium carrier gas 5 ml min⁻¹.

TABLE 8. Retention Indices and Salient Mass Spectral Features[a] of Selected Constituents of Pooled Newborn Urinary Steroids as Their O-*Iso*-pentyloxime Trimethylsilyl Ether (*i*-PO-TMS) Derivatives (See Fig. 26)

Fig. 26 designation	I	M[+]	Base peak	Prominent fragment ions[a]	Proposed structure
16α-HO-DHA[b] (1)	2976	533	129	446 (60) 356 (34) 266 (90) 145 (40)	3β,16α-dihydroxy-5-androsten-17-one
16α-HO-DHA[b] (2)	2993	533	129	446 (38) 356 (29) 266 (68) 145 (32)	3β,16α-dihydroxy-5-androsten-17-one
16-Keto-⁵A-diol	3076	533	129	446 (57) 356 (34) 266 (40) 230 (52) 160 (64)	3β,17β-dihydroxy-5-androsten-16-one
⁵P-triolone	3147	649	258	473 (26) 418 (37) 244 (67) 143 (38) 129 (59)	3β,11β,17α-trihydroxy-5-pregnen-20-one
16α-HO-Pregnenolone	3215	561	474	432 (7) 384 (13) 244 (49) 156 (53) 129 (46)	3β,16α-dihydroxy-5-pregnen-20-one

[a]Chromatographic and mass spectrometric conditions as noted in Fig. 26.
[b]*Syn*- and *anti*- isomers of the O-*iso*-pentyloxime derivative.

butyloxime *bis*-TMS ether, *m/e* 216) and provide further evidence of this genesis. These data and retention index correlations support the identification of the parent compound as $3\beta,17\beta$-dihydroxy-5-androsten-16-one.

A prominent GC peak at $I = 3147$ is indicated to be a pregnenetriolone *O-iso*-pentyloxime trimethylsilyl ether derivative by the mass of the molecular ion, *m/e* 649. This is further confirmed by an ion at *m/e* 472 $[M-87-90]^+$, while *m/e* 129 indicates a Δ^5-3-OTMS structure. The base peak, *m/e* 258, and a fragment ion *m/e* 244 are consistent with a 20-ketosteroid bearing a trimethylsilyloxy substituent in the region C-15 to C-21 and C-16 to C-21, respectively. These features, together with comparison of GC data with published results, permit the structure of the parent steroid to be assigned as $3\beta,11\beta,17\alpha$-trihydroxy-5-pregnen-20-one.

The ion at *m/e* 244 from the compound of $I = 3215$ is also consistent with a 20-ketosteroid containing a trimethylsilyloxy substituent in the region C-16–C-21 [i.e., in comparison with the simple 20-ketosteroid *O*-methyloxime: *m/e* $(100 + 56 + 88)$]. The molecular ion and the ion of *m/e* 129 support the Δ^5-3-OTMS structure, and the identity is confirmed by comparison with the authentic 16α-hydroxypregnenolone derivative.

Another type of structure that is thermally unstable is the β-hydroxyamine system present in many natural products such as epinephrine, ephedrine, serine, and sphingosine. The β-hydroxyphenylethylamines, in particular, tend to suffer degradation to benzaldehyde and its analogs:

$$\text{ArCH(OH)–CH}_2\text{(NHR)} \xrightarrow{\Delta} \text{ArCHO} + \text{CH}_3\text{NHR}$$

In this case, protection can be effected simply by acylation or trimethylsilylation; either process yields derivatives with very good gas-chromatographic properties. Figure 27 shows the separation of a mixture of

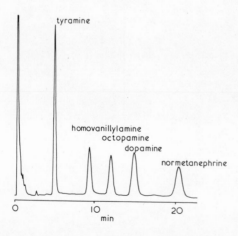

FIG. 27. Chromatographic separation of a mixture of fully acetylated phenylethylamines. Chromatographic conditions: 2 m × 3.5-mm-i.d. glass column packed with 7% F-60 and 1% EGSP-Z coated on 100–120 mesh Gas-Chrom P installed in a Barber-Colman Model 10 gas chromatograph equipped with flame ionization detector, temperature 215°C, argon carrier gas 50 ml min⁻¹. (Work by C.J.W. Brooks in Dr. E.C. Horning's laboratory, Baylor College of Medicine, 1963.)

phenylethylamines as their acetylated derivatives. The β-hydroxyamine groups in octopamine and normetanephrine are completely stabilized as the *O,N*-diacetyl derivatives, and acetylation also protects the easily oxidizable catechol group in dopamine. Useful correlations exist between retention times and structure; thus the introduction of a β-acetoxyl group (diacetyl tyramine → triacetyl octopamine; diacetyl homovanillylamine → triacetyl normetanephrine) increases the retention time by a factor of 2.6 ± 0.1 under the conditions indicated.

Mass spectra of two representative amides from the mixture separation shown in Fig. 27 are illustrated in Fig. 28. Molecular ions are weak under electron-impact conditions. In the case of triacetyl octopamine the major initial cleavages are benzylic, yielding $[M-59]^+$ and $[M-72]^+$ from losses of acetoxyl and acetamidomethyl, respectively. Other ions, with the exception of the acetylium ion (base peak, *m/e* 43), arise mainly from rearrangements. Triacetyldopamine affords a base peak at *m/e* 136 and a prominent ion at *m/e* 178, attributable to the following fragmentations:

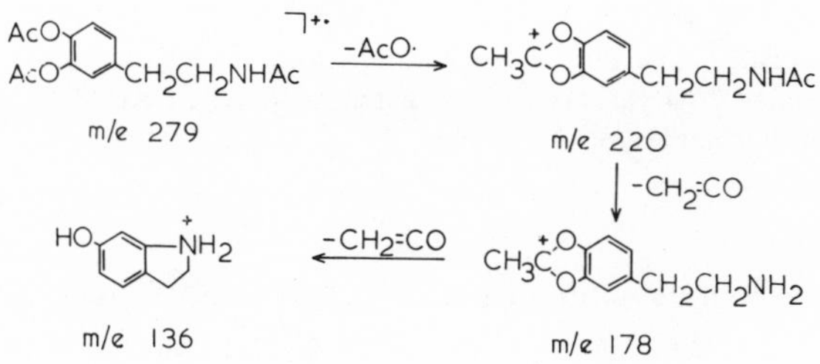

4.7. Applications of Selective Derivatization Reagents

Selective reagents have, in some respects, wider application in conjunction with gas-phase methods than in classical analytical chemistry. It is, for example, usually practicable to employ a large excess of reagent and thus to effect virtually quantitative formation of the product. This is useful for slow reactions, but expecially for equilibrium reactions where samples of the reaction mixture can be injected directly into the gas chromatograph.

One of the best examples is to be found in the reactions of amines with ketones. Thus acetone can conveniently be used simultaneously as a reagent and as a solvent; primary amines are converted into Schiff bases, whereas secondary amines—though they may form adducts in solution— are unaffected in their gas chromatographic properties:

$$RNH_2 + (CH_3)_2C{=}O \rightarrow RN{=}C(CH_3)_2 + H_2O$$
(Schiff base)

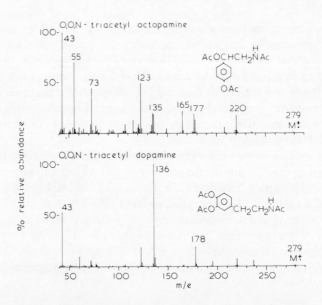

FIG. 28. Mass spectra of the fully acetylated dihydroxy phenylethylamines, octopamine (upper), and dopamine (lower). Mass spectrometer conditions: LKB 9000, source temperature 270°C, and electron energy 70 eV.

Secondary amines containing a β-hydroxy substituent can, of course, yield oxazolidines: these are distinguishable from Schiff bases by GC–MS.

Thus comparative studies of amines in inert solvents and in various ketonic solvents provide an experimentally simple means of differentiating the type and number of amine groups present.

An even more versatile class of selective reagents is provided by the boronic acids, $RB(OH)_2$ (cf. Table 5). These react with a wide range of bifunctional compounds in which the two reactive groups are in the 1:2 or 1:3 relationship, forming 5- and 6-membered cyclic boronates, respectively. (The formation of larger rings is also possible in favorable circumstances, e.g., from *cis*-cylohexane-1,4-diol.) An example of a 1,3-diol cyclic boronate has been noted on page 110. The formation of cyclic derivatives of this type has several advantages; e.g., (i) protection of the two reactive groups affords implicit evidence of their presence and relative disposition in the substrate; (ii) the derivative usually preserves the stereochemistry of the two reacting groups, and the rigidity introduced by ring formation enhances the differences in the GC–MS behavior of stereoisomers; (iii) the mass increment can be kept to a small value and can be adjusted by selection of any desired boronic acid; (iv) cyclic boronates have good gas chromatographic properties, and also possess considerable stability in mass spectrometry; molecular ions are often abundant under electron impact, while protonated molecular ions are usually the base peaks in isobutane chemical ionization MS; and (v) the boronate ring, in many instances, is stable enough to allow further derivatization of other functional groups present in the sample.

A simple illustration of the last-mentioned point is provided by Fig. 29. Methyl-β-D-glucopyranoside readily yields a 4,6-benzeneboronate: the 2,3-*trans*-diol grouping does not form a cyclic ester under the mild conditions used, and the presence of the free diol impairs the quality of the gas chromatography, as shown in trace (a). Trimethylsilylation of the boronate affords a derivative with excellent gas chromatographic behavior as seen in trace (b); the mass spectrum in this case shows no molecular ion, but the presence of the vicinal diol group not involved in boronate formation is clearly indicated by the ion of *m/e* 204, the structure of which is noted in Fig. 29.

Other important types of compounds for which cyclic boronates have proved very effective derivatives in GC–MS include sphingosines, prostaglandins of the F_α series, and arylethane-1,2-diols derived, e.g., from catecholamine metabolism.

In addition to the use of selective derivatization, it is valuable to carry out other selective transformations of substrates prior to GC–MS. For microchemical analysis it is desirable that the reagents used should be simple and the reactions essentially quantitative. Reduction of ketones with sodium borohydride would be a typical example. In some cases, a specific degradation of the sample may facilitate its analysis by GC–MS. Thus aldosterone is distinguished from other corticosteroids by periodate oxidation which affords a lactone with very good GC–MS properties:

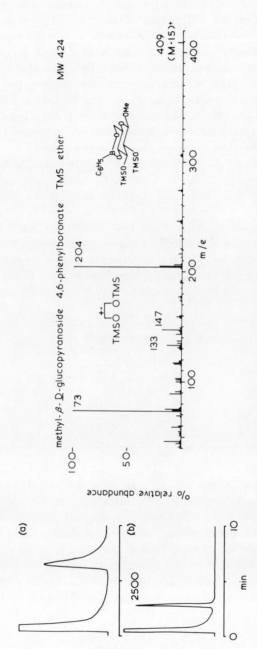

Fig. 29. Gas chromatographic peaks (left) of methyl-β-D-glucopyranoside 4,6-phenylboronate (a) and 4,6-phenylboronate trimethylsilyl (TMS) ether (b) and the mass spectrum of the fully derivatized compound (right). Chromatographic conditions: 3 m × 3.5-mm-i.d. glass column packed with 1% QF-1 coated on 100–120 mesh Gas-Chrom Q installed in a Varian Aerograph Model 204 gas chromatograph equipped with flame ionization detector, temperature 125°C, nitrogen carrier gas 40 ml min⁻¹. Mass spectrometer conditions: LKB 9000, GC inlet, source temperature 270°C, electron energy 70 eV. (Work by Dr. W. J. Reid.)

Recently, the application of enzymes as selective reagents has become feasible. For substrates in the steroid series, cholesterol oxidases derived from microorganisms are now available. These enzymes catalyze the aerobic oxidation of cholesterol and the rearrangement of the Δ^5 bond to yield 4-cholesten-3-one. The substrate specificity varies according to the enzyme source, but reaction can be effected with many 5α and Δ^5-3β-hydroxysteroids. This leads to many potential uses of the enzymes in the indirect analysis of mixtures of hydroxysteroids. Thus the separation of cholesterol from cholestanol by GC is difficult unless high-resolution columns are available. By treatment of the mixture of these sterols with cholesterol oxidase, they are converted into 4-cholesten-3-one and 5α-cholestan-3-one, respectively, and these products are easily separable, as shown in Fig. 30. Furthermore, the mass spectra of the ketones are quite different from each other and from those of the parent sterols, thus affording distinctive evidence of structure, as indicated in Fig. 31; the intense ion at *m/e* 124 is frequently observed in 4-en-3-ones.

4.8. Use of Isotope-Labeled Reagents

Mass spectrometry of isotopically labeled compounds is of value in numerous applications in quantitative and mechanistic studies in organic chemistry, pharmacology, and biochemistry. Isotope labeling has particular relevance to the investigation of fragmentation processes in mass spectrometry. A fragment ion may be formed by a single process or a combination of several processes. These may often be rationalized in terms of expected modes of breakdown of the decomposing ion; that is, specific bonds are

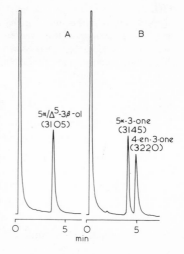

FIG. 30. Chromatogram of a mixture of cholesterol and 5α-cholestanol (A) and the same mixture after enzymic oxidation to 5α-cholestan-3-one and cholest-4-en-3-one. Retention indices are recorded on the chromatogram. Chromatographic conditions: 2 m × 3.5-mm-i.d. glass column packed with 1% OV-1 coated on 100–120 mesh Gas-Chrom Q installed in a Pye Series 104 gas chromatograph equipped with flame ionization detector, temperature 275°C, nitrogen carrier gas 40 ml min⁻¹. (Work by Dr. A. G. Smith.)

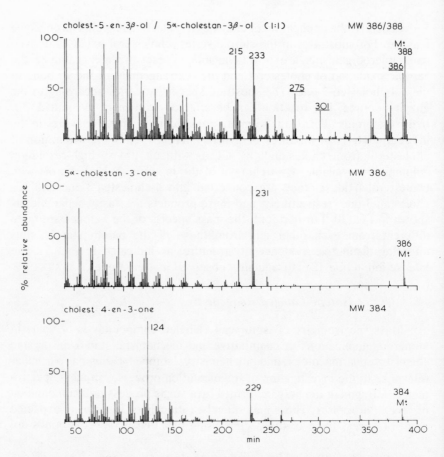

FIG. 31. Electron-impact (EI) mass spectra obtained by packed-column GC–MS analysis of the mixtures as shown in Fig. 30. In the spectrum of a 1:1 mixture of cholest-5-en-3β-ol and 5α-cholestan-3β-ol (top) the molecular ion and two of the prominent fragment ions associated with cholest-5-en-3β-ol are underlined. Mass spectrometer conditions: LKB 9000, source temperature 270°C, electron energy 70 eV..

formed and others broken, forming a stable ion and a stable neutral fragment. Isotope substitution in a molecule will result in corresponding shifts in mass of every fragment containing the isotope atom. Comparison of the spectra of the normal and labeled compound shows which peaks have shifted, and provides evidence for the fragmentation mechanism.

In mass spectra, characteristic patterns of natural isotopes (e.g., $^{10-11}B$, $^{35-37}Cl$, $^{79-81}Br$) may act as useful markers. Among the most useful isotopic labels, from a mass spectrometric point of view, are ^2H, ^{18}O, ^{13}C, and ^{15}N. Deuterium may be incorporated by exchange from ^2H_2O for

TABLE 9. Salient Features of the Mass Spectra[a] of 20β-Hydroxy-5α-pregnan-3-ones, 20β-Hydroxy-4-pregnen-3-ones, and Related Compounds

Steroid	MW[b]	[CH₃CHO–TMS]⁺ (base peak)	[M–CH₃CHO]⁺ (%)	[M–CH₃CHO–CH₃]⁺ (%)
20β-OTMS-5α-pregnan-3-one (I)	390	117	346 (9)	331 (2)
20β-OTMS-4-pregnen-3-one (II)	388	117	344 (8)	329 (3)
[3-¹⁸O]-20β-OTMS-5α-pregnan-3-one (Ia)[c]	392	117	348 (14)	333 (4)
[3-¹⁸O]-20β-OTMS-4-pregnen-3-one (IIa)[c]	390	117	346 (17)	331 (8)
20β-[¹⁸O]TMS-4-pregnen-3-one (IIb)[c]	390	119	344 (18)	329 (10)
20β-O[²H₉]TMS-5α-pregnan-3-one (Ic)	399	126	355 (33)	340 (9)
20β-O[²H₉]TMS-4-pregnen-3-one (IIc)	397	126	353 (27)	338 (10)
20β-OTMS-5α-pregnane (III)[d]	376	117	—	—
20β-OTMS-5α-pregnan-3-one-3-O-methyloxime (IV)[d]	419	117	375 (1)	—

[a] Mass spectrometer conditions: LKB 9000, source temperature 270°C, electron energy 20 eV.
[b] Molecular ion absent in electron-impact spectra, [M−15]⁺ present in low abundance demonstrating molecular species.
[c] Oxygen-16-containing ions not included.
[d] Electron energy 70 eV.

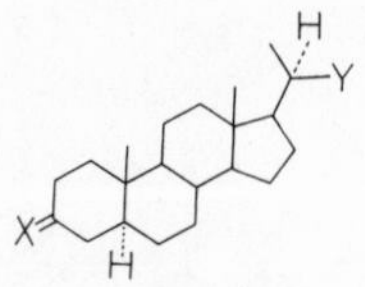

Structure	X	Y	Structure	X	Y
I	O	-OTMS	II	O	-OTMS
Ia	^{18}O	-OTMS	IIa	^{18}O	-OTMS
Ic	O	-O[^{2}H$_9$]TMS	IIb	O	-^{18}OTMS
III	-H$_2$	-OTMS	IIc	O	-O[^{2}H$_9$]TMS
IV	-NOCH$_3$	-OTMS			

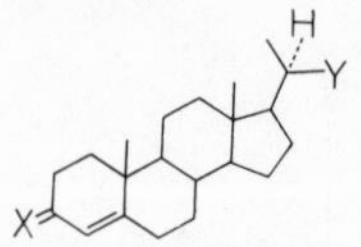

Fig. 32. Structures of analogs of 20β-hydroxy-5α- pregnan-3-one and 20β-hydroxy-4-pregnen-3-one referred to in Table 9.

active hydrogens in the mass spectrometer, in the gas chromatograph (Section 4.2), or *in vitro* prior to analysis. Deuterium may also be incorporated synthetically from a number of deuteriated intermediates [^{2}H$_6$]acetone, [^{2}H$_3$]acetic acid, [^{2}H$_4$]LiAlH$_4$, etc.) or by the preparation of derivatives. Deuterium labeling is particularly valuable in the investigation of site-specific hydrogen transfer reactions, which occur frequently in the mass spectra of organic compounds. Oxygen-18 is most readily obtained as H$_2$^{18}O and may be incorporated by exchange. Carbon-13 and nitrogen-15 are important in the investigation of mass spectrometric, chemical, and biochemical reactions. Their incorporation into organic compounds generally involves partial syntheses.

A simple example of the application of isotope labeling by the formation of derivatives is an investigation of a trimethylsilyl group migration during the mass spectrometry of TMS derivatives of 20β-hydroxy-5α-pregnan-3-one and 20β-hydroxy-4-pregnen-3-one (Fig. 32).

Spectra of these compounds were characterized by an intense ion at m/e 117, i.e., [CH$_3$CHOTMS]$^+$ (corresponding to cleavage adjacent to the trimethylsilyloxy group), no molecular ion, and few other ions of significant abundance. However, each spectrum included ions at [M−44]$^{+\cdot}$ and [M−59]$^+$ which appeared to correspond to the loss of CH$_3$CHO and CH$_3$CHO + CH$_3\cdot$ with retention of the TMS group. Isotope labeling of the 3-keto and 20-hydroxy oxygen atoms by ^{18}O, and of the TMS group with [^{2}H$_9$], was carried out to investigate these ions, and the mass spectrometric results are summarized in Table 9.

Oxygen-18 was incorporated at the 3-position in the two substrates by acid-catalyzed exchange in H$_2$^{18}O. Mass spectra of the derived TMS ethers (Ia and IIa) exhibited fragment ions at masses corresponding to

$[M-44]^{+\cdot}$ and $[M-59]^+$, indicating that the C-3 oxygen atom was not involved in these fragments. ^{18}O-labeling of the 20-hydroxyl group was achieved by acid-catalyzed exchange of the ketonic oxygen atom in 3β-hydroxy-5-pregnen-20-one, followed by lithium aluminum hydride reduction of the 20-^{18}O-ketone and selective enzymic oxidation of the Δ^5-3β-hydroxy group to the corresponding Δ^4-3-ketosteroid, by means of cholesterol oxidase. The TMS ether of this compound (IIb) showed masses of the fragments of interest corresponding to $[M-46]^{+\cdot}$ and $[M-61]^+$. The losses of ^{18}O demonstrate that the eliminated CH_3CHO moiety arises from the 20-keto oxygen atom together with C-20 and C-21, with concomitant migration of the trimethylsilyl group to the charge-retaining fragment. This was further confirmed by the preparation of the $[^2H_9]TMS$ derivative obtained with N,O-bis-$[^2H_9]$trimethylsilylacetamide. The 20β-O-$[^2H_9]TMS$ derivatives of compounds Ic and IIc showed base peaks shifted to m/e 126 $[CH_3CHO[^2H_9]TMS]^+$, and the fragment ions appeared at masses corresponding to $[M-44]^{+\cdot}$ and $[M-59]^+$ with the expected mass shifts of 9 amu. The 3-keto group is implicated as a terminus for the TMS migration by the absence of significant ions at $[M-44]^{+\cdot}$ and $[M-59]^+$ in 20β-O-TMS-5α-pregnane (II) and the analogous 3-methyloxime (III).

5. Exercises

1. A number of factors are considered in choosing a column for gas chromatography. Which is the most important for GC–MS?

2. Compare the various types of molecular separator used to interface a gas chromatograph and a mass spectrometer.

3. Why do we frequently use derivatives for GC–MS analyses?

4. The use of stable isotopes is discussed in Section 4.8. Can radioisotopes be used also?

5. Name one analytical technique which is more sensitive, selective, and versatile than GC–MS.

6. Suggested Reading

Gas Chromatography

L. S. ETTRE AND A. ZLATKIS, *The Practice of Gas Chromatography*, Interscience Publishers, New York, 1967.

H. M. McNAIR AND E. J. BONELLI, *Basic Gas Chromatography*, Varian Aerograph, 1969.

L. S. ETTRE, *Open Tubular Columns in Gas Chromatography*, Plenum Press, New York, 1965.

M. NOVOTNY AND A. ZLATKIS, Glass capillary columns and their significance in biochemical research, *Chromatogr. Rev.* **14**, 1–44 (1971).

Combined Gas Chromatography–Mass Spectrometry

W. H. McFADDEN, *Techniques of Combined Gas Chromatography–Mass Spectrometry: Applications in Organic Analysis,* Wiley-Interscience, New York, 1973.

J. T. WATSON, Mass spectrometry instrumentation, in *Biochemical Applications of Mass Spectrometry* (G. R. Waller, ed.), Wiley-Interscience, New York, 1972, pp. 23–49.

J. T. WATSON, *Introduction to Mass Spectrometry,* Raven Press, New York, 1976.

A. M. LAWSON AND G. H. DRAFFAN, Gas-liquid chromatography–mass spectrometry in biochemistry, pharmacology and toxicology, in *Progress in Medicinal Chemistry,* Vol. 12 (G. P. Ellis and G. B. West, eds.), North-Holland Publishing Company, Amsterdam, 1975, pp. 1–103.

C. J. W. BROOKS AND B. S. MIDDLEDITCH, Gas chromatography–mass spectrometry, in *Mass Spectrometry,* Vol. 5 (R. A. W. Johnstone, ed.) (Specialist Periodical Reports), The Chemical Society, London, 1979.

C. J. W. BROOKS (ed.), *Gas Chromatography–Mass Spectrometry Abstracts,* Vol. 10, P. P. R. M. Science and Technology Agency, London, 1979.

4

Selective Ion Monitoring

Edward C. M. Chen

1. Introduction

In Chapter 3, examples of the use of GC–MS were given in which mass spectra were recorded for individual components of mixtures separated by a gas chromatograph. The mass spectrometer may also be used as a specific detector for the gas chromatograph by tuning it to detect ions of a single m/e value. With minor modification to the instrument it is possible to continually record the abundances of ions of more than one m/e value by switching between them. This technique, under many guises, has been under parallel development almost since the inception of GC–MS in laboratories around the world. It is not surprising, then, that a multiplicity of terms has been coined to describe the method and its products. The term used in the chapter is "selective ion monitoring" (SIM) and the product is referred to as a "SIM profile." Other versions which have appeared in the literature include selective/selected/specific/single/multiple ion/mass monitoring/detection (there are 20 combinations there!) as well as combinations such as "multiple specific ion monitoring."

A related technique is one in which full mass spectra are recorded repetitively throughout a chromatogram, and a SIM profile is then reconstructed by computer. The term used by *Chemical Abstracts* to describe this technique is "mass fragmentography," with the product known as a "mass fragmentogram." This terminology is strictly inappropriate when molecular ions are considered, so "reconstructed SIM profile" is used in this chapter.

Edward C. M. Chen ● School of Sciences and Technologies, University of Houston at Clear Lake City, Houston, Texas 77058

The mass spectrometer is certainly the most expensive of gas chromatographic detectors, but its specificity is unsurpassed. SIM can also be performed under conditions which lead to greater sensitivity than any other detector.

At this point it is appropriate to define several terms related to the performance of SIM:

Noise. Long-term noise is a periodic excursion of the baseline which, in general, does not interfere with the detection of a chromatographic peak. Short-term noise is a relatively fast excursion of the signal within a limited range. The latter type of noise limits the detection of a peak.

Sensitivity. This is a measure of the response of a detector to a given compound. In a concentration-dependent detector the sensitivity is given by

$$S = \text{(peak area} \times \text{flow rate)/sample weight}$$

In a mass-flow-rate-dependent detector the sensitivity is given by

$$S = \text{peak area/sample weight}$$

Lower limit of detection. This refers to the smallest amount of a material which gives a signal distinguishable from that of the noise. An arbitrary signal-to-noise ratio of 2 is often used, although ratios of 5 or 10 are sometimes employed.

Linear dynamic range. The range of concentrations over which the sensitivity is constant.

Specificity. The relationship between the sensitivity to a given compound and the sensitivity to a potential interferent.

The simplest method of quantitation in GC is to use an instrument which is calibrated for each component of a mixture. The weight percent of a component (i) is given by

$$(\text{wt.\%})_i = 100 A_i S_i / \Sigma A S$$

where the A values are the areas of the GC peaks and the S values are the sensitivity factors for each compound. This procedure is used when all components of a mixture are known to be eluted from the column employed, such as in the analysis of an overhead distillate. More accurate quantitation can be effected for an individual component if a calibration curve is constructed for various concentrations of an authentic sample of the component, known as an "external standard." The most accurate results are obtained by coinjection of an "internal standard" with the sample, particularly if the standard is an isotopic analog of the sample (see p. 142).

When the GC–MS combination was first developed, the emphasis was placed upon the solution of the major shortcoming of gas chromatography, which is the lack of qualitative data. In 1961, however, early in the development of GC–MS, Henneberg employed SIM to characterize hydrocarbons eluting from a gas chromatographic column. The first biological application of SIM was reported by Sweeley, Elliot, Fries, and Ryhage in 1966. In the early 1970's, many SIM analyses of biological, medicinal, environmental, and industrial samples were reported, and the number of applications is expanding monthly. This is especially true since the advent of commercial GC–MS systems which have extended the sensitivities into the low picogram range. The discussion which follows is intended to give the reader a practical understanding of the procedure for developing analyses by SIM, so that he can contribute to the growth of this rapidly expanding field.

2. Operational Variables in Selective Ion Monitoring

The purpose of this section is to describe the specific equipment and operational variables pertinent to the SIM technique and to establish a basis for selecting the values of the adjustable variables in order to optimize the particular application. The use of SIM in quantitative analyses will also be compared with the use of other gas chromatographic methods. As before, it is convenient to consider three types of operational variables: the mass spectrometric, the gas chromatographic, and the combination. The combination variables are most easily discussed since they are essentially the same as for a standard GC–MS system. The key to the successful operation of such a unit is the separator or concentration device. These are described in more detail in Chapter 3. The property of such a device most important to SIM analysis is the lack of discrimination but, in general, there is nothing which can be modified short of changing the device. Another important point is that, in most separation devices, as much as 50% of the sample can be lost in the concentration device, so that an increase in sensitivity can be obtained by introducing the sample directly into the mass spectrometer. This has led to the development of GC–MS systems which can operate with relatively large flow rates of carrier gas entering the ion source, requiring an additional diffusion pump on the ion source, as when chemical ionization is used. The direct introduction of the effluent into the ion source eliminates the separator discrimination problem.

2.1. Mass Spectrometric Variables

When ions of a single m/e value are being monitored, there are no modifications that must be made to the mass spectrometer. However, as soon as multiple ions are being monitored, there must be additions to the mass analyzer and the recording section in order to accommodate the rapid switching and the multiple outputs. Since the major modifications are made in the analyzer section, this will be discussed first. There are three basic types of analyzers used in GC–MS systems: the magnetic sector, the quadrupole, and the time-of-flight instruments. The magnetic sector instruments are the most prevalent ones in laboratories today, but the quadrupole instruments are gaining in popularity. The time-of-flight instruments rank a poor third.

The basic equation describing the function of magnetic sector instruments is

$$m/e = R^2H^2/2V$$

where R is the radius of the sector, H is the magnetic field strength, and V is the accelerating voltage. The radius is fixed so that the m/e values must be varied by changing the voltage and/or the magnetic field. In one technique, the magnetic field is set to focus the ion of lowest m/e value of interest on the detector. Reduction of the accelerating voltage can bring ions of higher m/e value into focus. The range of m/e values is limited by the loss of sensitivity at lower accelerating voltages. Magnetic field switching has been accomplished by measuring the magnetic field and comparing it with a reference voltage corresponding to the preselected m/e value, but this is restricted by hysteresis of the magnet. Another technique developed for magnetic instruments is the combination of electric and magnetic switching where gross changes are made in the magnetic field and fine tuning is achieved with an auxiliary electromagnetic field. These modifications have resulted from the rapid growth of the SIM applications, and it is certain that further improvements designed to minimize the inherent limitations of the magnetic sector instruments will be forthcoming from the manufacturers of these instruments as the SIM applications continue to expand.

The separation of ions in the quadrupole instrument is obtained by varying an electric and a radiofrequency field applied to the four poles of the analyzer. Because these voltages can be varied rapidly and reproducibly, the quadrupole instrument is ideally suited for SIM analyses. Any ion within the nominal mass range of the instrument is accessible to SIM analysis. There are two ways of accomplishing this, one in which electronic hardware is used to switch to the various m/e values and send the signal to different output channels, and the other in which the ions are rapidly switched by a minicomputer which also stores or displays the data in real time. Up to as many as eight masses can be recorded simultaneously on some commercial instruments.

In the time-of-flight instrument, ions are alternately produced and accelerated and the m/e values are differentiated on the basis of their time of arrival at the detector. Thus the time-of-flight instrument is also well suited for SIM analysis since any portion of the mass spectrum can be monitored without altering from optimum the operating parameters of the instrument. This type of instrument has been used to monitor four ions simultaneously or to monitor 6 ions consecutively. Thus a maximum of 24 ions could be monitored sequentially by using 4 channels monitoring 6 ions each.

The simplest type of data collection system for SIM is a multichannel recorder. However, if successive scans of the mass spectra are to be recorded and a SIM profile obtained from these data at a later time, then a computer with disk storage is necessary. Some researchers have added this type of unit to existing GC–MS systems, and many of the new commercial instruments have this as an optional feature.

The ion source and the electron multiplier are the portions of a mass spectrometer which are the most likely to vary in performance from day to day. In addition, the ion source is the component which has the most adjustable variables. These variables are the same for SIM as for conventional mass spectrometry and, in general, will be the same values as for conventional mass spectrometry, except in the SIM applications which require very high sensitivity. The ion source can be an electron impact source, a chemical ionization source, a field ionization source, or a photoionization source. The major concern here is in the production of a relatively abundant ion in a reproducible and linear manner. The long history of research which has gone into the development of electron-impact ion sources for quantitative purposes supports their use in this case. In any analytical procedure, however, the reproducibility and linear dynamic range should be examined experimentally.

The variables which can be modified in the ion source to obtain greater sensitivity are the electron energy, the filament current, the source temperature, the source pressure, and the slit widths (for magnetic instruments) or peak widths (for quadrupole instruments). The optimum value of the electron energy is generally less than the standard 70 eV used for reproducibility (see p.136). An increase in filament current will increase sensitivity but decrease filament life. The temperature of the ion source will affect the fragmentation pattern of molecules. High pressures in the ion source will decrease sensitivity and induce ion–molecule reactions. An increase in peak width or slit width will increase sensitivity but decrease resolution. Probably the most significant factor in sensitivity is the cleanliness of the ion source and the detector. Contamination will generally reduce sensitivity markedly.

2.2. Gas Chromatographic Variables

In order to carry out effective GC–MS analyses, it is imperative that the chromatographic columns bleed very little. This restricts the type of columns which can be used and generally requires that the columns have liquid loadings in the range of 3% or less, and requires that the columns be well conditioned before use in the GC–MS system. In addition, the upper temperature limit for most columns is lower for GC–MS applications than for normal GC determinations. There is no restriction as to the type of column that can be used with GC–MS. Packed columns, SCOT columns, and capillary columns have all been used successfully. Both glass and metal columns have been used, but, in the case of trace analysis, glass is the preferred material in order to minimize adsorption losses.

In the applications of SIM which require high sensitivity, it is possible to sacrifice chromatographic resolution and compensate for this by using mass spectrometric selectivity. Thus the temperature of the column can be raised to give a sharper peak and a higher concentration at the peak if the compound being analyzed affords an ion at a characteristic *m/e* value. This would not be desirable if the characteristics of the compound are being determined for identification purposes. It is also possible to tolerate a greater amount of column bleed in these trace analyses so long as the bleed does not interfere with the ions being measured, although this will ultimately have the adverse effect of contaminating the ion source. In summary, almost any column developed for a specific gas chromatographic analysis can be used for SIM applications where the mass spectrometer is essentially a highly sensitive and selective detector.

3. Instrument Operation for Selective Ion Monitoring

The GC–MS system in use at the University of Houston at Clear Lake City is a Hewlett-Packard 5930A instrument. It consists of a HP-5700 dual column, temperature-programmed gas chromatograph with a membrane separator connecting it to the mass spectrometer, which is equipped with an HP-5933A data reduction system. The gas chromatograph does not have an auxiliary detector. The second column of the dual column instrument is usually connected directly to the mass spectrometer to introduce a chemical ionization reagent gas into the source. In this laboratory, however, this position has been used for capillary columns with helium as the carrier gas. The flow through the capillary columns is low enough to be allowed to enter the mass spectrometer directly. If chemical ionization is to be used, the reagent gas can be added to the effluent of the capillary column. The ion source is a dual electron-impact–chemical-ionization source which can be rapidly switched back and forth to either mode. In all other regards, the ion source is a typical quadrupole electron-impact source. The dodecapole mass analyzer is a modified quadrupole. Eight extra blades were added to the four basic circular rods in order to more nearly approximate the ideal hyperbolic surfaces. In addition, these blades can be used to compensate for changes on the surfaces of the rods due to contamination. The mass spectrometer is equipped with an electron multiplier and has an oscillographic recorder and a strip chart recorder for output. The hardware for multiple-ion recording directly from the mass spectrometer is not available on our instrument. The data system consists of an HP-2100S minicomputer with 16K of core memory. The minicomputer can also control the operation of the mass spectrometer. The system has a dual disk drive system, a paper tape reader, and a graphic display terminal. In addition, a high-speed plotter and a hard copy unit are available for ultimate data output.

3.1. Monitoring of a Finite Number of Chosen Ions

To analyze compounds present in the concentration range of parts per trillion, it is necessary to monitor a small number of ions. The ultimate limit of detection, which is in the picogram range, will be attained only if a single ion is measured but, in order to have a degree of certainty in identification, at least two ions should be monitored. To construct a total mass spectrum from the repeated application of this process, at least four ions at a time should be monitored with an overlapping ion in each cluster. In addition to the great sensitivity of this mode of operation, there are two other major advantages. First, there is no need to decide what sort of baseline correction is needed, and, second, there is a continuous output so that the operator can be assured that the analysis is proceeding properly. This is especially important when there is only a small amount of sample.

The Hewlett-Packard instrument has two programs for selective ion monitoring: SIM and MID. The former is more versatile than the latter. Both programs jump rapidly between *m/e* values, the SIM program having user-specified dwell times and the MID program having fixed dwell times. The SIM program records the data on a disk file and they are available for further data processing. It also allows for display of real-time chromatograms, permanent and temporary disk storage of raw and smoothed data, background subtraction, and GC peak area calculation with optional disk storage of the areas in a format compatible with further data processing routines. Only four *m/e* values can be monitored at any one time, but, in the SIM program, as many as five groups of four *m/e* values can be specified beforehand for use during the run. The operator can then choose any one of these groups of four *m/e* values at any time or the program can be used to start monitoring of a given group at a preset time. The dwell times can be varied from 50 to 32,767 msec. The data required for a routine run can be stored on a file for repeated use. Once the data have been obtained, areas can be calculated for the various GC peaks and can be normalized or stored for further data reduction. The major use of the MID program is for analytical procedures which have already been developed using the SIM program. Up to four *m/e* values are monitored sequentially and stored in temporary locations and are output on the Zeta plotter. The output scales are variable and are specified by the operator. The areas must be determined graphically and further data reduction done manually. This program is good only for the quantitation of a small number of GC peaks.

3.2. *Repetitive Scanning and SIM Profile Reconstruction*

In the original applications of GC–MS, the progress of the chromatogram was observed on a strip chart recorder, and a light beam oscillographic scan of the spectrum was initiated as the peaks of interest entered the mass spectrometer. As computers became less expensive and more available for use with GC–MS systems, the natural development was to scan the mass range repeatedly at a fixed time interval. The only step left then was to somehow reconstruct the SIM profile from the data stored on disk, which is a very easy problem for a minicomputer.

This technique can be used on both quantitative and qualitative analyses. It is possible to identify compound types by constructing the ion profiles of *m/e* values characteristic of a given compound type, such as *m/e* 91 for benzylic compounds (p.141). It is also possible by using this technique to obtain quantitative results for two compounds eluting simultaneously. The verification of analyses developed for other detectors can be carried out by this technique as long as the ultimate in sensitivity is not required. Thus the major difference in the two modes of operation is in degree and not in kind. In fact, if the number of *m/e* values being scanned is decreased to four, the repetitive scanning mode may approach the discrete ion mode in sensitivity.

To discuss the programs used in this mode of operation of SIM, we must consider both the programs designed to collect data and the programs designed to obtain the SIM profiles. There are two programs used to collect data, the MASYS program and the AQUIRE program. There are many features of these two programs used in other aspects of GC–MS applications which will not be discussed here. In the MASYS program, the *m/e* range and the rate of scan are specified. The data are stored in a specified disk file and are thus available for use in other programs. The AQUIRE program performs the same functions as the MASYS program as far as data acquistion is concerned, but has the additional capability of displaying on the output terminal two real-time SIM profiles. Three *m/e* values for these profiles can be specified beforehand and the operator can choose two for display at any time during the run. The program used to obtain SIM profiles from data stored on disk is the Reconstruct (RECON) program. The output device for this program is the Zeta plotter. As many as four *m/e* values can be specified by the operator, and the SIM profiles and TIC profile will be output. In addition, the retention time and scan number are plotted out. This allows the operator to then return to the MASYS program and obtain an output of the total mass spectrum. In the RECON program, different scale expansions can be specified for each *m/e* value. The profiles are normalized to the largest abundance of each of the specific ions. The areas of the various peaks may be obtained graphically or can be obtained from the TIC profile by using the MASYS program. Another program which can be used to obtain SIM files is the Spectrum Edit and Display program (SPEED). Only one profile at a time can be obtained by using this program. However, it has the advantage of being output on the display terminal, where peak areas can be obtained directly from the screen by using cursors. The execution of this program is quite rapid so that it is very valuable for survey applications of SIM. The SIM profiles are normalized to the largest abundance of the specific ion so that absolute concentrations must be obtained from the TIC profile, which is displayed along with the specific ion data.

4. *Procedures for the Development of SIM Analyses*

With the background just presented, it is possible to describe general procedures which are based upon the experience gained using the GC–MS system at the University of Houston at Clear Lake City. Basically, these procedures are described from the point of view of a novice in SIM analyses who has had considerable experience in the development of general analytical procedures, especially GC methods. In order to extend the scope of the applications beyond this laboratory, some examples given in the literature will also be presented.

It must be emphasized that these procedures are to be applied to the final analytical sample. There can and should be cleanup, derivatization, and/or concentration of the sample prior to analysis. The exact procedures for these operations will be discussed in the following section along with the survey of the SIM applications to various fields.

4.1. Analysis of Trace (Picogram) Quantities

The use of SIM in the quantitative and qualitative analysis of trace (picogram) quantities will be considered first. The major difference between this type of analysis and any other type of analysis using SIM is the extreme sensitivity which is required. Thus many of the steps in this procedure are just as applicable to the other types of SIM analyses.

The steps in the development of a procedure for the quantitative analysis of picogram amounts of material using SIM are as follows:

(i)   Decide upon a gas chromatographic column.
(ii) Obtain a mass spectrum of the compounds to be quantitated.
(iii) Determine the *m/e* value(s) to be monitored.
(iv) Choose and characterize a method of quantitation.
(v)   Run the sample on the GC–MS instrument using optimum conditions.
(vi) Check the quantitation procedure and calculate the results.

4.1.1. Hexachlorobenzene

The determination of hexachlorobenzene present in a waste stream will be taken as an example. In order to concentrate the organics, the aqueous stream was extracted with benzene and then the benzene was evaporated down to 0.5 ml. The concentration of hexachlorobenzene in the final sample was about 1–10 ppb, so that quantitation of picogram quantities was necessary.

The choice of a column was no problem in this case since the other organics did not have mass spectral ions that overlap the hexachlorobenzene ones, and since only one compound was being quantitated. Thus a 1 m × 3-mm-i.d. column containing 3% OV-101, with a 30-ml flow rate of helium, was used in order to get a sharp peak with a short retention time. In other cases, it may be necessary to use a capillary column to achieve a particular separation.

The mass spectrum of hexachlorobenzene was obtained by injecting a 100-ppb standard into the column and using the AQUIRE program. If a large sample of the pure material is not available, the spectrum would have to be taken from a reference collection and verified using a SIM procedure. It is always better to have the spectrum obtained on the particular instrument used in the analysis. The spectrum of hexachloro-benzene is given in Table 1 for the *m/e* range 100–300. If the ultimate sensitivity is not required, it would be valuable to monitor *m/e* 282, 284, 286, and 288. If, however, maximum sensitivity is required, then *m/e* 284 would be monitored continuously, as in this case.

For maximum sensitivity, the quantitation procedure of choice is the external standard method used in conjunction with a calibration curve. The calibration curve is prepared by injecting known amounts of the compounds over a range of concentrations and measuring the response. In addition, the lower limit of detection and the dynamic range can be determined. SIM profiles for *m/e* 284 are shown in Fig. 1 for the range of 10–10^5 pg. The dynamic range of 10^5 can be seen in Fig. 2, where the response is plotted versus the amount injected on a log–log scale. The unit slope is as expected and the deviations at the lower concentrations are probably due to loss of sample due to adsorption. The minimum detectable

TABLE 1.   Partial Mass Spectrum of
Hexachlorobenzene

m/e	Standard spectrum	Repeated SIM
	(percent of base peak)	
107	8.5	
115	6.7	
141	10.5	10
142	38.5	40
143	14.5	15
144	22.3	20
145	5.1	
146	4.3	
177	21.5	20
179	21.0	20
187	17.3	20
202	34.7	35
204	44.3	45
206	19.6	15
208	4.0	5
247	35.7	40
249	60.2	65
251	36.3	40
253	12.5	15
282	63.4	60
284	100.0	100
286	85.3	80
288	38.5	40
290	9.3	

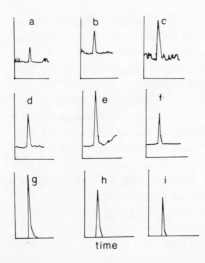

FIG. 1. SIM profiles (*m/e* 284) for hexachloro-benzene, using MID program: a, 10 pg (X10); b, 20 pg (X10); c, 40 pg (X10); d, 80 pg (X25); e, 100 pg (X25); f, 1 ng (X400); g, 2 ng (X400); h, 6 ng (X1,500); i, 20 ng (X6,000). See text for analytical conditions.

amount was 1.0 pg on column or 0.5 pg into the ion source. This latter figure was checked by introducing the sample directly into the ion source and bypassing the separator.

When a sample of unknown concentration is run on the instrument, at least one point of the calibration curve should be verified, preferably in the range of the unknown. In other words, the quantitation is actually based upon this external standard, and the calibration curve merely serves to establish the linear range. This is especially important when the sensitivity of the electron multiplier is varying, such as at the beginning or end of its life. The concentration of the unknown will be obtained from the ratio of the response of the unknown to the response of the standard.

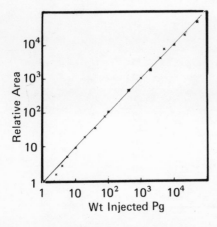

FIG. 2. Response curve for hexachloroben-zene. Data from Fig. 1 and additional analyses.

This can be corrected for nonlinearities in the extreme limit of detection, but the reproducibility in that range is often so poor that such corrections are within the limits of experimental error.

The use of SIM for qualitative analyses of trace components is merely the repeated application of the quantitative precedure. Because four *m/e* values are being monitored, the sensitivity will be lower by a factor of about 2. An example of the monitoring of four *m/e* values in the case of hexachlorobenzene is shown in Fig. 3. The scales for *m/e* 284 and 286 are one-half the scales for *m/e* 282 and 288. Another run, for example, would monitor *m/e* 247, 249, 251, and 282 so that the total mass spectrum could be constructed. This has been done and is given in Table 1, column 3. In the case of a very limited amount of material, a single scan with *m/e* 214, 249, 284, and 286 would suffice for a positive identification, since GC retention data would also be available for confirmation.

4.1.2. 2,4-D Butyl Esters

The analyses of trace quantities of several 2,4-D butyl esters was recently reported by Farwell *et al.* [*Anal. Chem.* **48,** 420 (1976)], using a SIM procedure on a HP 5930A GC–MS system. The limit of detection was in the 0.1–1.0 pg range, and the linear dynamic range was 10^4–10^5. In addition, a number of analyses were also carried out on both SIM and electron capture detection. For these particular compounds, the lower limit of detection for electron capture was in the 50-pg range. A plot of the electron capture results versus the SIM results is shown in Fig. 4. The ^{63}Ni electron capture results are in good agreement with the SIM results,

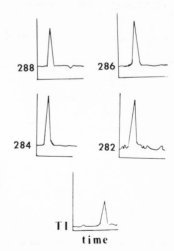

FIG. 3. TIC and SIM profiles (*m/e* 282, 284, 286, 288) for hexachlorobenzene, using MID program. 10 pg injection. Attenuation X2 for *m/e* 282, 288; X4 for *m/e* 284, 286. The mass spectrum obtained by "Repeated SIM" in Table 1 was obtained from these data and additional analyses at other *m/e* values.

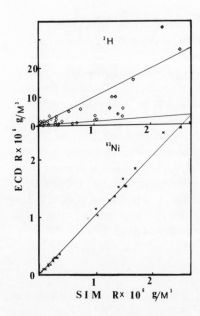

FIG. 4. Comparison of SIM response for *t*-butyl ester of 2,4-D with ^{63}Ni and ^{3}H electron capture GC responses.

but the ^{3}H electron capture results are quite spurious, often being a factor of 10 different. This could not have happened if an external standard were run with the unknown. This is an example of how SIM analyses can be used to check the reliability of other types of analyses. If SIM is to be used to check a nonchromatographic procedure, such as a fluorescence method, then a chromatographic procedure would have to be developed. However, this would still be worthwhile if the ultimate reliability of the original method is in question.

4.2. Analyses Using Reconstructed SIM Profiles

The original use of reconstructed SIM profiles was primarily for qualitative analyses as an aid to traditional GC–MS applications. Obviously, all of the applications for trace analyses can be applied to reconstructed SIM profiles as long as there is sufficient sensitivity. The maximum sensitivity is shown in the SIM profiles of hexachlorobenzene given in Fig. 5. A lower limit of detection is about 100 pg, which is about two orders of magnitude higher than in the trace analysis.

There are no particular instructions for the use of reconstructed SIM profiles since the data are acquired in the normal operation of a computer-controlled GC–MS system. For example, to use this procedure for functional group analyses, the only thing that is necessary is an ion characteristic of that functional group. An example of this is given in Fig.

6, where data from the chromatogram of a mixture of hydrocarbons are given. The aromatic compounds can be identified by the ion of *m/e* 91, the tropylium ion, and the aliphatic compounds are identified by the ion of *m/e* 57. This technique can be extended to any functional group to take full advantage of the selectivity of SIM.

4.2.1. Overhead Distillate Impurities

The reconstructed SIM profile can also be used to identify specific impurities in a rather pure compound. Here the mass spectrum of the major component must be known or, at least, it must be known that ions of certain *m/e* values are not in the spectrum of this compound so that, on the basis of the presence of other ions, the nature of the impurities can be inferred. An example is given in Fig. 7. An overhead distillate which is 98% pure has been chromatographed and subjected to SIM analysis. The

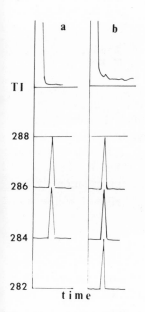

FIG. 5. Reconstructed TIC and SIM profiles (*m/e* 282, 284, 286, 288) for hexachlorobenzene, using AQUIRE and RECON programs. Injections of 100 pg (a) and 400 pg (b). Scans of *m/e* 50–300 in 10 sec.

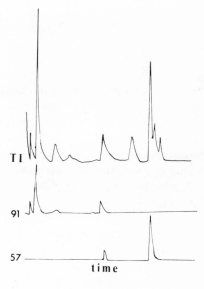

FIG. 6. Reconstructed TIC and SIM profiles (*m/e* 57, 91) for a hydrocarbon mixture, using AQUIRE and RECON programs. Scans of *m/e* 25–200 in 10 sec. 6 ft × ⅛-in.-o.d. column containing 3% Carbowax 20M, programmed from 100–200°C at 4°C min⁻¹, with carrier-gas flow rate of 30 ml min⁻¹.

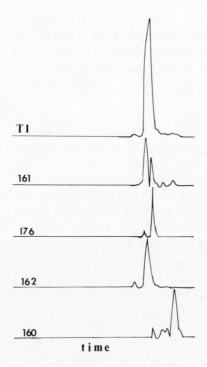

T I

161

176

162

160

t i m e

FIG. 7. Reconstructed TIC and SIM profiles (m/e 160, 161, 162, 176) for an overhead distillate containing 98% of a component with molecular weight 162, using AQUIRE and RECON programs. Scans of m/e 25–200 in 10 sec. Conditions as for Fig. 6, except column temperature isothermal at 170°C.

molecular weight of the compound is 162 and the majority of the remaining 2% of the sample is an isomer also of molecular weight 162. By using the RECON program with m/e values of 160, 176, and 161, it is possible to identify impurities affording ions of m/e 160 and 176. The loss of 15 amu from these ions to give fragment ions with m/e values of 145 and 161 indicates the presence of methyl groups. With this information, it was possible to develop a capillary column analysis of this stream with a flame ionization detector.

4.2.2. Isotope Labeling

The use of reconstructed SIM profiles for quantitative analyses is analogous to the use of any other gas chromatographic detector. However, there is one particular advantage of SIM which is overwhelming, that is, the ability to use isotopically substituted compounds as the internal standards, since the mass spectrometer can distinguish between such compounds. These compounds can be added to the system in rather large quantities and can even be carried through the derivatization or concentration procedure. The isotopes suitable for use are those of hydrogen, carbon, nitrogen, and oxygen. The most frequently used isotope is deuterium. Another important factor in the use of an isotopically labeled internal standard is the fact that the degree of ionization for the two compounds should be identical so that the relative response factor is unity. An application of this technique is shown in Fig. 8. The compound is 1,5-diacetyl-3,4,6-tri-*O*-methylaldononitrile. The isotopically labeled compound is the 1,5-dideuterioacetyl aldononitrile derivative. The attenuations of the ions of m/e 129 and 132 are the same.

Another example of the use of an isotopically substituted compound as an internal standard for quantitative analysis is the determination of the concentration of the peracetylated mannonitrile using an isotope of nitrogen, ^{15}N. A SIM profile for these compounds is shown in Fig. 9. This example emphasizes the need to correct the response of the internal standard for any interfering peaks. In this case, the molecular ion of the

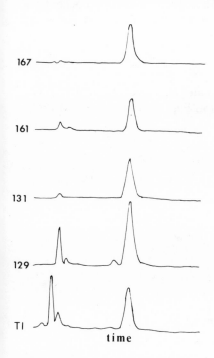

FIG. 8. TIC and SIM (m/e 129, 131, 161, 167) profiles for 1,5-diacetyl-3,4,6-tri-*O*-methylaldononitrile and the dideuterioacetyl analog, using MID program. 4 ft × ⅛-in.-o.d. column containing 3% neopentylglycol succinate, isothermal at 230°C, with carrier-gas flow rate of 60 ml min⁻¹.

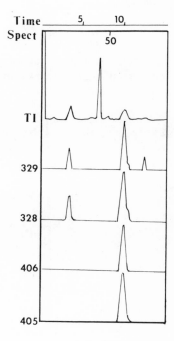

FIG. 9. Reconstructed TIC and SIM (m/e 328, 329, 405, 406) for peracetylated mannonitrile and the ^{15}N analog, using AQUIRE and RECON programs. NH₃ chemical ionization. Scans of m/e 50–450 in 20 sec.

¹⁵N compound and the normal isotopic [M+1] ¹³C peak overlap. Thus the response of the internal standard must be corrected for the natural abundance of ¹³C, which is 17.6%, based upon the normal peak being the base peak. This example also illustrates the use of a different type of ionization, ammonia chemical ionization. The 70-eV EI spectrum of peracetylated mannonitrile is compared to the ammonia CI spectrum in Fig. 10. The CI spectrum contains a ¹⁵N internal standard. The ion of *m/e* 405 corresponds to the [M+18] peak and the ion of *m/e* 328 corresponds to the [M+1−60] peak due to the loss of acetic acid from the [M+1] ion. These two ions represent over 90% of the total ionization so that it is possible to achieve better selectivity and sensitivity by using the chemical ionization mode. The ion of *m/e* 406 is due to the internal standard, and the relative values corrected for the natural isotope of carbon can be used for quantitation. This assumes that the concentration does not change as the mass spectrometer scans these two *m/e* values, which should certainly be true.

4.2.3. Other Internal Standards

Another type of analysis uses a naturally occurring material as an internal standard. An example is given in Fig. 11, where the SIM profiles

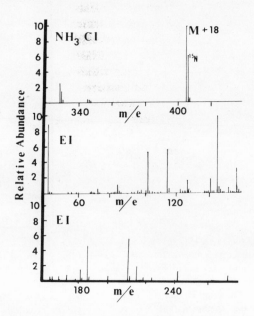

FIG. 10. NH₃ chemical ionization spectrum of a mixture of peracetylated mannonitrile and the ¹⁵N analog (upper) and 70-eV electron-impact spectrum of mannonitrile alone (lower).

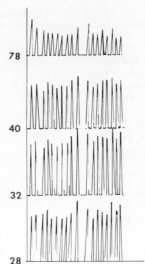

FIG. 11. SIM (*m/e* 28, 32, 40, 78) profiles for laboratory air, using MID program: *m/e* 78 at attenuation X1, others at attenuation X10⁴.

for benzene, argon, oxygen, and nitrogen are shown for multiple injections of 2 ml of air from a laboratory where benzene was being used as a solvent. The relative responses for benzene and argon were determined previously so that the benzene concentration could be determined by using argon as an internal standard. There is no separation of the components on the chromatographic column. This could be done with a flame ionization detector but an internal or external standard would have to be used.

4.3. Additional Considerations

The overhead distillate sample shown in Fig. 7 can also serve as an example of a normalization procedure for quantitation. The one prerequisite for this type of quantitation is that each component elutes from the chromatographic column and gives a characteristic response. This is generally true for any overhead distillate. If the relative responses of the components are not known then they can be assumed to be unity. This will generally not make much difference in the value of the major component.

The quantitative calculations which have just been discussed are summarized in Table 2. In general, the results using the SIM procedure are in agreement with the known concentrations of the components. In addition, all of the pertinent data for the calculations are given so that they can be repeated to insure an understanding of the procedure. However, it must be emphasized that SIM analyses should only be utilized if the extreme sensitivity and/or selectivity are absolutely necessary.

TABLE 2.   Quantitative Analyses with Selected Ion Monitoring

Analysis	Response Area	Response Height	Concentration (calculated)	Concentration (known)	Type
Aldononitrile					
m/e 129	7.60		2.38%	2.40%	Deuteriated int
m/e 132	4.75			4.80%	nal standa
m/e 161	7.75		2.43%		
m/e 167	4.75				
Mannonitrile					
m/e 338		1.50	1.50%	1.50%	N-15 internal
m/e 339		0.60		0.60%	standard
m/e 405		6.00	1.44%		
m/e 406		2.50			
Overhead distillate					
m/e 162 (meta)		0.10	1.60%		Area normal-
m/e 162 (para)		6.00	97.50%		ization
m/e 161 (MW=176)		0.01	0.20%		
m/e 145 (MW=160)		0.70	0.70%		
Benzene in air					
m/e 78		1.7×10^{-5}	10 ppm		Natural intern
m/e 40		3.5			standard

With high-resolution mass spectrometers, it is possible to obtain resolutions of 10,000–20,000 so that the exact elemental composition of ions with the same nominal m/e value can be determined. This extreme selectivity combined with the normal sensitivity of SIM represent the ultimate sophistication. Of course, the increase in cost and complexity makes the justification of the use of SIM, if at all, even more critical.

5.   Applications of SIM to Other Fields of Study

There are essentially no limitations as to the types of volatile compounds which can be analyzed by SIM. There are many different applications which have been reported in the literature, and it is well beyond the scope of this chapter to review each of these in detail.

The first extensive applications of SIM were in the medical, biomedical, and biochemical areas. SIM analyses in these areas cover all of the general types discussed earlier. One of the major areas of quantitative analysis is the determination of monomer sequences in biopolymers such as peptides, nucleotides, carbohydrates, and complex lipids. Studies of the monomers and small polymers have been carried out. In general, derivatives must be prepared in order to increase the volatility but some underivatized compounds have been studied. In general, isotopically substituted compounds using deuterium, ^{15}N, and ^{18}O have been employed as internal standards. Both electron impact and chemical ionization have been used. Ammonia chemical ionization is especially useful for multi-functional compounds. A number of studies have been carried out on feces, blood, urine, and other body fluids. In addition, GC–MS profiles of these are used to characterize various disorders. The use of SIM for functional group analysis adds a third dimension to this type of study.

The applications of SIM analyses to environmental problems is quite logical since the major pollutants are volatile organic compounds. Almost every type of SIM analysis has been carried out in this area. Aqueous samples from rivers, the ocean, sewage, and drinking water and all types of air samples have been studied. The types of compounds that have been studied include pesticides, herbicides, toxic impurities in herbicides, chlorinated hydrocarbons, freons, phthalates, aromatic and aliphatic hydrocarbons, combustion products, carcinogens, and the metabolic and photochemical degradation products of these. The use of isotopically labeled internal standards is not as common in this field. Chemical ionization, positive and negative electron impact, and atmospheric pressure ionization have all been utilized. One of the major problems in these studies is preconcentration. The large number of compounds present in these samples makes functional group analysis with SIM again very valuable.

The application of SIM to pharmaceuticals is closely related to the medical and biological applications, except that very specific compounds are being studied so that quantitative analyses play a greater role in the pharmaceutical studies. The types of compounds that have been studied include street drugs, antidepressants, hypnotics, barbiturates, anticonvulsants, antiepileptics, antitumor agents, anesthetics, antibiotics, analgesics, cardiac glycosides, antihypertensive drugs, narcotic antagonists, antimalarial drugs, sedatives, and muscle relaxants and their degradation products. One important type of analysis is to determine the nature and amount of drugs taken in overdoses. Another important type of study is that of the concentration and nature of drugs and their metabolites as a function of time. Deuteriated and ^{13}C and ^{15}N derivatives are routinely used as internal standards. Chemical ionization, electron-impact ionization, and atmospheric pressure ionization have been used in these studies.

Other applications include the study of foods, flavors, and fragrances, the study of petroleum and alternate sources of energy such as coal and shale oil, organic geochemistry, and the application of SIM analyses to the field of industrial analysis. Here again, all types of SIM analyses have been described. Specific quantitative analyses are important in the food, flavor, and fragrance areas, the study of organic geochemistry, and the development of industrial processes. Also, the use of functional group analysis is quite significant for all of these areas since, in general, complex organic mixtures are encountered. The use of high-resolution mass spectrometry has particular application to the study of coal and shale oil. Electron-impact ionization is the predominant mode of ionization for studies in these areas.

6. Summary

This chapter can be best summarized by tracing the path by which SIM analyses have gone through the "Seven Ages of an Analytical Procedure" as presented recently in an editorial by Laitinen [*Anal. Chem.* **45**, 2305 (1973)].

The first stage is conception, which, in the case of SIM, is rooted in the foundations of gas chromatography and mass spectrometry. The end of this stage is best represented by the birth of the separators following the marriage of GC and MS systems in the early 1960s.

In the second stage, experimental measurements are made to establish the validity of the principle. This stage was very short because of the vast GC and MS heritage.

The third step is characterized by the development of instrumentation to bring the method into the hands of the nonspecialist. The milestone for SIM analyses in this stage is the development of the integrated GC–MS and data system at an achievable price. This occurred in the late 1960s and early 1970s with the development of several generations of systems designed strictly for GC–MS applications.

The fourth phase is the detailed study of the principles and mechanisms using the improved instrumentation in order to gain general acceptance for the technique. In the case of SIM analyses, the advantages of sensitivity and selectivity were so obvious that this stage was essentially bypassed.

In the fifth age, the technique is applied to an ever widening scope of areas with appropriate modifications in procedure. The SIM procedures are well into this phase as evidenced by applications in the medical and biomedical, pharmaceutical, environmental, industrial, geochemical, and flavor and fragrances fields.

The sixth stage is the application of well-established procedures to new as well as old problems. The SIM analyses are at this stage as evidenced by the standard procedures for drugs and pollutants that are being established by various governmental agencies using SIM and by the fact that many applications of SIM are no longer being published as separate items but are being incorporated as a portion of the experimental procedure. This chapter is, in fact, testimony to the arrival of SIM analyses at this stage of development.

The final stage is a period of senescence when other methods of greater speed, economy, convenience, sensitivity, or selectivity, etc., surpass the method under consideration. This situation is not even in sight for SIM. The factors of speed and convenience are generally equivalent for SIM analyses. The only weak points for SIM. are economy and maintenance. Thus, if these factors are improved, as they have been in the past, SIM analyses will be the ultimate analytical method for volatile compounds, until a better method passes through the "Seven Ages."

7. Exercises

1. Compare the sensitivities attainable by monitoring a single *m/e* value throughout a chromatogram and by repetitive scanning and reconstruction of SIM profiles.

2. In GC–MS we normally generate a "total-ion" chromatogram and can produce "single-ion" SIM profiles. Are there occasions when it would be useful to obtain a chromatogram derived from ions within a limited *m/e* range?

3. Discuss the factors which lead to the selection of internal standards for SIM.

4. Discuss the factors which lead to the selection of appropriate ions for trace analysis by SIM.

5. In analyzing a mixture of fatty acids as methyl esters, which ions would be most useful in SIM for the selective detection of:

   (a) all saturated straight-chain methyl esters, and

   (b) methyl stearate?

8. Suggested Reading

E. COSTA AND B. HOMSTEDT (eds.), *Advances in Biochemical Pharmacology,* Vol. 7, Raven Press, New York, 1973.

J. W. EICHELBERGER, L. E. HARRIS, AND W. L. BUDDE, *Anal. Chem.* **46**, 227 (1974).

F. C. FAULKNER, B. J. SWEETMAN, AND J. T. WATSON, *Appl. Spectrosc. Revs.* **10**, 51 (1975).

C. FENSELAU, *Appl. Spectrosc.* **28**, 305 (1974).

B. HOLMSTEDT AND L. PALMER, *Adv. Biochem. Psychopharmacol.* **7**, 1 (1973).

M. G. HORNING, J. NOWLIN, K. LERTRATANANGKOON, R. N. STILLWELL, W. G. STILLWELL, AND R. M. HILL, *Clin. Chem.* **19**, 845 (1973).

B. S. MIDDLEDITCH AND D. M. DESIDERIO, *Anal. Chem.* **45**, 806 (1973).

Concentration Techniques for Volatile Samples

Albert Zlatkis and Henry Shanfield

1. Introduction

The direct analysis of trace concentrations of many organic compounds stretches the very limits of, or completely defies, even the most sensitive analytical equipment. For many areas of research, substances of interest are often at or below the part per billion level.

A good example is the isolation and identification of compounds which have profound physiological effects at very low concentrations. Another growing area of interest is the analysis of volatiles from biological fluids (e.g., urine) to seek distinctive differences between "normals" and those afflicted by disease. More recently, the awareness has grown that minute concentrations of chemical pollutants can have far-reaching effects as health hazards, further underscoring the need for reliable analytical techniques. Flavor and odorant studies should also be considered.

At the part per billion level or lower, it is almost always necessary to use some cumulative or concentrating technique to obtain measurable amounts of sought-after compounds. Ideally, we wish to eliminate as much as possible of unwanted "background" compounds (usually water or air) while accumulating the desired substances quantitatively. For most techniques, the result is a compromise of these two goals. Several approaches come to mind readily: fractional distillation, freeze concentration, zone

Albert Zlatkis and Henry Shanfield • Department of Chemistry, University of Houston, Houston, Texas 77004

melting, solvent extraction, chromatographic techniques, and adsorption. Recent work in using adsorbents is summarized in this chapter.

2. Selective Adsorption

A wide variety of adsorbents have been investigated as selective trapping materials for trace organics. These range from strong adsorbents like activated charcoal to relatively weak polymeric adsorbents. Desorption from strong adsorbents involves either solvent extraction or heating to temperatures which often result in chemical changes. Weak adsorbents, on the other hand, will allow many desired substances to escape.

For most analytical situations it is convenient to sample at ambient temperatures, without the necessity for cold traps. We prefer to use adsorbents which are efficient collectors for many compounds of interest at room temperature, and which do not introduce chemical change or artifacts during thermal desorption.

2.1. Selective Adsorption Using Tenax-GC Polymer

Tenax-GC, a porous polymer of 2,6-diphenyl-*p*-phenyl oxide, has proven to be a versatile selective adsorbent. It has a modest surface area of 18 m^2 g^{-1} and has a low retention for low-molecular-weight polar substances, especially water. Higher-molecular-weight compounds, having relatively low polarity, are trapped and desorbed thermally with high efficiency. This is illustrated in Table 1, which lists the trapping capacity and desorption characteristics of Tenax-GC for various organic compounds. This is evaluated by injecting a fixed amount of each compound on the Tenax-GC and eluting the trap with 0.5, 1.5, and 5.0 liters of pure nitrogen. The amount retained is determined chromatographically by desorption at 300°C. At this temperature Tenax-GC does not contribute detectable artifacts, due to its unusual thermal stability.

Tenax-GC is like any other chromatographic stationary phase and must be evaluated from the point of view of partitioning of a compound between adsorbent and carrier gas. As a consequence, the results in Table 1 apply to the specific amount of Tenax-GC employed in these experiments (0.28 g). That is to say, the "breakthrough" volume is directly proportional to the amount of adsorbent. Table 1 shows that for molecules with molecular weights as high as that of *n*-octadecane, adsorption and desorption are reversible and complete. Low-molecular-weight polar compounds (e.g., methanols, ethanol, and acetone) have low retention and relatively small breakthrough volumes. Water is not noticeably adsorbed.

TABLE 1. Recovery (%) of Organic Compounds from Tenax-GC After Adsorption at Room Temperature

Compound	Volume of N_2 passed through trap (liters)		
	0.5	1.5	5.0
Methanol	1	0	0
Ethanol	1	0	0
Methyl chloride	3	1	0
Acetone	68	2	0
Chloroform	100	84	5
Diethylamine	80	50	1
Isobutanol	100	95	16
n-Pentane	100	50	9
Cyclohexane	100	50	9
n-Hexane	100	100	20
Ethyl acetate	100	100	35
n-Butanol	100	100	35
Toluene	100	100	100
C_7–C_{12} alkanes, alkenes	100	100	100
Styrene, ethylbenzene	100	100	100
Xylenes	100	100	100
Pyridine	100	100	100
Chlorophenols	100	100	100
n-C_{13},C_{14},C_{15},C_{16},C_{17},C_{18} alkanes	100	100	100

Figure 1 illustrates a typical system for trapping compounds from ambient air or any other source of volatile organics. The air or other carrier gas is drawn through the trapping tubes, which may be arranged in parallel or series. Typically, a sampling tube will be of about 4-mm i.d. and 12-cm length. One-quarter gram of Tenax-GC occupies a length of about 9 cm, and is held in place by two plugs of glass wool. Flow rates are maintained at about 30 ml min^{-1} to allow satisfactory contact time with the Tenax-GC. For many applications a total of 5 liters of air or carrier gas

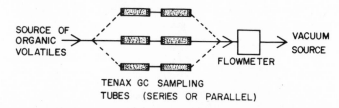

FIG. 1. Schematic diagram of typical system for trapping organic volatiles with Tenax-GC.

is adequate for 0.25 g of adsorbent, where substances are present in the gas at the fraction of a part per million level.

Desorption is carried out at 300°C by rapidly heating the Tenax-GC trap while passing an inert gas through it, then cold-trapping the volatiles ahead of a chromatographic column. Subsequently, the cold trap is flash heated and the contents analyzed by chromatographic techniques.

Selected examples of the application of this technique will be presented.

2.1.1. Air Pollution Studies

Tenax-GC porous polymer adsorbent has been successfully applied to air pollution studies where it offers convenience and simplicity. Figure 2 illustrates the system employed with regulating devices, and the sampling tubes containing Tenax-GC adsorbent. Sampling tubes were designed to interface with the injection port of gas chromatographic equipment, as is shown in Fig. 3.

The tubes containing the adsorbent were preconditioned at 350°C in a stream of pure nitrogen, then allowed to cool in screw-capped test tubes. Fig. 3 also shows the heater surrounding the Tenax-GC tube for desorption of volatiles at about 300°C into a cryogenic precolumn trap.

Analysis was carried out in a high-resolution capillary column. Several hundred compounds were observed, of which nearly one hundred were identified by mass spectrometry. Figure 4 is a typical chromatogram of a sample of 200 liters of urban air. The reproducibility for successive samplings is excellent.

2.1.2. Volatiles of Biological Origin

The analysis of volatile constituents derived from biological sources is a continuing and extensive area of medical research. Tenax-GC has

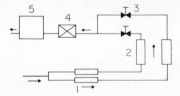

FIG. 2. Schematic diagram of trapping system components: 1, traps; 2, rotameters; 3, needle valves; 4, pump; 5, flowmeter.

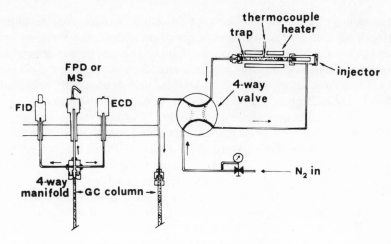

FIG. 3. Injection, analytical, and detection systems (FID, flame ionizaton detector; FPD, flame photometric detector; ECD, electron-capture detector).

proven to be a versatile adsorbent for trapping such volatiles, particularly those above C_6 in molecular weight. These compounds are commonly "buried" in a background of water, for which Tenax-GC has no affinity.

The volatile metabolites in urine have been examined using this adsorbent, with interest centering on the disease *diabetes mellitus*. One

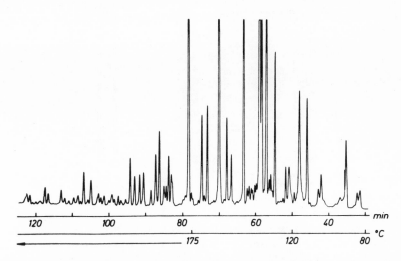

FIG. 4. Partial chromatogram of volatile organic compounds in Houston, Texas air.

common approach used has been the so-called "head-space" technique. A stream of inert gas is passed over the urine sample, the volatiles being swept out with the vapors above the liquid. The carrier gas then passes through a Tenax-GC adsorbent tube. This is subsequently desorbed at 300°C. The volatiles are trapped cryogenically, then ultimately analyzed in a gas chromatograph. Figure 5 shows the system schematically.

In a particular investigation, a pooled "normal" urine sample was divided into ten 200-ml portions and refrigerated at 4°C for storage. The volatiles from each aliquot were then collected for 1 hr on Tenax-GC adsorbent as previously described, the urine being kept at 100°C (see Fig. 5). The analytical results for three such samples analyzed three days apart are shown in Fig. 6. The reproducibility is quite good. Subsequent data have shown that the same reproducibility is achieved over a period of weeks when the Tenax-GC sample tube is stored at room temperature in screw-capped tubes. The same technique has been applied to other situations, e.g., human breath, and serum. A highly advantageous aspect of Tenax-GC sampling is the flexibility of shipment and storage of samples taken at diverse geographical locations. Interlaboratory comparisons are thereby greatly facilitated.

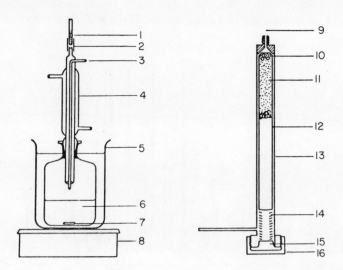

FIG. 5. Head-space volatiles sampler and Tenax-GC trap adapted for chromatography: 1, trap insert with Tenax-GC; 2, Teflon sleeve; 3, helium inlet; 4, condenser; 5, boiling water bath; 6, sample; 7, magnetic stirring bar; 8, hot-plate stirrer; 9, precolumn inlet; 10, silanized glass wool; 11, Tenax-GC; 12, trap insert; 13, modified injection part; 14, spring;

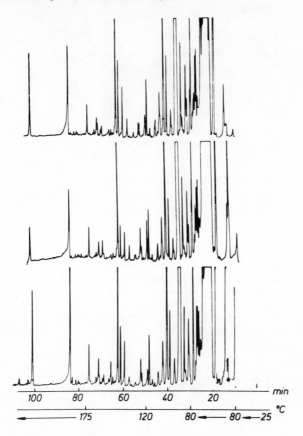

FIG. 6. Volatile organic compounds in a single sample of normal human urine, analyzed on three consecutive days.

2.1.3. Miscellaneous Applications

Two other examples of the application of Tenax-GC adsorbent will be noted here. The first of these is a study of exhaled tobacco smoke. Standard reference cigarettes were smoked under the standard conditions used by industry, and 3.5 liter air samples containing the exhaled smoke were passed through Tenax-GC adsorbent. For comparison, cigarette smoke samples were taken directly with a simple sampling device shown in Fig. 7. The sampling tubes were desorbed in the usual manner, and the gases analyzed by GC–MS. The reproducibility of sample profiles were surprisingly good.

Another area of research to which Tenax-GC has been applied is that

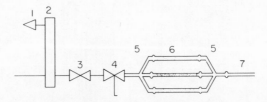

FIG. 7. Sampler for a small fraction of a standardized puff of fresh cigarette smoke: 1, vacuum source; 2, flowmeter; 3, shutoff valve; 4, fine metering valve; 5, flow dividers and sample tube holders; 6, adsorbent tubes and bypass tube; 7, cigarette.

TABLE 2. Recovery (%) of Organic Compounds from Carbopack B After Adsorption at Room Temperature

	Volume of N_2 passed through trap (liters)		
Compound	0.5	1.5	5.0
Methanol	3	0	0
Ethanol	3	0	0
Methyl chloride	1	0	0
Acetone	5	1	0
Chloroform	100	54	1
Diethylamine	100	80	50
Isobutanol	100	100	25
n-Pentane	100	100	100
Cyclohexane	100	100	100
n-Hexane	100	100	100
Ethyl acetate	100	100	100
n-Butanol	100	100	100
Benzene	100	100	100
Toluene	100	100	100
C_7–C_{12} alkanes, alkenes	100	100	100
Styrene, ethylbenzene	100	100	100
Xylenes	100	100	100
Pyridine	100	100	100
n-C_{13} alkane	65	65	65
n-C_{14} alkane	46	46	46
n-C_{15} alkane	25	25	25
n-C_{16} alkane	8	8	8
n-C_{17} alkane	1	1	1
n-C_{18} alkane	0	0	0

of determining the trace contaminants in water for a pollution study. Here again the hydrophobicity of Tenax-GC works to great advantage. As usual, significant adsorption occurs for compounds above C_6.

2.2. Selective Adsorption Using Carbopack B

Carbopack B, a graphitized carbon black, is a moderately strong adsorbent. This thermally stable substance is valuable both as a versatile column material for chromatography and as a trapping material, particularly for analysis of atmospheric pollutants. It suffers from the disadvantage of irreversibly adsorbing high-molecular-weight compounds (over 200). Table 2 lists the adsorption–desorption characteristics of the same compounds listed in Table 1 and evaluated by the same procedure used for Tenax-GC.

Only one particular air pollution study will be mentioned here. Carbopack B traps were used for sampling organic volatiles in a chemical plant where solvents are used during synthesis operations. A 1-liter air sample was taken in the area of the plant where workers usually spend most of their working day, thus providing a representative sample of inhaled air.

Particular emphasis was placed on the search for aromatic solvents. Analysis following desorption was by GC–MS, 2 hr after sampling.

The conclusion of this study was that Carbopack B trapping and desorption gave reliable and producible results. It is especially useful for compounds in the range of C_4 through C_8. Above this molecular weight, irreversible adsorption becomes increasingly prominent, and Tenax-GC becomes the preferred trapping material.

3. Exercises

1. If you were using a Tenax trap to collect air pollutants for quantitative analysis, how could you determine whether the trap was saturated, leading to "breakthrough" of some components?

2. In most GC–MS analyses it is best to use an internal standard for quantitation. Why is this usually not done using the procedures described in this chapter?

3. If you were using a Tenax trap to collect a pheromone for identification, how could you determine whether it is degraded:

   (a) during the adsorption/desorption, or

   (b) during gas chromatography?

4. Suggested Reading

A. ZLATKIS, H. A. LICHTENSTEIN, AND A. TISHBEE, Concentration and analysis of trace volatile organics in gases and biological fluids with a new solid adsorbent, *Chromatographia* **6**, 67–70 (1973).

W. BERTSCH, R. C. CHANG, AND A. ZLATKIS, The determination of organic volatiles in air pollution studies: characterization of profiles, *J. Chromatogr. Sci.* **12**, 175–182 (1974).

J. C. FLETCHER AND A. ZLATKIS, Analysis of Volatile Organic Compounds, U.S. Patent Application Number 450,504, Filed March 12, 1974.

A. ZLATKIS, H. A. LICHTENSTEIN, A. TISHBEE, W. BERTSCH, F. SHUNBO, AND H. M. LIEBICH, Concentration and analysis of volatile urinary metabolites, *J. Chromatogr. Sci.* **11**, 299–302 (1973).

H. M. LIEBICH, O. AL-BABBILI, A. ZLATKIS, AND K. KIM, Gas-chromatographic and mass-spectrometric detection of low-molecular-wieght aliphatic alcohols in urine of normal individuals and patients with diabetes mellitus, *Clin. Chem.* **21**, 1294–2396 (1975).

P. CICCIOLI, G. BERTONI, E. BRANCELEONI, R. FRATAR-CANGELI, AND F. BRUNER, Evaluation of organic pollutants in the open air and atmospheres in industrial sites using graphitized carbon black traps and gas chromatographic–mass spectrometric analysis with specific detectors, *J. Chromatogr.* **126**, 757–770 (1976).

G. HOLZER, J. ORÓ, AND W. BERTSCH, Gas chromatographic–mass spectrometric evaluation of exhaled tobacco smoke, *J. Chromatogr.* **126**, 771–785 (1976).

Automatic Data Processing

Richard N. Stillwell

1. Introduction

Computers have been applied to all aspects of mass spectrometry from control of the instrument and acquisition of data to automated compound identification. This chapter will focus on the factors to be considered in interfacing a computer to a mass spectrometer and in designing a system for interactive data storage and display, and will touch on some of the other applications of computer technology.

2. Instrument Control and Data Acquisition

Packaged "data systems" are available for most mass spectrometers either from the manufacturer of the spectrometer or from independent vendors. Rather than describe any particular system or systems, I shall discuss some of the factors which one would consider in designing such a system. Although few mass spectroscopists now build their own data systems, an understanding of how they work is essential in choosing and using such a system.

It is first necessary to consider the characteristics of the instrument we are going to control and of the signal we are going to measure. A mass spectrometer is a device for generating and separating ions. It will be assumed that from the point of view of the data system the ions are

---

Richard N. Stillwell ● Institute for Lipid Research, Baylor College of Medicine, Houston, Texas 77030

separated in time, rather than space. (Even with a photoplate instrument, in which the ions of different mass are recorded simultaneously, the plate is subsequently scanned over a period of time. If, as seems probable, an instrument is built in the future in which each element of a linear detector is directly connected to a memory element of a computer, the computer will undoubtedly scan the memory sequentially.) The data system, then, will measure ions of different m/e values at different times, or, to put it another way, at any given time it will be measuring the flux of ions having a distribution of m/e values about some central value, and a range of distribution which is a function of the resolution of the mass spectrometer.

If a mass spectrometer is set to measure ions of a given m/e value, and a sample is admitted to the ion source inlet at a given rate, then the ratio of the output signal to the rate of introduction of the sample is the sensitivity of the mass spectrometer. Sensitivity, of course, is one of the parameters that sells mass spectrometers, and manufacturers' brochures say a great deal about it, but seldom in very specific terms. A typical brochure, for example, refers to "very high sensitivity," "unequalled sensitivity," "outstanding sensitivity," or presents a normalized spectrum labeled "200 picograms." Before going on to consider the design of a system to process the output signal we should examine the sensitivity question in a more quantitative way.

The sensitivity of a mass spectrometer as defined above is the product of two terms:

$$S = T_{ms} \times G_a$$

where T_{ms} is the transmission efficiency or throughput of the mass spectrometer (i.e., the number of ions striking the detector per molecule admitted to the inlet system) and G_a is the gain of the amplifier system. Since G_a can be increased to any desired value, it might seem that the sensitivity of the mass spectrometer could be made as high as we please. However, the noise level will increase with the gain, so the information content of the output signal will not increase, and information is what we really want. So it is necessary to redefine sensitivity as *the ratio of the output signal to the rate of introduction of the sample at a signal-to-noise ratio of 3:1* (or 6:1, or whatever we think is an acceptable value). Note that this definition, like the first one, implies a steady-state situation, and gives a value measured in units such as amperes per picogram per second. We can, of course, integrate over time and obtain a value in, say, coulombs per picogram; this is valid for the first definition but not for the second since, as we shall see, the noise in the system is not strictly proportional to the signal.

The amplifier gain G_a can in fact be eliminated as a factor of interest.

By using an electron multiplier (the usual mass spectrometer detector) with a very fast amplifier and a discriminator which blocks signals below a certain level, it is possible to count individual ions striking the detector. In this case, G_a is identically 1, and the sensitivity of the mass spectrometer is simply its throughput, and can be expressed in units of ions/molecule. I do not intend to discuss the factors affecting the throughput of a mass spectrometer, but shall present the results of an experiment on a typical instrument.

The mass spectrometer used for the experiment was an LKB 9000. This is a single-focusing, magnetically scanned instrument with a gas chromatographic introduction system using a jet-orifice separator (Chapter 2). The instrument was old but well maintained, and was tuned up to a resolution of better than 600 and focused on $m/e = 372.4$, corresponding to M^+ for cholestane. It was equipped with an ion-counting interface and small computer system similar to the one which is described later in this chapter. The computer was programmed to read and clear the counter once a second and to sum the counts between times specified by the user. A 100-ng sample of cholestane was injected into the gas chromatographic column. This was a large enough sample that loss on the column or by adsorption in the inlet system and source should be insignificant. The peak width was about 30 sec. The maximum count rate observed was 2.3×10^5 ions/sec, almost 100 times the background count, which was 300 ions/sec between injections with the inlet valve open. The total counts measured were 7.9×10^6. For the molecular ion of cholestane, then, the throughput of this mass spectrometer was 7.9×10^4 ions/ng, or about one molecular ion per 20 million molecules of cholestane injected. Since the molecular ion carries about 4% of the total ion current of the electron-impact spectrum, the total throughput is one ion per 800,000 molecules. It is worth remembering this figure as an order-of-magnitude number, say, one-in-a-million, for the typical modern instrument. The SuperSpec 999 might put out an ion for every hundred thousand molecules on its best day (at a resolution of 50), but on a bad day it might take 10 million molecules to get an ion.

An experiment such as this one allows us to calculate not only the sensitivity of the mass spectrometer and related quantities such as the smallest detectable sample, but also parameters we need to design a data system. As we have seen, we really need to qualify our definition of sensitivity further by specifying the particular ion of the particular compound we want to measure. After all, we can measure only one ion at a time unless we reduce the resolution, and that defeats the whole purpose of the mass spectrometer, which is to *separate* ions. If, then, we want to detect the presence of cholestane in a sample by monitoring the molecular ion, and we insist on a confidence level of 95% (note that this is simply detection, not quantification), then we must have a S/N ratio of 2. If the background signal is 300 ions/sec the noise level of the background is $300^{1/2}$ or about 17. The signal due to the sample must be at least 35 ions/ sec. If the throughput of the mass spectrometer is 80 ions/pg, then the rate of sample introduction must be very nearly 0.5 pg/sec. If we have a GC inlet system and the column parameters give a peak width of 20 sec, then (approximating the GC peak as a triangle) a total sample of 5 pg will give a detectable signal at the apex of the peak. This is in good agreement with another experiment which was run on a different mass spectrometer of the same model in our laboratory a couple of years earlier. This instrument had the standard analog amplifier and was interfaced to a computer through an analog-to-digital converter. The signal was smoothed by summing a series of 2000 measurements and treating the result as a single value; this is essentially the same as increasing both the amplifier gain and the time constant of the filter circuit (*vide infra*). In this experiment, an injected sample of 20 pg of cholestane gave a peak about $^1/_5$ of the background level, with a S/N ratio of about 6.

From the above discussion, it can be seen that there are two possible ways to increase the sensitivity of the system: (1) reduce the background, or (2) increase the throughput of the mass spectrometer. Of these the second will be the more effective, since the signal is directly proportional

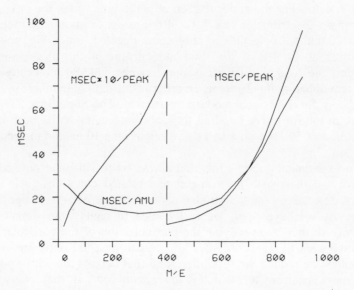

FIG. 1. LKB scan rate: msec/amu and msec/peak width (with a 10:1 scale change) as a function of *m/e* at resolution 700.

to the throughput, while the background noise is only proportional to the square root of the average background.

Before going on to show how the sensitivity of the mass spectrometer affects the design of the data system, it is interesting to calculate the minimum sample size for a scanned spectrum. Setting the scan speed knob of the LKB 9000 to 6 gives about a 6-sec scan from *m/e* 20 to 400. Figure 1 shows the experimentally determined rate of scan as a function of mass at a resolution of 700. The scan rate is fairly constant up to about *m/e* 600 but, since at constant resolution the peak width is proportional to *m/e* value, the time spent on each peak increases sharply as the spectrometer scans to higher masses. At the molecular ion of cholestane (*m/e* 372), the peak width is approximately 7 msec. A sample size of 1 ng in a 20-sec GC peak gives a maximum sample flow into the system of 100 pg/sec. This, in turn, provides a signal of 8000 counts/sec or 8 counts/msec at the detector. This is large enough that we can ignore the background count rate of 300 counts/sec (all of these figures, of course, are very approximate). The total count across a 7-msec mass spectrum peak is 56. The *noise due to the signal* is $S^{1/2} = 7$. The intensity of the peak then is 56 ± 7, about a 15% standard deviation. Probably, we would not want much less precision than this in our measurements. One nanogram, then, is approximately the lower limit of sample size for scanned spectra in this instrument.

3. The Minicomputer

Having examined the characteristics of the signal we are going to process, let us now look at the characteristics of the instrument which will process them. We will assume that this is to be a minicomputer connected to the mass spectrometer through an interface which we will construct in Section 5. Let us start with some definitions.

A *bit* is the smallest unit of information which a computer can manipulate. Apart from its ordinary English meaning, the word suggests *bi*nary dig*it*, the basic unit in the binary number system, having a value of either 0 or 1. Since digital electronics are most easily and reliably constructed with circuits and other elements having two alternate states (conducting or nonconducting, charged or uncharged, etc.), computers always use the binary number system, just as ten-fingered people use the decimal system.

A *byte* is a small group of bits (usually 6 or 8) handled together as a unit. Since communication between the computer and its master, the user, in binary notation would be exceedingly inconvenient for the latter, a *code* is used to relate byte values to the alphanumeric characters recognized by the user. One such code is the ASCII code, a 7-bit code in which A is 1000001, a is 1100001, etc. If ASCII text is stored one character per 8-bit byte, the 8th bit is simply unused, or used for a check bit. In order to print a value on a typewriter, the computer converts its internal value to ASCII code and transmits this over a wire (one bit at a time) to the typewriter terminal, which then translates the ASCII code to the appropriate keystroke.

A *word* is a longer collection of bits, generally an integral number of bytes, which is used to hold an integer value. A 12-bit computer such as the PDP-8 can directly handle numbers in the range of 0 to $2^{12}-1$, or 4095. If one bit is used for the sign of the number, the range is ± 2047. Numbers outside of this range can be handled in *multiple precision,* in which two or more words are used to store each number, but arithmetic then involves more operations.

A *register* is the hardware necessary to contain one byte or one word (or sometimes one or a few bits). A computer consists of a *processor* (containing a few registers, arithmetic and logic units, and timing circuits), *memory* (containing many registers, each of which can be referred to by an *address*), and *peripherals*, controlled by the processor via an *I/O* (input–output) *bus*.

A *program* is a series of words stored in memory which act as *instructions* to the processor. One register in the processor is the *program counter*, which always points to the next instruction to be executed. The processor fetches the word pointed to, interprets the value in it as an instruction, increments the program counter to point to the next word, and executes the indicated action. The action, at least in a minicomputer, will always be a simple one; for example, "store the contents of accumulator 2 (a processor register) in memory at address X," or "if the result of the last operation was negative, add Y to the program counter." It is the presence of conditional instructions such as the second one which distinguish a computer from a calculator; the computer can test the data it is operating on and perform different operations depending on the result.

Creating a program is simplified by the use of an *assembler*, a program which reads shorthand forms of instruction descriptions such as those just given and translates them into instruction values. Such shorthand forms generally consist of a *label*, an *opcode*, one or more *operands*, and a *comment*. The label assigns a symbolic name to the address at which the instruction will be stored, and is used only if some other instruction refers to it. The opcode is the thing to do, and the operands are the things to do it to. The comment should be present on almost every line of an assembly language program, and should explain why the action is being done. For example, the two instructions given above would appear in PDP-11 assembly language as

      SAVDIF: MOV R2,DIFF ;save the difference for later use
            BMI DNSLP   ;negative means downward slope

and somewhere in the program, DIFF and DNSLP would appear in label fields. The assembler would translate these two lines to values such as

$$0\ 0\ 0\ 1\ 0\ 0\ 0\ 0\ 1\ 0\ 0\ 1\ 1\ 1\ 1\ 1$$
$$1\ 1\ 1\ 1\ 1\ 1\ 0\ 0\ 1\ 1\ 1\ 1\ 0\ 0\ 1\ 0$$
$$1\ 0\ 0\ 0\ 0\ 0\ 0\ 1\ 0\ 0\ 1\ 1\ 0\ 0\ 1\ 0$$

in three successive memory words. The actual values would depend on the values of DIFF and DNSLP.

4. Interrupts and Asynchronous Processing

Every computer must, of course, have means of controlling and responding to its peripherals (the terminal, mass spectrometer interface, disk, etc.). Generally this is provided by means of a set of *I/O instructions*. For example, on the PDP-8 family of minicomputers, a character is printed on the terminal by the following sequence:

```
TSF            ;has terminal finished last
               character?
JMP  .-1       ;no—go back and test again
TAD  CHAR      ;yes—load character in
               accumulator
TLS            ;transmit it to the terminal and
               clear flag
```

The *I/O* instruction TSF tests a flag which is set by the terminal interface when the terminal has completed printing a character; if the flag is set, the program counter is incremented by one, which causes the next instruction to be skipped. (Recall that the PC was incremented by the processor after the TSF was fetched for execution.) Operating in this way, however, is wasteful of processor time. The loop TSF; JMP .-1 takes 7.5 μsec on a PDP-8/I, while a Teletype takes 100 msec to print each character. If the processor has other things to do, it can just check the flag periodically and continue processing if it is not set, but this adds a great deal of complication to the program.

A better solution is to use the *interrupt facility* of the computer. An instruction in the startup code enables the interrupt. Thereafter, whenever the terminal flag goes up, the processor completes its current instruction, saves critical registers such as the PC, and transfers to a predetermined location in memory. There are assembled instructions to clear the flag and perform other appropriate actions: e.g., to determine if there is another character to be printed and, if so, to transmit it to the printer. Then the registers are restored by a "return from interrupt" instruction and processing resumes where it left off. Servicing devices in this way is called *asynchronous* processing because the interrupt code is executed at unpredictable times with respect to the main program.

If data are to be transferred at a very high rate, as for example to or from a disk, interrupt processing would consume most of the processor's capability, or might not even be able to keep up with the data. Devices such as disk and tape units are therefore controlled through a *direct memory access* (DMA) interface (also called data break, channel transfer, etc.). Here the processor starts an I/O operation, either in the main program or in interrupt code, specifying the starting address in memory of the data to be transferred, the number of words to be transferred, the direction of transfer, etc. Then the DMA controller, which can be considered an auxiliary, special-purpose processor, performs the transfer, halting the main processor temporarily each time the device is ready to transfer a word (or byte), transferring the word to or from memory, incrementing its memory pointer, and decrementing its counter. When all the data have been transferred, the controller signals the processor by means of an interrupt. Just as the main program is not "aware" of interrupt processing, the main processor is not aware of DMA transfers, except that programs take a little longer to execute because of the *cycle stealing* by the DMA controller.

5. Interfacing

Now that we have examined the characteristics of the signal to be processed and the general features of minicomputers, let us pick a particular mass spectrometer and a particular computer and see how to design an interface between them. The mass spectrometer is the LKB 9000 which was referred to in Section 2. The minicomputer we shall use is the Digital Equipment Corp. PDP-11/10. This computer has a 16-bit word length and can support up to 28K (28 × 1024) words of memory. Memory is addressed by bytes. It is a two-address machine, meaning that for operations such as "add" and "move" both the source and the destination are specified in the instruction (as opposed to a single-address machine in which one operand is always the accumulator). It has six general registers (R0–R5) which can be used as accumulators, pointers, etc., a stack pointer (SP) whose use is discussed below, and a program counter (PC). SP and PC are addressed by instructions just like R0–R5; this makes for powerful and flexible addressing mechanisms. In addition to memory there is an "external page" of registers which is addressed just like memory locations but which actually consists of I/O registers controlling the peripherals. Instead of loading an accumulator with a character (in ASCII code) and issuing an I/O instruction to transmit it to a terminal, the character is simply "MOV"ed to the terminal output register just as if to location in memory.

The external page also contains a register called the *program status word* (PSW) which contains the *condition codes* resulting from the last operation performed (zero, negative, carry, overflow) and the priority at which the processor is operating (which determines what devices can interrupt).

Interrupt processing works as follows: when a device sets its flag to request service (say, the user has struck a key on his terminal), then *if* the program has enabled the keyboard interrupt by setting a bit in its external page register, and *if* the processor is operating at a priority lower than the fixed priority assigned to the keyboard, the current instruction is completed and the PC and PSW are "pushed on to the stack." This means that each in turn is copied to memory at the location pointed to by the SP after the SP has been decremented by 2. The SP must always point to an area of memory reserved for this kind of "last in, first out" storage. It works just like a stack of plates in a cafeteria, where customers remove plates in the reverse of the order in which they are stacked by the kitchen helper. After the PC and PSW have been stacked, they are reloaded from two words at a fixed pair of locations in memory; this "interrupt vector" has been set up by the initialization code. The processor now starts executing instructions from a part of memory and at a priority determined by the interrupt vector; the priority is chosen so that the device cannot interrupt again before the first interrupt has been processed. The interrupt routine can use the stack to save any of the general-purpose registers it needs to use itself. When it has finished processing the interrupt, it "pops" the saved values from the stack back into the registers, and executes an RTI (return from interrupt) instruction which "pops" the old PSW and PC, thereby resuming the interrupted code at the former priority and with the condition codes the same as before the interrupt.

6. Timing

The rate at which the mass spectrometer signal is to be sampled has already entered into our calculations of sensitivity of the instrument (Section 2). Now we must consider it from the point of view of the data acquisition system. For the moment let us consider the signal to be a continuously varying voltage which can be sampled as often as we like without loss of precision; this is nearly the case for the spectrum of a large sample (say, 1 μg). To determine the worst-case timing requirement for our data system, we tune our mass spectrometer for its best resolution and take a scan at the maximum speed we expect to use. Suppose we find that the narrowest peaks (those at the low end of the mass range) are about 0.5 msec wide at the base and are nearly triangular. In order to be sure of sampling the peak when it is within 10% of its maximum value, we would have to take 10 samples on each side of the peak; this means a sample interval of 25 μsec. Actually, although the peaks are unlikely to be true Gaussian curves, they will be somewhat rounded on top so the requirement will not be quite so severe; on the other hand, we may want intensity measurements to better than 10% accuracy. How many instructions can our computer execute in 25 μsec? Instruction time in the PDP-11 family depends on the instruction being executed and the mode of addressing of the operands. If we take an instruction to move a general register to memory (in assembly language: MOV RO,MEM) as typical, this instruction takes about 7 μsec, so we could execute only three or four instructions per sample. If we had enough memory, we could store successive intensity values, but clearly we could not detect peaks on the fly, let alone do any other processing during a scan.

There are, indeed, data systems which use a simple analog-to-digital converter and detect peaks in software, but (a) they do not meet the 25-μsec requirement, so peak intensities at high scan rates are not very accurate, and (b) they do not permit any other use of the computer during a scan. If our system is not to be limited in these two ways, we must reduce the load on the central processor by building some "intelligence" into our interface. If the interface were capable of detecting the occurrence of a peak and notifying the computer when it occurs, the sampling rate of the computer would be reduced to the rate of arrival of peaks. Going back to our test scan, we might find that this was about 5 msec between peaks at one peak per mass unit, at the fastest part of the scan. At 7 μsec per instruction, about 700 instructions can be executed between peaks. In addition, since peak detection is now a hardware function, all the computer has to do is to store the mass and intensity values presented to it by the interface. Hence there will be plenty of processing power left over for other operations, such as displaying or plotting previous spectra. Since our computer will not be serving a large number of simultaneous users, there seems to be no point in going further, for example, to place the whole spectrum in memory by means of a DMA transfer (see p. 168) before interrupting the processor.

7. Designing the Interface

We are now ready to design the mass spectrometer–computer interface. We start by considering it as a "black box" and specifying what its inputs and outputs are to be. Then we draw a functional diagram showing what the logic circuits and storage registers inside the box should do, and finally we select the actual components (which for the most part will be standard integrated circuits) and lay out the printed circuit boards and wiring.

There will be one input from the computer to the interface: the rate at which the interface is to sample the signal from the mass spectrometer. There will be three input signals from the mass spectrometer: (1) the pulses from the electron multiplier, (2) the digital value from the mass marker, and (3) a signal from the scan relay (on during a scan; off at the end of the scan). The output signals from the interface to the computer will be: (1) an interrupt signal when a peak has been detected, (2) the maximum ion count for the peak, (3) the mass marker reading at the top of the peak, and (4) an interrupt signal at the end of the scan. Obviously, other signals could be specified: for example, we might want to drive the magnet under computer control. But the ones just listed will do for now.

The next step is to lay out the interface in functional terms. This has been done in Fig. 2. Each box represents a register or logic unit; arrows represent the control and data paths between them. The whole thing is controlled by a clock, which puts out a train of pulses at intervals determined by loading its counter register from the computer. Each time the clock "ticks" the following actions take place:

(1) the contents of CNT2 are shifted to CNT3;

(2) the contents of CNT1 are shifted to CNT2, and

(2b) the contents of MM1 are shifted to MM2;

(3) the contents of the ion counter are shifted to CNT1, and

(3b) the contents of the mass marker register are shifted to MM2;

(4) the comparator generates an interrupt signal if $(CNT2) > (CNT1)$ and $(CNT2 > CNT3)$ and the scan relay is closed, and

(4b) CNT0 is reset to 0.

'n order for the comparator to detect a peak, it must have access to the ion count for three successive intervals; thus CNT1, CNT2, and CNT3 are a "history" of the count rate at successively earlier times. Since the mass marker is asynchronous with respect to the clock in the interface, it is necessary to gate the mass value through a series of registers in parallel with the ion count, so that when the computer reads the ion count from CNT2, which contains the maximum value, it can also read the correspond-

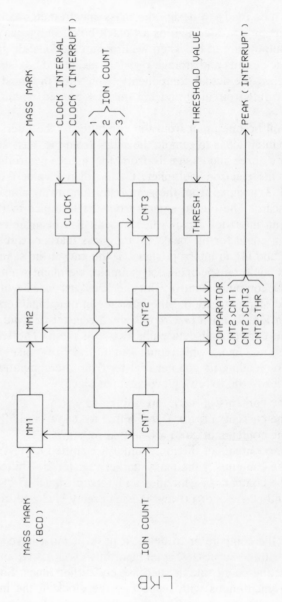

FIG. 2. Block diagram of mass-spectrometer–computer interface.

ing *m/e* value. Also, since the computer will not be monitoring the mass marker continuously, it must be told when the scan is ended. The scan relay gives this information, but the level change must be converted into a short pulse so that it will not continue to interrupt until the next scan starts. The "1-shot" is a logic element which converts a level change into a pulse.

A real interface would almost certainly be somewhat more complex than the one in Fig. 2. The reader can undoubtedly think of many useful modifications: for example, a computer-settable threshold register so that interrupts only occur if the value in CNT2 exceeds the threshold. Also, we have ignored the fact that the comparator will detect peaks and cause interrupts during the back scan as well as the forward scan, so we will want to add a gate to "and" the scan-on signal from the relay with the comparator interrupt output.

The next step in the project is to redraw Fig. 2 in terms of the logical functions of the components we are actually going to use to construct the interface. There is available at low cost a wide variety of integrated circuits in uniform packages containing, for example, a 16-bit counter, an 8-bit comparator, or four 2-input NAND gates (whose output is "true" if one or both inputs is "false"). Sometimes it will be necessary to use two or three IC's to implement a box on the overall logic diagram; sometimes one IC can be used for several boxes in different parts of the diagram. All this must be worked out, and the various connections drawn in as pin-to-pin connections. Then a printed-circuit board is laid out (with the fewest possible crossovers between connections) and etched, the IC's are soldered in place, and the board is mounted in a frame or box with a power supply and connected by cables to the computer and the mass spectrometer.

This is probably the most economical way to build the interface, either as a single unit or (if you are going into the business) by the dozen. Other approaches, however, are possible and may be desirable. If you expect that these interfaces will sell like pocket calculators, you could get Texas Instruments or Intel to build the whole interface (except the power supply, probably) as a single large-scale integrated circuit (LSI). On the other hand, if you were going to build only a single interface and expected that some changes might have to be made as the result of experience with it, you could buy a general-purpose interface board from the computer manufacturer and build the interface on it using a wire-wrap tool (this wraps a wire tightly around a square pin so that the corners of the pin cut into the wire and make a good contact). Or, if the interface were going to be a good deal more complicated than this one, it could be built with a microprocessor. This is a computer CPU, generally of small word size and with a limited instruction set, built on a single LSI chip. It would need a program which would probably be written and debugged (by simulation) on a larger computer, then "burned" into a read-only memory (ROM). Here again timing would be critical, since microprocessors are generally slower than minicomputers.

8. Designing the Software

Now we have assembled the hardware of our system. Let us assume it consists of the mass spectrometer, the interface, the computer, a disk drive to hold programs and data, a second disk or a tape unit for storage and transfer of data, a CRT (cathode ray tube) terminal with graphics capabilities, and an electrostatic printer–plotter for hard-copy output. The next job, and it is a big one, is to write the programs to run all this hardware and to interact with the user.

First we must choose (or, if absolutely necessary, write) the operating system. This is a program which directly controls the hardware, handles interrupts, and allows the user to request the applications programs. The applications programs in turn rely on the operating system to handle I/O and create, open, and close files on disk. A good operating system simplifies programming enormously: for example, if the user instructs the data acquisition program to create a new data file called RUN37, the program just passes this request on to the operating system, which allocates the necessary blocks on disk, makes an entry in the disk index, and notes that the file is empty. Later on, the program will pass data to the operating system to place in the file, and eventually will request that the file be closed. Another program can then request the operating system to open RUN37 and read data back from it. Neither program needs to know which disk blocks have been allocated for the file: in fact, the operating system should not allow applications programs to access the disk directly, in order to prevent conflicting use of disk blocks.

Since fast response to interrupts by the mass spectrometer interface is required, the operating system must be a real-time system. This means that the device or operation of highest priority gets serviced first, and continues to get service until it is finished with its current operation. This is opposed to a time-sharing system which allocates a certain amount of time to each user in turn. The two modes of operation are not compatible except in the most elaborate operating systems on computers larger than the one we are contemplating here. For this reason, if you want to run programs in BASIC or play Space Wars on your mass spectrometer data system, you will have to shut down the mass spectrometer operations and probably load a different program disk. Assuming that a suitable real-time system is available from the computer manufacturer or an independent software house, you will need to write a device handler which the operating system can call to control the mass spectrometer interface. The documentation which comes with the system should tell you how to do this, but writing and debugging device handlers is tricky business, partly because of their asynchronous nature, which can lead to problems such as conflicts in register usage, and partly because they run "inside" the operating system and therefore are not restricted by the system's protection schemes.

Next you will have to decide on a file format for the mass spectrometric data. This will be limited by the formats available under the operating system, and also by the languages you plan to use for the various programs, since not all file structures available under the operating system may be available to a FORTRAN or BASIC program. The simplest scheme is to place the data from one GC–MS run in one sequential file, with each mass spectrum occupying one logical record or unit of data in the file. (The length of a logical record need not be commensurate with the length of a physical record or disk block, and indeed may vary from record to record.) The sequential format conserves both computer code and disk space, especially if the blocks are linked together by pointers so that they may be scattered about the disk wherever space is available. It has the grave disadvantage, however, that a given spectrum can only be accessed by starting from the beginning of the file (or from a known point in it) and reading forward one logical record at a time. Also, to construct an ion current profile, either for the total ion current or for a set of discrete ions, every record in the file must be read. This is quite time consuming, even from disk.

A common solution is to use some variety of indexed file structure. Such a file has space reserved in it, or in an associated file, for an index which contains an entry for each logical record. The entry consists of the block number in which the record starts and (unless each record starts at the beginning of a new block) the offset within the block at which it starts. Creating and accessing an indexed file is obviously more complicated than creating and accessing a sequential file. Moreover, the blocks of the file cannot be linked, but must either be contiguous on the disk or be mapped. A mapped file has another index (a map) to relate file blocks to actual disk blocks. An indexed file allows reasonably direct access to any given spectrum (the operating system must read first the index, then the map, then the record itself), but is even more inefficient than a sequential file for recovering ion current profiles.

If reconstruction of ion current profiles is important to you (and you are not taking full advantage of the GC–MS–COM system if you do not use them), there are two choices. One is to construct, after a GC–MS run is complete, an inverted file containing the ion current profiles. Since this requires reading the original data file once for each ion current profile (or a few profiles, if the computer has sufficient memory), inverting a file is very time-consuming with a small computer. What is more, you now have two files, which we can assume are indexed files, each of which requires perhaps a hundred thousand bytes of disk storage. This is a substantial fraction of the 2.5 million bytes available on the most commonly used type of minicomputer disk. Another approach is to create a semi-inverted file, in which each record contains intensity data for a range of m/e values (say 50) from a range of spectra (say 20): for example, record 135 might contain the intensities for m/e 300–349 from scans 161–180. With such a file, retrieving a given spectrum or profile requires (1) calculating which records contain segments of the spectrum or profile, (2) finding and reading the records, and (3) assembling the spectrum or profile from the segments. If the file contains data for 200 scans with a mass range of 500 amu, either operation requires ten disk accesses. Since all records are the same length, the operating system can calculate block numbers directly and indexing is not required (although mapping may be). Such a file is called a direct access file.

The format in which data are stored in the records of the file is the next consideration. We saw in Section 2 that 1 ng of cholestane gives 56 counts for the molecular ion in a full-scan spectrum. Let us assume we want to be able to handle samples up to 5 μg, and allow another factor of 50 for more intense peaks and slower scan rates. Therefore we will have a maximum of $56 \times 5000 \times 50$ or 14,000,000 counts in any peak. The next higher power of 2 is 2^{24}, so out interface must have 24-bit counters. However, there is no need to store 24 bits of intensity data. If we need 1% precision, or, say, 1 part in 128, we need only seven bits for the mantissa of the intensity, and five bits (2^5 is greater than 24) for the characteristic, a total of 12 bits. Data stored in this way, with a characteristic (i.e., exponent) and mantissa (proper fraction) are called floating-point data. (All minicomputers for scientific work can handle floating-point numbers either in hardware or with special subroutines. Normally, however, they will use 8–12 bits for the characteristic, 23–56 bits for the mantissa, and one bit for the sign.) Or, if we think the floating-point format is too cumbersome, we can scale the intensities so that the largest value for a run has no more than perhaps 14 significant bits, giving a range of 1–16,384. If our computer has a word length of 16 bits, this leaves two bits for flags; we will see possible uses for these bits as we go along. If we decide to use the semi-inverted file format, we must use one word for each intensity value. If a run consists of 200 scans with a mass range of 500 amu, the file will occupy 100,000 words (200,000 bytes) of storage. (This could be reduced slightly by trimming off zero intensities at the ends of groups of spectra, at some additional cost in access overhead.) If we use indexed files (perhaps with an inverted file for ion current profiles) we can save some space by not storing zero intensities (or those below some threshold). One way to do this is to use a bit map. If the mass range is 500 amu, for example, we reserve 512 bits (32 words) at the beginning of the record as a map showing which masses have nonzero intensities. These are followed by the nonzero intensity values. To determine the intensity for a given mass we isolate the corresponding bit from the map. If it is 0, the intensity is 0. If it is 1, we count the number of 1 bits up to that point. If this number is n, the nth intensity belongs to the given mass. (In the limit, intensities are saved to only 1 significant bit; i.e., the peak is either present or absent. In this case, the bit map *is* the spectrum. This is an efficient and often adequate representation for library searching: see Chapters 7–9.)

In general, savings in storage must be paid for by increased complexity of program code and computation time. There is no "best" solution to the data format problem; each application will have a different optimum compromise.

Now we must decide what facilities the system should provide. We will want a set of file operations to list the directory, copy, and delete files; these often come with the operating system as a standard utility. We need programs to calibrate, tune, and test the mass spectrometer and interface. Test programs are often neglected; it is pleasanter to be optimistic and assume everything will work as it should, but in the real world this seldom happens, and a good set of test programs can be a great help in diagnosing problems. We need a data acquisition program, of course, or possibly one program for acquisition of scanned data and another which is optimized for selected ion detection for quantitative work. The latter should allow you to monitor enough ions for your application; if it will only do four ions, sooner or later you will need five, and will have to run each sample twice. It may be desirable to be able to switch to a different set of ions at a preset time, or at an interrupt command from the user. We need programs to display spectra and (multiple) ion current profiles. If we are going to be doing quantitative work then a program to integrate peak areas and calculate concentrations using internal standards is very useful (and seldom provided). For the identification of minor components in mixtures, such as minor drug metabolites, there are programs for resolving overlapping GC peaks, either by simply flagging ions which have a local intensity maximum at a given scan, or by more complex curve-fitting procedures. The identification of spectra by comparison with reference libraries is covered in the next three chapters; the data system can provide an interface to the search systems which are available through time-sharing services. Even though these systems are very comprehensive, it is useful to be able to maintain and search your own mass spectrum library, since you can add spectra which you are not ready to publish or which are not of good enough quality to submit to the reference collections.

We should now design the user interface, or the way the data system communicates with the user. All too often this is left up to the programmer, who naturally designs the dialog from the program's point of view, not the user's. There are essentially two types of user interface. One is to program the computer to display on the terminal a series of questions to which a simple answer is to be given: "yes," "no," a number, a file name. This is the easiest type of interface for the user to learn, since there is very little that needs to be remembered. It is, however, rather tedious for the experienced user, although with a fast CRT terminal the number of questions can be cut down by using a multiple-choice format. The other approach is to let the user type in a command in some standardized command language, and to default anything not specified to something reasonable (such as what is was before the command was typed). Although this approach requires the user to learn the conventions of the command language, it is easier to use in the long run. It does, of course, require a much more complex program to interpret the user's command than to go from question to question.

All that is left to do now is to write the programs. Three good programmers who can work together effectively should be able to do this in three or four years. Once they have started work, of course, you are stuck with the basic design until you can afford to start over from scratch, although you can always make minor modifications. (A minor modification is one which introduces no more than ten new problems.) I make these remarks partly to try to convince you not to build your own system, even though it would undoubtedly be far superior to any other if it worked, and partly to show why GC–MS–COM analytical systems are pretty much the same today as they were five years ago. Even so, the capabilities of the analytical system are greater than the sum of the capabilities of the component parts, and the data system should be considered not an accessory but an integral part of the system.

9. Exercises

1. Can the computer do anything that the mass spectroscopist cannot do manually?

2. Are there disadvantages to using a computer?

3. How does using a GC–MS–COM system differ from using a gas chromatograph, a mass spectrometer, and a computer which are not physically converted?

4. What part of a GC–MS instrument is analogous to the MS–COM interface discussed in this chapter? What are the significant differences between the two interfaces?

7

Collections of Mass Spectral Data

Brian S. Middleditch

1. Introduction

The newcomer to mass spectrometry, until he has mastered the art of interpreting spectra, leans heavily upon comparisons of his "unknown" spectra with those obtained by others. Even when a spectrum has been interpreted from first principles, it is usually desirable to confirm an identity by spectral comparisons. Fortunately, there are several central repositories for mass spectral data which make such data available on a continuous basis or which publish printed collections of data periodically. Also, individual groups of mass spectroscopists have compiled more specialized data tabulations. There is considerable overlap between the various data collections, so one does not need to subscribe to all of them.

The following lists are not comprehensive. No attempt has been made, for example, to list data collections supplied by manufacturers of mass spectral data handling systems. In any case, such collections are usually derived from those described.

2. Open-Ended Data Collections

The main repository for mass spectral data is the Mass Spectrometry Data Centre (Chapter 8). These data are accessible through the Mass Spectral Search System (Chapter 9) and are provided to other collections

Brian S. Middleditch ● Department of Biophysical Sciences, University of Houston, Houston, Texas 77004

on a regular basis. There is particularly close collaboration with the Environmental Protection Agency and National Institutes of Health collections.

Environmental Protection Agency. In an effort to provide a larger mass spectral data base for identification of materials found in the environment, the EPA is collecting such data on a voluntary basis. All data submitted are considered in the public domain and are not to be reproduced in any form for profit.

The EPA also assists in the administration of the Mass Spectral Search System and is in the process of publishing (on a nonprofit basis) an inexpensive yet comprehensive compilation of mass spectral data.

Further details may be obtained from:

> Dr. Stephen R. Heller
> EPA, PM-218
> 401 M Street, SW
> Washington, D.C. 20460

The Athens Environmental Research Laboratory of the EPA maintains a separate collection of spectra of common water pollutants. By 1978, 1489 spectra ordered by molecular weight were on file. Data can be transcribed onto your 9-track magnetic tape.

Further details may be obtained from:

> Dr. W. M. Shackelford
> Athens Environmental Research Laboratory
> EPA
> Athens, Georgia 30605

Food and Drug Administration. The FDA maintains and regularly updates a "Mass Spectral Data Compilation of Pesticides and Industrial Chemicals" for use in surveillance analysis. EI, CI, FI, and FD spectra are included. A 100-page printed version was distributed by the FDA in 1976, and an enlarged version is to be published shortly by Heyden and Son.

Further details may be obtained from:

> Dr. Thomas Cairns
> FDA
> Los Angeles District
> Los Angeles, California 90015

Karolinska Institutet. One of the activities of the World Health Organization Research Training Center on Human Reproduction is the assembly of mass spectral data for steroids and steroid derivatives suitable for GC–MS analyses. About 2000 spectra have been contributed, primarily by Drs. H. Adlercreutz, C. J. W. Brooks, E. C. Horning, M. G. Horning, D. H. Smith, G. Spiteller, C. C. Sweeley, and R. Vihko. Further details may be obtained from:

> Dr. Jan Sjövall
> Kemiska Institutionen
> Karolinska Institutet
> Sölnavagen 1
> S-104 01 Stockholm 60
> Sweden

Mass Spectral Search System. This system is described in detail in Chapter 9.

Mass Spectrometry Data Centre. See Chapter 8 for information on the Centre.

National Institutes of Health. Dr. Stephen Heller and others at the NIH, in 1971, pioneered development of an innovative mass spectral search system. Several thousand spectra of biologically important compounds were stored in a PDP-10 computer in Bethesda, which was accessible by teletype over telephone lines. It is now used only for in-house and intragovernment work. Nongovernmental users may obtain access to a parallel collection through the Mass Spectral Search System. Further details may be obtained from:

> Dr. Henry M. Fales
> National Heart and Lung Institute
> Bethesda, Maryland 20014

Thermodynamics Research Center. The TRC is responsible for the collection, evaluation, and distribution of mass spectral data for:

(i) The American Petroleum Institute Research Project 44. The API 44 catalog of selected mass spectral data commenced publication in loose-leaf form in 1947. It now contains a total of 2687 data sheets in eight volumes; six volumes in standard columnar tabulation and two volumes in matrix format. These data were measured on pure hydrocarbons and related sulfur and nitrogen compounds. This compilation contains a molecular formula index arranged in ascending C–H order with corresponding compound name, molecular mass, serial number, and contributing laboratory. A second index is arranged in numerical order of serial numbers.

(ii) The TRC Data Project (formerly the Manufacturing Chemists' Association Research Project). The TRCDP compilation dates from 1959 and is published in a format similar to that of the API 44 catalog. It now contains 569 data sheets in two volumes, one in columnar tabulation and one in matrix array. These volumes contain spectra of organic compounds other than C–H–N–S and on simple inorganic compounds. This compilation is indexed in the same manner as the API series.

A comprehensive index of all API 44 and TRC serial publications is available. This master index has two sections—one arranged alphabetically by compound name and the other by molecular formula in the Chemical Abstracts system. Wiswesser Line Notations have been provided for all the compounds and can be used for computer indexing and retrieval.

Contributed spectra of fully characterized pure samples are subjected to review before publication. The spectral data from both compilations are available on loose-leaf sheets from:

> Data Distribution Office
> F. E. Box 130
> College Station, Texas 77843

Further details may be obtained from:

> Dr. Bruno J. Zwolinski
> Department of Chemistry
> Texas A & M University
> College Station, Texas 77843

University of Gothenberg.   This collection, maintained by Dr. Sixten Abrahamsson (and, formerly, the late Dr. Einar Stenhagen), with the collaboration of Dr. F. W. McLafferty, forms the basis of the printed compilations *Atlas of Mass Spectral Data* and *Archives of Mass Spectral Data*. A parallel data collection at Cornell University is used for the Probability Based Matching system, accessible through the TYMNET network in a manner similar to that of Mass Spectral Search System.

Further details may be obtained from:

> Dr. Sixten Abrahamsson
> Crystallography Group
> MRC Unit for Molecular Structure Analysis
> Medicinaregatan 9
> S-400 33 Goteborg 33
> Sweden

or from:

> Dr. F. W. McLafferty
> Department of Chemistry
> Cornell University
> Ithaca, New York 14853

3. Published Data Collections

Some of these collections are no longer available since they were distributed in limited quantities on an informal basis. It is suggested that those who wish to obtain access to such collections contact persons who were members of the American Society for Mass Spectrometry (formerly, Committee E-14 of the American Society for Testing and Materials) at the time that the compilations were distributed.

Dow Uncertified Mass Spectral Data. Edited by Dr. R. S. Gohlke, Published by the Dow Chemical Company in cooperation with the ASTM E-14 Subcommittee IV, 1963, 320 pages. This collection, no longer available, contains uncertified mass spectral data in tabular form, ordered by molecular weight, for more than 2000 organic compounds, from the Dow collection of mass spectra.

Compilation of Mass Spectral Data. A. Cornu and R. Massot, Heyden and Son, Ltd., London, 1966, 617 pages. First Supplement: 1967, 138 pages. Second Supplement: 1971, 161 pages. This compilation is a computer listing now covering the spectra of 7000 compounds. They have been sorted into four different sections to facilitate easy reference. Prefaced by a table of the origins of the spectra, the four parts are arranged in the following order:

(i) By reference number: giving the name of the compound, the reference number in the original collection, molecular weight, reference value, listing of the ten strongest peaks and their relative abundances, and the molecular formula of the compound.

(ii) By molecular weight: giving the identical information as in (i), but listing the compounds by increasing molecular weight.

(iii) By molecular formula: in order of C, H, D, Br, Cl, F, I, O, P, S, Si, and others.

(iv) By fragment ion values: In order of increasing m/e value: this section is three times as large as the others, and the ten peaks are listed in the order of the first, second, and third most intense.

These data are also available on punched cards and on magnetic tape, together with a retrieval program.

ASTM Index of Mass Spectral Data. American Society for Testing and Materials, Philadelphia, 1969, 632 pages. This volume contains a "six-peak" index of approximately 8000 uncertified mass spectra which comprise the University of Gothenberg Collection (as of the time of publication) and others assembled by the ASTM E-14 Subcommittee IV. It contains seven indexes based on molecular weight, most intense peak, and second to sixth most intense peaks.

Atlas of Mass Spectral Data. Edited by E. Stenhagen, S. Abrahamsson, and F. W. McLafferty, Interscience, New York, 1969, 3 volumes.

This publication contains tabulated mass spectral data for over 6000 organic compounds from many sources, including those in the API 44 catalog, the ASTM E-14 uncertified data collection, and the Dow compilation. Spectra are listed in order of molecular weight (Vol. I: 16–142; Vol. II: 142–213; Vol. III: 214–703), and there is a complete molecular formula index in each volume. The Atlas is also available on magnetic tape.

Mass Spectrometry of Biologically Important Aromatic Acids. C. M. Williams, A. H. Porter, and M. Greer, University of Florida Press, 1969. This specialized collection contains data for 118 compounds.

Archives of Mass Spectral Data. Edited by E. Stenhagen, S. Abrahamsson, and F. W. McLafferty, Interscience, New York, 1970–1972 (quarterly). This periodical was intended to supplement the *Atlas of Mass Spectral Data* by publishing refereed spectra in tabulated form and line diagrams. Eleven hundred spectra appeared during a three-year period, but publication has now ceased.

Identification of Endogenous Urinary Metabolites by Gas Chromatography–Mass Spectrometry: A Collection of Mass Spectral Data. S. P. Markey, H. A. Thobhani, and K. B. Hammond, B. F. Stolinsky Research Laboratory, Department of Pediatrics, University of Colorado Medical Center, Denver, 1972, 260 pages. The first 26 pages of this booklet are devoted to discussions of screening for organic acidurias, GC–MS, and the interpretation of mass spectra. The major portion of the booklet contains mass spectral data for derivatives of 190 endogenous urinary metabolites and seven-peak indexes ordered by molecular weight, most intense peak, second most intense peak, and third most intense peak. A line diagram, tabulated data, and structural formula are provided for each compound.

Mass Spectra of Drugs. Assembled, printed, and distributed by the MIT Mass Spectrometry Laboratory, under the sponsorship of ASMS Committee VI, with financial support from ASMS and the NIH, 1972, 190 pages. This is a collection of tabulated mass spectral data and line diagrams for 376 compounds, contributed by the Institute for Lipid Research, Baylor College of Medicine, the Department of Chemistry, Massachusetts Institute of Technology, and the National Heart and Lung Institute, NIH. Most of the compounds included are drugs, drug metabolites, and their derivatives, but a few artifacts and other substances frequently encountered in body fluids are also included.

Separate eight-peak and alphabetical indexes have been distributed by MIT.

Gas Chromatography and Mass Spectrometry of Selected C_{19} and C_{21} Steroids. R. H. Thompson, Jr., N. D. Young, J. E. Harten, T. A. Springer, R. Vihko, and C. C. Sweeley, Michigan State University, East Lansing, Michigan 1973, 244 pages. This is a collection of tabulated data and line diagrams for 214 spectra of C_{19} and C_{21} steroids (free and derivatized) at 22.5 and 70 eV. They are listed in order of molecular weight. A base peak index contains mass and abundance data for the six most intense peaks in each spectrum. There is also an index arranged according to carbon and oxygen content. Gas chromatographic retention data (on 3% QF-1, relative to cholestane) are listed for 68 steroid trimethylsilyl ethers.

Applications of Gas Chromatography–Mass Spectrometry to the Investigation of Human Disease: The Proceedings of a Workshop. Edited by O. A. Mamer, W. J. Mitchell, and C. R. Scriver, McGill University, Montreal, 1974, 314 pages. The first 240 pages of this monograph summarize current practical aspects of the application of GC–MS to metabolic disorders. An appendix, compiled by Dr. S. P. Markey and his former co-workers at the University of Colorado Medical Center, contains spectral data for 397 human metabolites or derivatives. An alphabetical listing is followed by a seven-peak index arranged in molecular weight order and additional seven-peak indexes based on most intense peak and second to seventh most intense peaks.

Final Report on the Rapid Identification of Drugs from Mass Spectra. Batelle Columbus Laboratories, Columbus, Ohio, 1974, 42 pages plus appendix. The appendix of this report to the National Institute on Drug Abuse contains major peaks from the methane chemical ionization and electron impact mass spectra of approximately 400 drugs, drug metabolites, and other compounds commonly found in body fluids.

Mass Spectra of Compounds of Biological Interest. Edited by S. P. Markey, W. G. Urban, and S. P. Levine, National Technical Information Service, Springfield, Virginia, Report No. TID-26553, 1974, 2 volumes. This compilation, prepared in collaboration with ASMS Committee VI, includes line diagrams and tabulated data for more than 2000 compounds, including steroids, sugars, bases, prostaglandins, amino acids, aromatic acids, keto acids, hydroxy acids, and fatty acids, and their trimethylsilyl, *O*-methyloxime, and acetyl derivatives. Spectra were contributed by M. Couch, D. M. Desiderio, A. Duffield, M. G. Horning, E. Jellum, M. Levenberg, O. A. Mamer, S. P. Markey, J. A. McCloskey, B. S. Middleditch, G. Petersson, W. T. Rainey, Jr., V. N. Reinhold, W. Sherman, D. H. Smith, C. C. Sweeley, B. R. Webster, and C. M. Williams. Volume I contains indexes, and volume II contains the spectral data. Spectra are arranged in order of molecular weight and are indexed alphabetically and by the most intense peak every 50 amu.

Registry of Mass Spectral Data. E. Stenhagen, S. Abrahamsson, and F. W. McLafferty, John Wiley and Son, New York, 1974, 4 volumes. This publication is based upon the Gothenberg/Cornell collection. From a data base of more than 25,000 spectra, 18,806 spectra of different compounds were selected for the printed version. Each spectrum was examined for errors and inconsistencies and is presented as a line diagram. Spectra are listed in order of molecular weight and there is a molecular formula index. Each spectrum is accompanied by a source code, usually a reference to the original literature. One magnetic tape version of the Registry contains 23,879 spectra of 18,806 different compounds, and an enlarged version contains 30,476 spectra without duplicates.

Catalog of the Mass Spectra of Pesticides. J. Freudenthal and L. G. Gramberg, National Institute of Public Health, Bilthoven, The Netherlands, 1975. Contains mass spectra (EI, 70 eV) of 294 pesticides. Line diagrams and tabulated data are included, together with indexes for the ten most intense peaks, compound name, and molecular formula.

Mass Spectra of Aromatic Acids and Amines. M. W. Couch and C. M. Williams, Veterans Administration Hospital and Department of Radiology, University of Florida, Gainesville, Florida, 1977. Mass spectral data for 180 compounds are contained in this compilation. Line diagrams are given for each compound, together with some high-resolution data and retention index values on OV-17 and SE-30. Mass spectral fragmentation routes substantiated by ion kinetic energy spectroscopy are presented. There are indexes of compound name, retention index, molecular weight, molecular formula, and the eight most abundant peaks.

Mass Spectra of Compounds of Biochemical Interest. C. E. Folsome, Laboratory for Exobiology, Department of Microbiology, University of Hawaii at Manoa, Honolulu, Hawaii 96822, 1977. Mass spectra of 495 compounds, mainly purines, pyrimidines, amino acids, and organic acids, and of their TMS derivatives are included. Also available on magnetic tape.

Mass Spectra of Organic Compounds. B. H. Kennett, K. E. Murray, F. B. Whitfield, G. Stanley, J. Shipton, P. A. Bannister, and K. Shaw, Commonwealth Scientific and Industrial Research Organization, Division of Food Research, North Ryde, N.S.W. 2112, Australia, 1977. Six volumes, each containing 75 spectra in line diagram and tabulated form, have been published and more are planned. Compounds encountered during studies of food flavors and atmospheric pollution are included.

Reference Guide to Mass Spectra of Insecticides, Herbicides, and Fungicides and Metabolites. Edited by S. I. M. Skinner and R. Greenhalgh, Agriculture Canada, Ottawa, Ontario KIA OC6, Canada, 1977, 136 pages. Mass spectra of 250 compounds are given as line diagrams. There is a five-peak index and indexes of common and trade names are also included.

Drugs Used in Horse Racing. Race Track Supervision, Agriculture Canada, Ottawa, Ontario, KIA OC6, Canada, 1978. Line diagrams and tabulated data for mass spectra of 300 legal and prohibited drugs are contained in this collection. GC, IR, TLC, and UV data are also given. There is an index of common names.

Handbook of Clinical Toxicology. I. Sunshine, CRC Press, Cleveland, Ohio 44128, 1978. A total of about 1000 EI and CI spectra are presented as line diagrams. Most compounds are drugs, metabolites, and by-products. There is an eight-peak index.

Mycotoxins Mass Spectral Data Bank. Association of Official Analytical Chemists, Washington, D. C. 20044, 1978. 104 mass spectra recorded by the FDA are included. There is an alphabetic listing, and one in order of increasing molecular weight.

4. Donations of Spectra

Most of the spectra in the collections mentioned were donated by persons working in about a dozen laboratories. A major criticism of these data collections is that they are insufficiently comprehensive. The obvious solution to this problem is for many more mass spectroscopists to donate additional spectra. Indeed, those who use these compilations have a moral obligation to contribute to their expansion.

5. Exercise

If you have recorded spectra which are not included in a published compilation, send them to the Mass Spectrometry Data Centre.

6. Suggested Reading

B. S. MIDDLEDITCH AND J. A. MCCLOSKEY, *A Guide to Collections of Mass Spectral Data,* Baylor College of Medicine, Houston, Texas, 1974. This guide, prepared for ASMS Committee VI, provided the basis for this chapter.

8

The Mass Spectrometry Data Centre

H. D. M. Jager, David C. Maxwell, and Andrew McCormick

1. History and Organization

The latter part of the 1950s marked the beginning of a rapid rise in the use of mass spectrometry as an analytical technique in several branches of science and as a subject of research. Until 1965 British mass spectrometrists kept in touch with developments by various informal methods, including the production of lists of published articles. Scientists of the United Kingdom Atomic Energy Authority, particularly the Atomic Energy Research Establishment at Harwell and the Atomic Weapons Research Establishment (AWRE) at Aldermaston, were especially active in these endeavors. These methods soon became inadequate to cope with this rapidly expanding subject so, after much discussion, a group of mass spectrometrists, representing industrial, academic, and government scientists, approached the British Government suggesting the establishment of a Data Centre.

After discussion of the subject and the appointment of a committee to assess requirements, the government department concerned, the Office for Scientific and Technical Information (OSTI), accepted the proposals and the Mass Spectrometry Data Centre (MSDC) was established in May 1966 at AWRE. This venue was selected as it already housed one of the largest mass spectrometry groups in the country and was able to provide library, computing, printing, and accounting services.

H. D. M. Jager, David C. Maxwell, and Andrew McCormick ● Atomic Weapons Research Establishment, Aldermaston, Reading, Berkshire RG7 4PR, England

The MSDC was operated and managed by AWRE staff with responsibility to, and financial backing from, OSTI. In the 10 years of the MSDC's existence, this sponsorship was first shared with, and finally taken over completely by, the Ministry of Technology (now the Department of Industry, DOI). AWRE has itself now been transferred from the United Kingdom Atomic Energy Authority to the Ministry of Defence.* From the start the MSDC has had the valuable guidance of an Advisory Committee, consisting of representative mass spectrometrists from government, industry, and the universities, which meets regularly to discuss progress and advise on future policy. It should be pointed out that while the British Government is willing to sponsor research activities, including such as undertaken by the MSDC, any purely service operations stemming from these activities should be run as far as possible on a cost recovery basis. Sponsorship of the MSDC is thus limited to meeting the difference between running costs and income from sale of products, and a fairly close surveillance is kept by DOI to ensure that routine services of MSDC are appropriately priced.

Close liaison has always been maintained with other national mass spectrometry groups, particularly in Europe and the USA. The establishment of the Centre was announced at the American Society for Testing and Materials Committee E-14 mass spectrometry meeting in 1966. Reports and papers have been given at major meetings since 1966, and an informal international advisory group has been convened at these meetings to discuss MSDC activities and proposals in relation to the needs of the mass spectrometry community. In addition, the MSDC has discussions directly with individual users of its services and other cooperating bodies, whose opinions are noted and if sufficiently supported, may influence future policy.

2. Activities of the Centre

The MSDC engages in three principal activities:

(i) compilation of the monthly *Mass Spectrometry Bulletin,*
(ii) collection and dissemination of mass spectra, and
(iii) assistance in compound identification by mass spectrometry.

---

*Since this paper was written the United Kingdom Chemical Information Service, a branch of the Chemical Society, has taken over management of the MSDC on behalf of the Department of Industry. The new address is: The Mass Spectrometry Data Centre, UKCIS, The University, Nottingham NG7 2RD, United Kingdom.

2.1. The Mass Spectrometry Bulletin

The *Bulletin* is designed to provide a complete current awareness service and contains a list of references selected from current literature, including reports, conference proceedings, patents, new books, etc. Each reference includes the full title (translated into English where necessary), names of all authors, full journal reference and CODEN (or report number and place of origin), and selected descriptors which supplement the information given in the title. The descriptors also form the basis for the subject index. An asterisk before the title indicates that the article contains no fewer than six relative ion abundances from a low-resolution spectrum in either numerical or graphical form. Provision is also made for comments which cannot be conveyed by the standard descriptors. Articles are grouped under eight main subject headings, as shown in Fig. 1.

2.1.1. Indexes

The Subject Index.   This index was originally based on a thesaurus of 242 terms. These were selected after consultation between the MSDC and its Advisory Committee and representatives of the German Physical Society. To cope with more recent developments, terms have been added and a few have been deleted from the list. The present total is 289 (see Fig. 2). Subject Index terms appear for each entry in the printed *Bulletin* in alphabetical order after the source references. In the Subject Index itself, each term is followed by the number of each entry containing this term as shown in Fig. 3.

The Compounds Classification Index.   This index lists article numbers under one or more of 88 group classifications, which indicate the types of compounds studied in the article numbers listed. Combination of these classifications leads to greater selectivity in retrieval. For example, Fig. 4 under the "chlorine" classification shows 13 article numbers referring to nitrogen heterocycles, only one of which occurs also under phosphorus.

The Elements Index.   This lists article numbers for elements to which specific reference is made.

The General Index.   Originally called the Materials Index, the General Index was an open index listing the various types of materials studied in specific articles, as well as materials and elements used as catalysts, filaments, surfaces, targets, etc. This has recently been expanded and subdivided to list ions, compounds, conference sites, geological and extraterrestrial terms and materials, alloys and systems, drug types, medical and pharmacological items, and species, etc.

		PAGE
CONTENTS LIST		i
NOTES ON MEETINGS, CONFERENCES		ii
CURRENT LITERATURE REFERENCES		
SECTION 1	INSTRUMENT DESIGN AND TECHNIQUES ANALYSERS, DATA PROCESSING DETECTION, ELECTRONICS, INSTRUMENTS, PARTICLE OPTICS, SAMPLE HANDLING, SOURCE, VACUUM, OTHER TECHNIQUES	33
SECTION 2	ISOTOPIC ANALYSIS, PRECISION MASS MEASUREMENT, ISOTOPE SEPARATION AND AGE DETERMINATION ETC	34
SECTION 3	CHEMICAL ANALYSIS	35
SECTION 4	ORGANIC CHEMISTRY GENERAL, BIOLOGICAL, ORGANO-METALLIC, INORGANIC, FOSSIL FUELS	35
SECTION 5	ATOMIC AND MOLECULAR PROCESSES GENERAL, ELECTRON IMPACT PROCESSES, PHOTON EFFECTS, HEAVY PARTICLE COLLISIONS, NEUTRAL SCATTERING	42
SECTION 6	SURFACE PHENOMENA AND SOLID STATE STUDIES	48
SECTION 7	THERMODYNAMICS AND REACTION KINETICS	49
SECTION 8	OTHER REFERENCES	50
SUBJECT INDEX		51
COMPOUNDS CLASSIFICATION		54
ELEMENTS INDEX		56
GENERAL INDEX		57
AUTHOR INDEX		60

FIG. 1. Contents page of the *Mass Spectrometry Bulletin*.

The Author Index. This lists all authors in alphabetical order with the relevant article numbers.

2.1.2. *Selection of Material for the Bulletin*

Mass spectrometry is used in connection with work in a wide range of scientific disciplines and, in consequence, relevant articles may be found in a large variety of scientific journals. In fact, over 300 primary journals contain a significant percentage of articles of interest and a similar number will contain at least one or two items of interest each year.

Before starting on the collection of references for the *Bulletin,* statistical studies were made to determine the numbers of articles on mass spectrometry appearing in the various journals during a limited period. As a result of this study, about 150 journals available in the Aldermaston Library were selected for visual scanning, and a further ten abstract journals were chosen to give some coverage of the balance of the world's technical literature. During subsequent years the number of journals scanned visually has been subject to frequent review in the light of retrieval statistics, and the number covered has gradually increased to the current figure of 240, of which about 50 are obtained from the British Lending Library.

Visual Scanning Methods. Scanning methods vary widely, depending on the type of journal, its subject, and the language in which it is printed. The great majority are scanned and indexed without translation from the original language. This is possible because mass spectrometry is a fairly new and specialized technique, developed mainly by English-speaking scientists and, in consequence, many of the key words and terms used in foreign languages are based on English roots and are therefore easily recognized. The almost universal use of the symbol *m/e* to indicate mass/charge ratio is also helpful. In addition, many of the foreign language journals contain English abstracts and/or titles and these, combined with a study of the text and figures, usually permit a satisfactory assessment of the content of the article to be made. In case of difficulty, selected passages are translated by experts in the various languages.

Originally, it was intended that all journals selected for visual scanning would be scanned page by page, but in practice, this has been found unnecessary. Many of the physics journals can be scanned on the basis of titles, or titles plus abstracts, with only an occasional reference to the full text. Organic chemistry journals, on the other hand, require more detailed examination because mass spectrometry is often only one of several analytical methods used for the determination of the structure or composition of an organic compound, and there may be no mention of its use in either title or abstract.

Many of the articles in organic chemistry journals mention that use has been made of mass spectrometry in the course of the work being reported, but the amount of detail given is considered to be insufficient to warrant inclusion in the *Bulletin.* If there is reason to believe that a full spectrum has been obtained for a specific compound, a letter is sent to the senior author asking for full details of available spectra. Besides providing a significant source of new spectra for the MSDC collection, this practice gives useful publicity for the Data Centre's activities.

MASS SPECTROMETRY DATA CENTRE

BULLETIN CODING SHEET № 2

TEMP 70	74	ART 80	NO BULLETIN ARTICLE NO
0 0 1 3	4 1	6 4 8 9 5	1 0 / 8 8 4

PUNCH ON EVERY CARD

SUBJECT INDEX

PUNCH COLUMNS 13-16

Term	Code	Term	Code
ABERRA	0040	INST PERF	3320
ABSORP	0060	INSTRS	3360
ABUN SENS	0080	ION COUNT	3400
ACCURACY	0120	IC RES	3420
ADSORP	0160	ION-E CON	3440
ADSORP	0200	IEL SPECT	3460
AGE DETN	0240	ION IMPCT	3480
EUROPE	0241	ION IMPL	3520
USSR	0242	IKES	3540
ASIA	0243	I-M REACT	3560
AFRICA	0244	ION PAIR	3580
AUSTRALIA	0245	ION TRAP	3590
N AMER	0246	IONIZAT	3600
S AMER	0247	ION EFF	3620
ANTARCTICA	0248	I P	3640
ISLANDS	0249	IONS	3660
MULT LOC	0250	ISOT ANAL	3680
OCEAN BEDS	0251	ISOT DILN	3720
ANALYSER	0280	ISOT EFF	3740
ANAL ELEC	0320	ISOT FRAC	3760
APP POT	0360	ISOT SEP	3800
ARCHAE	0400	K E	3840
AT BEAM	0440	K IN IF	3880
AT WT	0480	KIN REA	3920
AUGER PR	0540	KINO CELL	3960
AUTO-ION	0560	LAB NO	4000
AUTOMAT	0600	LASER	4040
BACKGR	0640	LEAK DET	4080
B F SPECT	0660	LIB MATCH	4100
BEAM PROF	0680	LIFETIMES	4120
BIBLIO	0690	LEEI	4160
BIOCHEM (circled)	0710	LRPD (circled)	4200
BOOK	0760	MAG ANAL	4240
BRAN RAT	0800	MAG FO	4280
CALL SAM	0840	MAG FPI	4320
CALI FUM	0880	MASS DISC	4360
CATALYSIS	0920	MASS FRAG (circled)	4380
CHANNG	0960	MASS MEAS	4400
CHARA CURV	1000	MATRIX EF	4480
CHARGE DN	1040	MEDICAL (circled)	4500
CHARGE EX	1080	MEMORY	4520
CHEM BEN	1120	MERG BEAM	4540
CHEM ION	1140	METAST	4560
CHEM REAC	1160	META STUD	4580
CLUSTERS	1180	MICROPROB	4600
COINC METH	1260	MINERALOG	4640
COLL ELEC	1280	MOBILITY	4680
COMBUST	1320	MOD TECH	4700
COMP PROG	1370	MOL BEAM	4720
CONF	1380	MOL IONS	4800
COOL SOLN	1400	MOL SEP	4840
COSMOLOGY	1440	MOL VEL SEL	4890
CONDENS	1480	MEEI	4920
XED BEAMS	1520	MONOPOLE	4960
XEO FIELD	1540	MOV ELEC	5000
X SECN	1560	MULT PN IN	5020
CRUCIBLE	1600	MULT DET	5040
DATA PROC	1640	MULT C ION	5080
DC DISC	1680	MULT SPECS	5120
DECONVOL	1700	NAT 1 AB	5160
DESORPT	1720	NAT PROD	5180
DETACHMT	1740	NEG IONS	5200
DETECTION	1760	NEUT PI	5240
DETN LIM	1800	NEUT REAC	5280
DETR OPT	1840	NEUT REAC	5320
DETR RESP	1880	NUCL PHYS	5360
DEU ANAL	1900	ODOUR	5370
DEU LAB	1920	OL DATA PROC	5375
DIFFUSION	1960	OPT SPECT	5380
DIR EFF	2000	OSCILL STR	5390
DISSOC	2040	OTH ANAL	5400
DISS ION	2060	OTH APPL	5440
DF SPECT	2080	OTH DET	5480
DOU INLET	2120	OTH PIF	5520
ELEC DETN	2200	OTH TECH	5560
EPP 1	2240	OTH SPEC	5600
ELN CAP	2300	OUTGAS	5640
EEL SPECT	2310	PALAE	5680
ELN IMP	2320	PARAB TS	5720
ELN MULT	2360	PART OPT	5760
ELECT SPEC	2380	PM ADV	5800
ELINCS	2400	PENN ION	5810
ELECTRONS	2440	PHARMACOL (circled)	5815
ELSPRAY MS	2460	PHOT EL SP	5820
ES ANAL	2480	PHOT PL	5840
ESFO	2520	PHOT-ION	5880
ELEM LIST	2530	PHOTOLYS	5920
ELEM MAP	2540	PHOT IMPACT	5940
ENER DIST	2560	PLASMA	5960
EM LEV	2580	PLAS GC/MS	5980
EVAP	2600	POLARIZN	6040
EEC REACT	2640	POLLUTION	6050
EX ITN	2680	POWDER T	6060
EXO PART	2720	PRECISION	6120
F DESORPT	2740	PREDISSOC	6140
FD IONIZ	2760	PRESS EFF	6160
FISS PROD	2800	PRETREAT	6200
FLAVOUR	2820	PROBE INL	6220
FLOW AFTER	2830	PROC CONT	6240
FRACN	2840	PROC HMPD	6280
FRAG MECH	2880	PROT AFF	6320
FREE RAD	2960	PULS SOU	6340
GAS ANAL	3000	PUMPING	6360
GC/MS	3020	PURIFICN	6400
GAS DISC	3040	PYROLYSIS	6440
GEOCHEM	3080	QUADRUP (circled)	6480
GEOCHEM	3100	Q A OH	6520
GEOLOGY	3120	QE THEORY	6540
GEOTHERM	3140	QUENCHING	6550
HALF LIFE	3160	RAD CHEM	6560
HEAT INL	3200	RANGE	6640
HOMOGEN	3240	RATE CON	6680
HPMS	3260	REARR	6720
HRPD	3280	RECOMB	6760
		REFLECTION	6800
		REP ANAL	6840
		RG ANAL	6880
		RESLN	6920
		RESONANCE	6960
		RET POT MEAS	6980
		REVIEW	7000
		RF DIS	7010
		RF SPECT	7020
		RYD STATE	7030
		SAMP ENR	7040
		SAMP HAND	7080

SAMP PREP 7120	SCAN PROB 7160	SCATTER 7200	SCAT ELECT 7205	SCAT IONS 7210	SCIN DET 7280	SCRAMBLE 7300	SECO EMSN 7320	SEC EM EL 7330	SEC EM IONS 7340	SENSIT 7360	SHOCK TUBE 7400	SIN CRYST 7440	SS DETR 7480	SOURCE 7520	SOU ELEC 7560	SPACE CH 7600	SPACE RES 7640	SSMS 7660	SPEC COR 7680
SPUT 7720	STOP POWER 7740	STRIPPING 7750	STRUC IDENT 7765	SUBLIM 7800	SURFACE 7840	SURF ION 7860	TEMP EFF 7920	THEORL 7940	THEOR MS 7960	THERM ANAL 7980	TD QUANT 8040	TLC/MS 8060	TOF SPECT 8080	TRMS 8120	TRACE ANAL 8160	TRANSF REACT 8180	TRANS PR 8200	TRANSIT 8220	U/L ATM 8240
VACUUM 8280	VAC DISC 8320	VAC LOCK 8360	VAC VALV 8400	VEL DISTR 8410	VIB 8420	WK FUNCTN 8440	YIELD 8480			MASS SPEC 4440									

PUNCH 19–22																		Cont. 70	Type
																		0 0	2 3
Sat 1000	= 1010	≡ 1020	Allc 1030	Alic + 1040	Ph 1050	Ph + 1060	Ph/Al 1070	Other 1080	hN 1090	hO 1100	hS 1110	hOth 1120	OH 1130	-O- 1140	Acel 1150	Ketl 1160	CHO 1170	CO 1180	COOH 1190
Anh 1200	Est 1210	MeEst 1220	Salt 1230	Ltne 1240	COhal 1250	CONH 1260	Ltam 1270	OthO 1280	NH_2 1290	CN 1300	NO_2 1310	NN 1320	OthN 1330	SH 1340	C=S 1350	$-SO_2$ 1360	CSSH 1370	OthS 1380	F 1390
Cl 1400	Br 1410	I 1420	P-3 1430	P-5 1440	TMS 1450	Si 1460	Bate 1470	B 1480	Oth 1490	I A 1500	I B 1510	II A 1520	II B 1530	III A 1540	Lanth 1550	Actin 1560	III B 1570	IV A 1580	IV B 1590
V A 1600	VI A 1610	VII A 1620	VIII 1630	M-CO 1640	Alk 1650	Am Ac 1660	Sug 1670	Flav 1680	Myco 1690	Nucl 1700	Ster 1710	1-T 1720	1 5-T 1730	2-T 1740	3-T 1750	4-T 1760	Isop T 1770	Carot 1780	Porph 1790
Chlor 1800	P-Lip 1810	Gibb 1812	Prost 1814	Oth 1820	Polym 1830	D-lab 1840	λ-lab 1850												

FIG. 2. Coding sheet 2 used to prepare the entry shown in Fig. 6 with relevant Subject Index thesaurus terms and Compounds Classification codes circled.

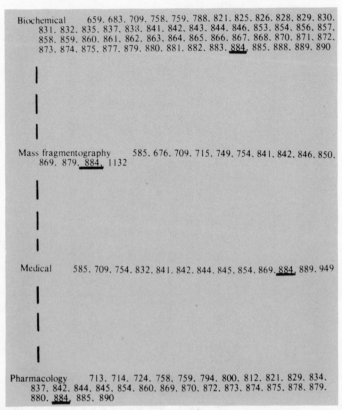

Biochemical   659. 683. 709. 758. 759. 788. 821. 825. 826. 828. 829. 830.
     831. 832. 835. 837. 838. 841. 842. 843. 844. 846. 853. 854. 856. 857.
     858. 859. 860. 861. 862. 863. 864. 865. 866. 867. 868. 870. 871. 872.
     873. 874. 875. 877. 879. 880. 881. 882. 883. 884. 885. 888. 889. 890

Mass fragmentography   585. 676. 709. 715. 749. 754. 841. 842. 846. 850.
     869. 879. 884. 1132

Medical   585. 709. 754. 832. 841. 842. 844. 845. 854. 869. 884. 889. 949

Pharmacology   713. 714. 724. 758. 759. 794. 800. 812. 821. 829. 834.
     837. 842. 844. 845. 854. 860. 869. 870. 872. 873. 874. 875. 878. 879.
     880. 884. 885. 890

FIG. 3. Extract from the Subject Index of the *Mass Spectrometry Bulletin* showing entries for the article of Fig. 6.

Computer-Based Search Services. Since work was started in 1966, some use has been made of computer-based search services for the retrieval of material from sources not scanned by visual methods and from abstract journals such as *Nuclear Science Abstracts.* Since 1970 the use of such services has been confined to those which are available locally at AWRE.

A study completed by Searle in 1970 confirmed earlier work which indicated that computer-based services using relatively simple and inexpensive profiles were unable to retrieve more than about 50% of the material obtained by visual scanning methods.

On several occasions during the past few years, claims have been made of better than 90% retrieval by computer techniques, but in all the cases examined it was found that sophisticated, and therefore expensive, profiles were used on a limited data base, and that the subject was one having a specific and well-defined terminology, such as high-energy physics. A check of 60 items picked at random from articles selected for inclusion in the organic chemistry section of the *Bulletin* showed that only 26 (or less than 44%) contained any mention of mass spectrometry (or relevant terms) in the title or abstract. (Articles which contained no abstract were omitted.) A computer search of these articles based on titles, or titles and abstracts, would therefore be using a data base containing less than 44% of the relevant material. One based on the titles alone would probably retrieve only about 20% of the organic chemistry articles selected for the *Bulletin*.

The use of computers allows a large increase in the size of the data base which can be searched, but the number of items retrieved has been

Halogens
Fluorine
    Acyclic   662, 786
    Alicyclic   817
    Aromatic   750
    Heterocyclic (nitrogen)   874, 879
    Other natural products   841
Chlorine
    Alicyclic   702, 705, 706, 732
    Aromatic   676, 750, 751
    Heterocyclic (nitrogen)   672, 710, 716, 719, 722, 729, 789, 791, 794, 872, 874, 875, 894
    Heterocyclic (oxygen)   724, 749
    Heterocyclic (other)   796, 804, 805
    Steroid   774
    Other natural products   843, 850, 851, 853
Bromine
    Alicyclic   818, 819
    Aromatic   680
    Heterocyclic (nitrogen)   660, 746, 752, 789, 791, 807, 875
Iodine
    Acyclic   744
    Aromatic   901
Other non—metallic elements
Phosphorus — trivalent
    Alicyclic   903
    Aromatic   893, 898, 900
    Heterocyclic (nitrogen)   752, 820
    Heterocyclic (oxygen)   896
Phosphorus — pentavalent
    Heterocyclic (nitrogen)   672, 872, 874
    Heterocyclic (oxygen)   668
    Heterocyclic (other)   768

FIG. 4. Extract from the Compounds Classification Index of the *Mass Spectrometry Bulletin*.

relatively small and has rarely exceeded about 6% of the total number of articles included and indexed in the *Bulletin*.

2.1.3. Use of the Computer for Elimination of Duplicate Items

Because of the considerable overlap between the data bases searched by various methods, it is inevitable that the same item should often be retrieved by several different methods. By making it possible to provide up-to-date listings of items which have already been noted and indexed for inclusion in the *Bulletin*, the computer serves as a valuable aid toward ensuring that a minimum of duplication occurs. Three types of listings are used, as detailed below:

(1) A monthly listing of items ready for insertion giving accession number, first author's name and initials, and the first 50 characters of the title of the article or report, in alphabetical order of authors' names.

(2) A cumulative monthly listing, similar to the above, but listing all items included in the six latest issues of the *Bulletin*, with the addition of the *Bulletin* reference number.

(3) Every third month a cumulative listing of all items included in the most recent three issues of the *Bulletin* is compiled with the items arranged in alphabetical source order. Each entry gives the *Bulletin* article number, the accession number, and the first 94 characters of the source reference. A six-month cumulative listing is also produced, in June and December.

Unfortunately, there is a time lag between the location of an article and completion of processing for computer sorting, so that there are always some articles which cannot be checked by mechanical methods. These have to be checked by visual search of accession lists and journal order records.

2.1.4. Production and Printing

Printed articles are passed on by the scanner to the indexers, who complete the specially prepared coding sheet illustrated in Figs. 2 and 5. Some expertise is required for this, as the relevant subject index terms (Fig. 2) are usually not explicitly stated in the article: a good working knowledge of structure and nomenclature in organic chemistry is essential. Standard 80-column punched cards are prepared from the coding sheets, and each week a batch of these is put through a computer program which arranges them in order and allocates a weekly reference number. Certain checks for punching errors and omissions are made and a weekly printout is produced. Errors noted on this are coded and punched onto correction cards. Every four weeks the batches are combined, together with any necessary correction cards, and a printout of the complete *Bulletin*

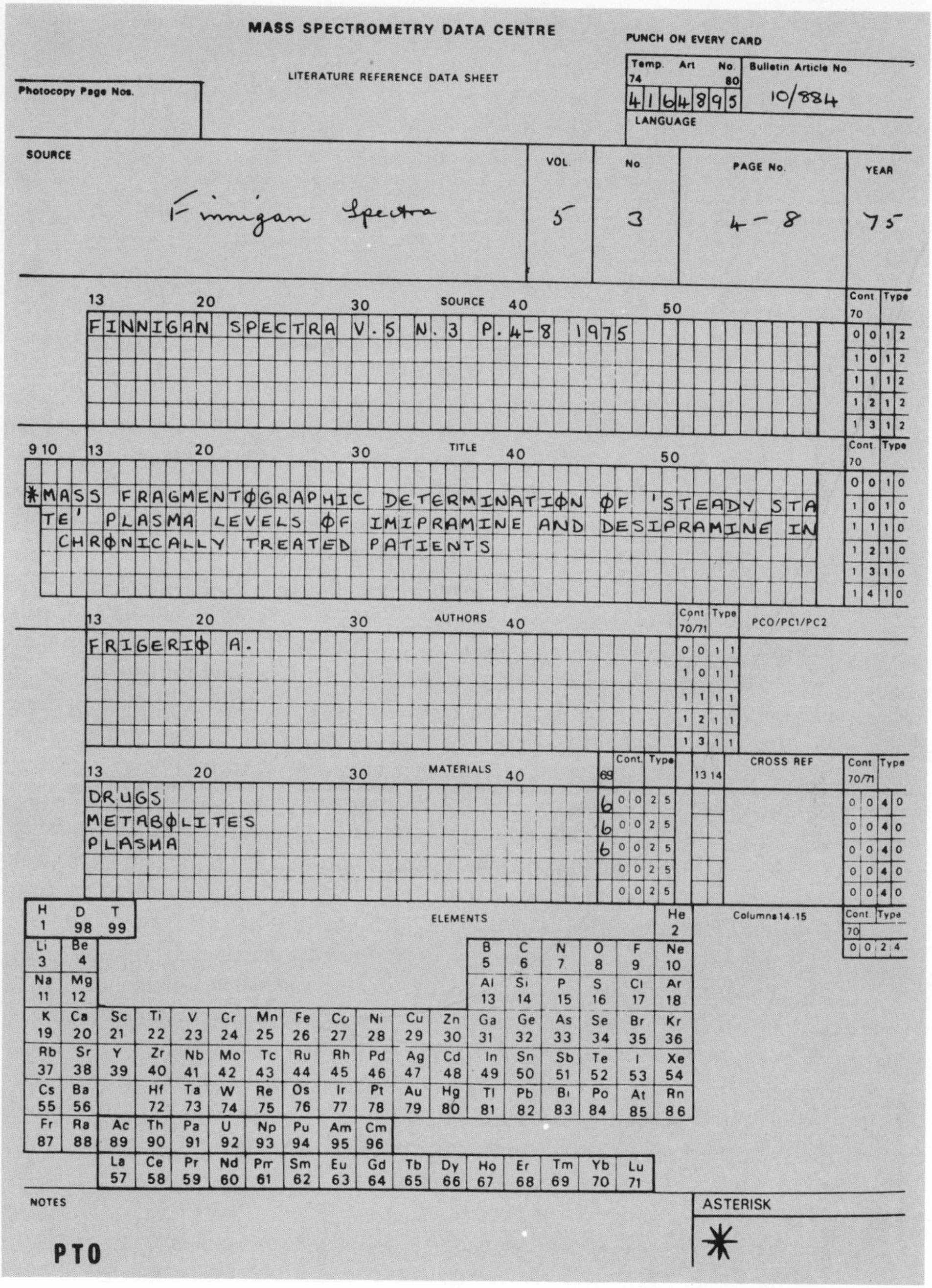

MASS SPECTROMETRY DATA CENTRE

LITERATURE REFERENCE DATA SHEET

PUNCH ON EVERY CARD

Photocopy Page Nos.

Temp. 74	Art	No. 80	Bulletin Article No
4 1 6 4 8 9 5			10/884

LANGUAGE

SOURCE: Finnigan Spectra

VOL.	No.	PAGE No.	YEAR
5	3	4 - 8	75

SOURCE (cols 13–50):
FINNIGAN SPECTRA V. 5 N. 3 P. 4-8 1975

TITLE (cols 9–50):
*MASS FRAGMENTOGRAPHIC DETERMINATION OF 'STEADY STA
TE' PLASMA LEVELS OF IMIPRAMINE AND DESIPRAMINE IN
CHRONICALLY TREATED PATIENTS

AUTHORS (col 13):
FRIGERIO A.

PCO/PC1/PC2

MATERIALS (col 13):
DRUGS
METABOLITES
PLASMA

CROSS REF

ELEMENTS

H 1	D 98	T 99											He 2
Li 3	Be 4							B 5	C 6	N 7	O 8	F 9	Ne 10
Na 11	Mg 12							Al 13	Si 14	P 15	S 16	Cl 17	Ar 18
K 19	Ca 20	Sc 21	Ti 22	V 23	Cr 24	Mn 25	Fe 26	Co 27	Ni 28	Cu 29	Zn 30	Ga 31	Ge 32	As 33	Se 34	Br 35	Kr 36
Rb 37	Sr 38	Y 39	Zr 40	Nb 41	Mo 42	Tc 43	Ru 44	Rh 45	Pd 46	Ag 47	Cd 48	In 49	Sn 50	Sb 51	Te 52	I 53	Xe 54
Cs 55	Ba 56		Hf 72	Ta 73	W 74	Re 75	Os 76	Ir 77	Pt 78	Au 79	Hg 80	Tl 81	Pb 82	Bi 83	Po 84	At 85	Rn 86
Fr 87	Ra 88	Ac 89	Th 90	Pa 91	U 92	Np 93	Pu 94	Am 95	Cm 96								
		La 57	Ce 58	Pr 59	Nd 60	Pm 61	Sm 62	Eu 63	Gd 64	Tb 65	Dy 66	Ho 67	Er 68	Tm 69	Yb 70	Lu 71	

NOTES

PTO

ASTERISK ✱

FIG. 5. Coding sheet 1 used in preparation of the example shown in Fig. 6. Entries in the Materials section appear in the General Index of the printed *Mass Spectrometry Bulletin*. The first two digits of the temporary article number result in the article appearing in section 4 (Organic Chemistry), subsection 1 (Biological).

Agr. Biol. Chem. V.39 N.8 P.1687-8 1975
Biochemical. Mass spectra. Natural products. Structure identification. Tribolium castaneum
884   *   Mass fragmentographic determination of 'steady state' plasma levels of imipramine and desipramine in chronically treated patients
Frigerio A.
Finnigan Spectra V.5 N.3 P.4-8 19/5
Biochemical. Low resolving power data. Mass fragmentography. Medical. Pharmacology. Quadrupole. Sensitivity. Drugs. Metabolites. Plasma
885   Alkaloids of Lupinus formosus. Structures and spectral properties
Fitch W.L.

FIG. 6. An entry from the February 1976 issue of the *Mass Spectrometry Bulletin*.

contents is produced. This is further scrutinized and any necessary corrections made, again by the correction card method, and the process repeated, if necessary, until a satisfactory product is obtained. Finally, a series of sort programs is used to arrange the articles in section and subsection order and to produce the indexes.

For the first four years, the *Bulletin* was produced by photoreduction of computer printout. This was not very satisfactory because of the restriction to capital letters, which are not easy to read, and because of the variable quality of computer print. Printing was therefore transferred to computerized typesetting using the PHOTON 713 phototypesetter. This involved the writing of a complicated program to produce the information in a form which, through the PHOTON, would give an acceptable printed output. Variable character width, line justification, different fonts, and capital and lower case are now all available to produce the easily readable version illustrated in Fig. 6, which shows the entry for the article whose code sheet appears as Figs. 2 and 5. As the PHOTON equipment is now obsolete, Her Majesty's Stationery Office has rewritten the phototypesetting software so that a more efficient modern typesetter can be used, but the appearance of the finished product will be virtually unchanged.

Introductory material, such as covers, content pages, announcement of forthcoming meetings, etc., is typeset in the ordinary way.

Every year the indexes of the previous 12 issues are combined and published as a cumulative Annual Index. An example of the cumulative Author Index is shown in Fig. 7. The *Bulletin* from its inception was

Mickiewicz M.   6808, 6809
Micovic I.V.   5646
Middleditch B.S.   2176,   2232,
   4723, 5777, 6205, 6291
Middleton D.   203
Middleton S.   3795, 5639
Midha K.K.   2224, 3350, 3351,
   3362

FIG. 7. An extract from the Author Index of the 1975 Annual Cumulative Index of the *Mass Spectrometry Bulletin*.

designed with computer retrieval in mind and is available also in magnetic tape form for this purpose.

The total number of citations in the first nine volumes is 57,051 and there are currently some 700 regular subscribers in 40 different countries. The *Bulletin* is thus the principal current awareness tool for practicing mass spectrometrists and features regularly as an invaluable aid to authors of review articles. Coverage is generally praised, but the criticism is made from time to time that the contents of published articles are too scantily described. However, to change the format to that of an abstracting journal would both slow down production and enormously increase costs and, therefore, it is unlikely that any departure will be made from the present compromise position of using descriptive terms to indicate the scope and content of each article.

2.2. Collection and Dissemination of Mass Spectra

2.2.1. Full Spectra

Before the advent of the MSDC, reference mass spectra were available in printed form in three different collections only, issued, respectively, by the American Petroleum Institute Project 44 (API), the American Society for Testing and Materials, and the Dow Chemical Company (Chapter 7). With assistance from the Massachusetts Institute of Technology and the Jet Propulsion Laboratory, these were keypunched and transferred to magnetic tape in what has become the standard format for the MSDC, which provides for contributor's name, compound name, molecular weight and formula, instrument conditions, and compound classification codes. At this time the continuing practice of soliciting additional spectra was also begun. Requests are sent to all authors of articles appearing in the *Bulletin* which contain, or suggest that there might be available, new spectral data (see also p. 193).

Spectra, on receipt, are subjected to some simple quality checks, e.g., agreement of molecular weight and formula with the spectral data, and are then keypunched along with the relevant compound classification codes. A printout of the resulting magnetic tape version is further checked for errors and, when it is thought that a satisfactory transcription of the data has been achieved, a copy is returned to the originator for checking. Finally, the tape of new spectra is checked against the existing collection for possible duplications and is then added as an update. Updates are issued in batches of 100 spectra, plus any additional spectra which may in the meantime have been added to the API and other collections. A file of errors reported by customers is maintained, and the whole collection is from time to time upgraded. By late 1977, 10,000 spectra contributed directly to the MSDC had been processed. These, together with the API and other collections mentioned above, bring the total number of complete spectra available from the MSDC on magnetic tape to 16,900, and there are currently some 4000 others being processed. This library of reference spectra (commonly referred to as the "Aldermaston Library") may be purchased by individual laboratories for their own use or is available in a reformated version as part of the compound identification facilities offered by several laboratory computer system suppliers.

The first 7000 MSDC spectra were also issued as data sheets giving, in tabular and line diagram form, the same information as the tape version, but with the addition of a structural formula. With the growing use of computers by customers and the escalating costs of printing and distribution, it has been found necessary to discontinue the sheet version.

2.2.2. Abbreviated ("Eight-Peak") Spectra

In cooperation with ICI (Organics) Ltd., the Centre produced, in 1971, an *Index* of the eight major peaks in each of 17,000 mass spectra. The file was compiled from the MSDC collection, some 2000 ICI spectra, and about 5000 extracted from the published literature. In addition to the spectral data, molecular weight, formula, name, and source were included for each compound. The *Index* was arranged in three tables as follows:

   (i) arranged in molecular weight order, subordered on formula;
  (ii) arranged according to molecular weight, subordered on m/e value of the most abundant ion; and
 (iii) arranged according to a particular m/e value occurring, respectively, as the first, second, and third most abundant ion.

The *Index* proved to be very successful as a manual method of compound identification, so a second, larger, edition was published in 1974, containing data from 31,101 spectra. For the second edition, the opportunity was taken of correcting errors in the first edition and of using phototypesetting as for the *Bulletin*. Examples from Tables 1 and 3 of this much more readable format are given in Figs. 8 and 9.

2.3. Assistance in Compound Identification by Mass Spectrometry

It has always been the policy of the MSDC not only to supply information and data, but to assist wherever possible in the efficient exploitation of its resources. To this end one of the first computer-based compound identification schemes was devised in cooperation with the

MASS TO CHARGE RATIOS								INTENSITIES								Parent	C	H	O	N	Cl	Br	F	S	P	B	Si	X	M.W.	Lit Ref.	No	COMPOUND NAME
114	63	57	115	50	81	75	94	100	18	8	8	5	4	4	4		6	4	–	–	–	–	2	–	–	–	–	–	114	52895	M2152	1,3-difluorobenzene
77	79	51	39	50	78	27	114	100	95	70	50	45	45	32	26		6	7	–	–	1	–	–	–	–	–	–	–	114		X 0510	Vinyl chloroprene
44	38	40	72	71	114	46	70	100	68	66	46	30	23	20	15		6	10	–	–	–	–	–	1	–	–	–	–	114		X 0323	Diallyl sulfide
73	45	41	72	114	39	71	99	100	98	88	87	83	82	60	55		6	10	–	–	–	–	–	1	–	–	–	–	114	53907	M6160	Diallyl sulphide
81	114	80	41	39	27	79	45	100	82	64	50	43	43	39	34		6	10	–	–	–	–	–	1	–	–	–	–	114		X 1539	Cyclohexene sulfide
85	86	114	80	39	27	45	60	100	61	57	56	46	43	41	34		6	10	–	–	–	–	–	1	–	–	–	–	114		Y 1879	7-thiabicyclo[2.2.1]heptane
43	99	71	114	57	15	27	42	100	27	13	9	6	5	3	3		6	10	2	–	–	–	–	–	–	–	–	–	114		Z 0374	2,5-hexanedione
71	71	41	27	44	39	114		100	77	32	23	9	9	7	5		6	10	2	–	–	–	–	–	–	–	–	–	114		Z 0375	Vinyl butyrate
55	42	84	41	56	28	70	27	100	76	37	35	34	19	16	14	13.31	6	10	2	–	–	–	–	–	–	–	–	–	114		Z 0388	6-hydroxyhexanoic acid lactone
45	86	27	43	29	31	41	44	100	38	33	32	20	17	16	16	0.14*	6	10	2	–	–	–	–	–	–	–	–	–	114		Z 0377	1,2-bis(vinyloxy)ethane

FIG. 8. An extract from Table 1 of the *Eight Peak Index of Mass Spectra*, 2nd edition. The numbers in the "parent" column are relative intensities of parent ions which do not occur as one of the eight major peaks. The "Lit. Ref." entry gives volume (first digit) and article number in the *Mass Spectrometry Bulletin*.

1868   [82]

m/e	MASS TO CHARGE RATIOS	M W	INTENSITIES	Parent	C	H	O	N	Cl	Br	F	S	P	B	D	Si	X	COMPOUND NAME	Lit.Ref.	No
82	163 135 134 109 108 120	163	100 76 61 40 34 18 16		7	9	-	5	-	-	-	-	-	-	-	-	-	5,7,8-trimethyltetrazolo(1 5-c)pyrimidine	51536	M5035
82	168 84 30 56 112 125 44	168	100 70 42 28 26 26 21 19		10	20	-	2	-	-	-	-	-	-	-	-	-	4,4'-bipiperidyl		C 0743
82	182 110 95 124 167 138	182	100 35 23 20 6 5 5		10	14	3	-	-	-	-	-	-	-	-	-	-	3-exo,8,8-trimethyl-7-oxabicyclo (3,3,0) octa-2,6-dione	43378	M3667
82	184 44 54 156 102 128 51	184	100 99 66 58 56 55 54 28		12	8	2	-	-	-	-	-	-	-	-	-	-	2-phenyl benzoquinone		D 0720
82	224 268 81 69 155 83 56	492	100 54 52 30 27 26 26 21	0.00	25	36	4	-	-	-	-	-	-	-	-	-	-	Methyl-9,10-diisonadeuteriotrimethylsilyloxy)-octadecanoate	24777	L 4126
82	330 83 148 331 248 149 332	330	100 34 34 19 18 9 7 4		20	26	4	-	-	-	-	-	-	-	-	-	-	Di-o-cyclohexyl phthalate	50863	M6195
82	357 282 277 375 202 374 358	375	100 99 95 63 59 52 47 32		19	9	2	1	-	-	-	-	-	-	18	2	-	Salicyl-anilide-di-(perdeuterio-tms)	55255	M7113
18	82 17 84 83 111 29 47	164	100 22 22 14 8 8 7 7	0.00	2	3	2	-	3	-	-	-	-	-	-	-	-	Chloral hydrate		Z 1086
41	82 27 83 43 54 29 55	111	100 68 62 52 50 44 38 37	0.34	7	13	-	1	-	-	-	-	-	-	-	-	-	Nor-heptanenitrile		X 1941
41	82 43 54 55 83 29 39	111	100 69 56 46 40 38 31 22	0.00	7	13	-	1	-	-	-	-	-	-	-	-	-	Nor-hexyl cyanide	22289	L 1774
41	82 43 83 54 55 27 29	111	100 95 63 58 56 51 30 28	0.70	7	13	-	1	-	-	-	-	-	-	-	-	-	Nor-hexyl nitrile	(1)	L 0009
82	82 55 57 67 83 39 54	129	100 93 89 73 53 51 45 35	0.23	6	11	2	1	-	-	-	-	-	-	-	-	-	Nitro cyclohexane		C 0186

FIG. 9. An extract from Table 3 of the *Eight-Peak Index of Mass Spectra* showing entries for *m/e* 82 as the most intense and then as second most intense peak in the spectrum. This index is particularly useful for cases in which no molecular ion is detected.

Unilever Research Laboratory at Sharnbrook, Bedfordshire, England. This is still available from the MSDC as a series of programs suitable for IBM 360 and 370 computers. The programs enable the user to convert the MSDC magnetic tape file of spectra into a format on a magnetic disk more suitable for searching, as well as the input and search programs themselves. Various search options are available using either "eight-peak" spectra or spectra reduced in the sense of having only the n (usually 2) largest peaks in each interval of m (usually 14) mass units in the comparison file, as is the commonly adopted procedure for most search systems. Output consists of the ten nearest matches to the input "unknown" spectrum, giving molecular weights, formulas, names, and a similarity index.

A system such as this has several disadvantages. It is necessary to have access to a large computer, and the level of usage has to be such as to justify maintaining a large data file permanently on disk. The alternative of batching up work and regenerating the search files for each session is time consuming and prohibitively expensive. Further, it is often necessary to have identification performed immediately after obtaining the spectrum, which is usually impractical with a batch-processing system, and there is the added expense and irritation of system maintenance and updating of the reference library.

To overcome the difficulties mentioned above, the MSDC has collaborated with the U.S. National Institutes of Health and the U.S. Environmental Protection Agency in making available a time-sharing system which is easily accessible by ordinary telephone from most cities in Europe and North America. This system is described in detail in Chapter 9. This interactive method of searching has several advantages. It is very easy to use, is continuously (and immediately) available, and is relatively inexpensive in its use of computer processing time; the user does not have to concern himself with maintenance and updating.

The success of this central system, with costs shared by many users, has encouraged the extension of available services beyond compound identification from mass spectra alone. Perhaps the most interesting development, as far as users of mass spectrometry are concerned, will be computer searching of the *Mass Spectrometry Bulletin*.* The *Bulletin* was designed with this aim in view and now, ten years later, the idea is on the point of coming to fruition (though it should be added that Biemann's group at MIT has had an in-house system for searching *Bulletin* tapes for some years now). Programming and testing the *Bulletin* retrieval system at the NIH is nearing completion, and there remains only the reaching of an agreement with the time-sharing vendor before it is publicly available. An example of a typical search is shown in Fig. 10. After logging in, the

*Note added in proof. The *Bulletin* search is now an integral part of MSSS (see Chapter 9). The data base is updated annually on completion by MSDC of each volume of the *Bulletin*.

.MSBULL

BULLETIN SEARCH SYSTEM

PLEASE TYPE YOUR INITIALS
USER:VAA
TO LIST OPTIONS, TYPE OPT
USER RESPONSE:OPT
TO SEARCH ON SUBJECT CODES, TYPE CODE
TO LIST THE SUBJECT CODES, TYPE SLIST
TO SEARCH ON COMPOUND CODES, TYPE COMP
TO LIST THE COMPOUND CODES, TYPE CLIST
TO SEARCH FOR ELEMENTS, TYPE ELE
TO EXIT FROM THE PROGRAM, TYPE EXIT

USER RESPONSE:CODE
SUBJECT CODE SEARCH
TYPE SUBJECT CODE
CR TO EXIT, 1 FOR LIST OF REFERENCES

CODE: 1140
       # REFS   CODES
         184   1140

NEXT CODE: 710
       # REFS   CODES
          44   1140 710

NEXT CODE: 5180
       # REFS   CODES
          32   1140 710 5180
NEXT CODE: 2320
       # REFS   CODES
          15   1140 710 5180 2320

NEXT CODE: 7765
       # REFS   CODES
           8   1140 710 5180 2320 7765
NEXT CODE: 1
    901435   CHARACTERIZATION OF DIPEPTIDES BY ELECTRON IMPACT AND
              CHEMICAL IONIZATION MASS SPECTROMETRY   SCHIER G.M.
              HALPERN B.   MILNE G.W.A.   BIOMED. MASS SPECTROM. V.1 N.4
              P.212-8 1974

    901460   THE ANTIBIOTIC XK-41 COMPLEX. PART 2. STRUCTURAL IDENTIFICATI(
              EGAN R.S.   MUELLER S.L.   MITSCHER L.A.   KAWAMOTO I.   OKACH
              KATO H.   YAMAMOTO S.   TAKASAWA S.   NARA T.   J. ANTIBIOT. (T(
              V.27 N.7 P.544-54 1974
    902021   CHEMICAL IONIZATION MASS SPECTRA OF AMINO ACID DERIVATIVES
              HUNT D.F.   STAFFORD G.   DEVINE C. 26TH SOUTHEASTERN
              REGIONAL ACS MTG., NORFOLK SCOPE, VA., USA 23-25 OCT, 1974 ABS
              N.274 1974

FIG. 10. An example (truncated) of a literature search, using the *Mass Spectrometry
Bulletin*, Vol. 9, on magnetic disk and the interactive software developed at the NIH.

user, whose initials are V.A.A., asks to see a list of options available for searching the *Bulletin* data base. He chooses to search by subject code, continually narrowing the scope of his search to single out papers of special interest. In response to his first request, he finds that there are 184 references to papers on CIMS (code 1140: See Fig. 2), but only eight papers refer to the use of both EI and CIMS in the identification of naturally occurring biochemicals. The first three references printed are also shown in Fig. 10.

3. Suggested Reading

Mass Spectrometry Bulletin, published by Her Majesty's Stationery Office, London, monthly.

Eight-Peak Index of Mass Spectra, Her Majesty's Stationery Office, London, 2nd ed., 1974.

9

The Mass Spectral Search System

Stephen R. Heller

1. Introduction

With the mass spectrum of a pure compound in hand, only one additional but major step is required to make a correct identification: the interpretation of the data. Interpretation may be made by applying the theory of mass spectra, and the rules of fragmentation of ions in the gas phase. This process is tedious, and it is difficult to sustain the necessary deductive reasoning process for the long periods of time required to make a large number of identifications. Lack of sufficient knowledge about the details for the fragmentation process further limits the effectiveness of this approach in interpreting some spectra. Instead, the mass spectrometrist may take advantage of the collections of reference mass spectra that have been accumulated in recent years. Empirical methods have been developed for searching a file of reference mass spectra to find a similar or exact match of an experimental mass spectrum.

Any empirical search and match system has two fundamental components:

(i) a data base that is nothing more than an organized collection of reference data, and

(ii) the system or approach that a user takes to search the data base.

Stephen R. Heller • U. S. Environmental Protection Agency, MIDSD, PM-218, Washington, D.C. 20460

Manual searching of printed data bases was explored first, and some elaborate indexing schemes were developed to facilitate the user's search. Nevertheless, all manual search systems are rather slow, subject to human error, and mentally fatiguing. It is also difficult and expensive to update a printed data base since the index usually requires complete revision. The application of computers overcomes many of these problems, but computerized search systems are constrained by the size and validity of the data base, the thoroughness of the searching algorithms, and the cost of using the system.

2. History and Organization

In 1971 the EPA undertook the operational development of a computerized search system patterned after an approach developed by Biemann and his associates at the Massachusetts Institute of Technology. A significant feature of this system is that data are transmitted over conventional voice-grade telephone lines directly from a minicomputer to a program running in a large-scale remote time-sharing computer.

The remote computer has access to the data base, conducts a search for a match based on the transmitted mass and abundance data, and sends the results back to the minicomputer in a matter of seconds. A major advantage of this system is that the names of compounds whose spectra are similar to the spectrum of the unknown are automatically printed at the user's terminal. Furthermore, they are printed in order of the similarity of their spectra to the spectrum of the unknown. The degree of similarity is measured by a numerical value on a scale of 0 to 1 that is included on the printout. Since the whole operation is relatively automatic, probable identification can be made without full-time interaction with a highly trained spectrometrist.

A typical search of this type is shown in Fig. 1: user responses are underlined for clarity. Mass and abundance data for seven salient ions (m/e 15, 5%, etc.) are input, and 16 spectra are found to be similar to that of the "unknown." The names of the best five are printed. Ethyl isocyanide, with a similarity index of 0.999, most resembles the unknown.

The major part of the original data base used by the EPA was acquired from the Mass Spectrometry Data Centre (see Chapter 8). This data base was augmented by 600 EPA pollutant spectra.

About the same time the Division of Computer Research and Technology (DCRT) and National Heart, Lung and Blood Institute (NHLBI) of the National Institutes of Health (NIH) implemented a matching system that, from the user's point of view, was somewhat different. The data base selected was a slightly updated MSDC file, but the data entry and the

```
OPTION? KB

COMPLETE SPECTRUM SEARCH

MAIN FILE (Y OR N)? Y
INPUT 2 TITLE LINES FIRST
DEMO FOR CHAPTER 9
FIGURE 9.1
INPUT
15,5
28,93
29,100
39,8
40,29
54,33
55,78
0,0
DATA OK? YES
SEARCHING HAS BEGUN
15000 SPECTRA SEARCHED
```

SI	REGN	QI	MW	MF	NAME
.999	624-79-3	659	55	C3H5N	ETHANE, ISOCYANO -(9CI); ETHYL ISOCYANIDE (8CI); ETHYL CARBYLAMINE; ETHYL ISONITRILE
.218	17758-50-8	308	110	C5H6N2O	PYRIMIDINE, 5-METHYL-, 1-OXIDE (8CI9CI)
.168	110-65-6	634	86	C4H6O2	2-BUTYNE-1,4-DIOL (8CI9CI); BIS (HYDROXYMETHYL) ACETYLENE; 1,4-BUTYNEDIOL; 1,4-DIHYDROXY-2-BUTYNE; 2-BUTYNEDIOL
.153	74-80-6	646	104	C5H12O2	HYDROPEROXIDE, PENTYL (9CI); PENTYL HYDROPEROXIDE (8CI); N-AMYL HYDROPEROXIDE; AMYL HYDROPEROXIDE
.139	3984-19-8	654	100	C5H8O2	P-DIOXANE, METHYLENE- (8CI)
.129	16681-77-9	446	84	C2H4N4	1H-TETRAZOLE, 1-METHYL- (8CI9CI); N-METHYLTETRAZOLE; 1-METHYL-1H-TETRAZOLE; 1-METHYLTETRAZOLE
.119	764-81-8	655	132	C7H16O2	HEPTYL HYDROPEROXIDE (8CI)
.101	4040-81-7	654	100	C5H8O2	5H-1,4-DIOXEPIN, 2,3-DIHYDRO- (8CI)

```
ANOTHER SEARCH (Y/N) ? N
```

FIG. 1. Typical Biemann-type search performed using the MSSS.

search procedures were different. The user enters mass and abundance data, one pair at a time, from an inexpensive keyboard/printer terminal that is interfaced to a conventional voice-grade telephone line. This terminal has no obligatory connection to a GC–MS minicomputer, which allows a large group of users of noncomputerized GC–MS systems to test and evaluate the spectrum matching system. The user-entered data are transmitted to a large remote time-sharing computer that has access to the data base. The search is conducted and the number of spectra in the file having the mass/abundance pair is returned to the user in a matter of seconds. By a repetition of this procedure the number of spectra that fit can be minimized until the choice is among a small number of spectra. The user then requests the names of these compounds to be printed at his terminal.

This type of search is illustrated in Fig. 2, again with user responses underlined. The user finds first that 1008 spectra have the ion of m/e 85 in high relative abundance. The search progresses by providing additional m/e values (128, 29, 69) for ions in the spectrum of the "unknown," together with the corresponding relative abundances, until only six of the 1008 spectra remain. All six are printed out, giving identification numbers, molecular weights, molecular formulas, and names. No similarity index is provided.

An important feature of the NIH System is that the user can search the data base with information other than mass and abundance data. Spectra can be retrieved based on molecular weights, partial or complete molecular formulas, mass losses from the molecular ion, classification codes, and combinations of all of these. Furthermore, complete spectra can be typed out, displayed, or plotted (Fig. 3) at the user's terminal. Since the user must impose his judgment in entering data, this system is oriented to the experienced user. The flexibility of this system permits its use in situations where a good match is not available in the data base. Spectra can be retrieved that have features similar to the experimental spectrum, and these provide clues to the identity of the unknown. The NIH system has found wide use in many government laboratories, including the EPA. In addition, a number of private, industrial, and university laboratories have used the system.

3. The Mass Spectral Search System

With the development and refinement of these systems it became apparent that a consolidation of the two systems would be economical and beneficial. The EPA in conjunction with the NHLI, The Food and Drug Administration (FDA), and the MSDC is supporting the consolidation of the systems into an international Mass Spectral Search System (MSSS), which is part of a larger NIH/EPA Chemical Information System (CIS). The goal of this merger is to provide a user-oriented, flexible, and self-supporting MSSS for the worldwide mass spectrometry community. The entire system is designed to encourage experimentation in the expectation that a better and more useful system will evolve. Significant advantages of

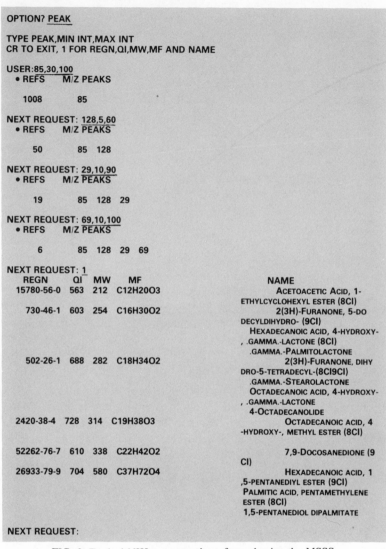

```
OPTION? PEAK

TYPE PEAK,MIN INT,MAX INT
CR TO EXIT, 1 FOR REGN,QI,MW,MF AND NAME

USER:85,30,100
 • REFS    M/Z PEAKS

   1008       85

NEXT REQUEST: 128,5,60
 • REFS    M/Z PEAKS

    50       85  128

NEXT REQUEST: 29,10,90
 • REFS    M/Z PEAKS

    19       85  128  29

NEXT REQUEST: 69,10,100
 • REFS    M/Z PEAKS

     6       85  128  29  69

NEXT REQUEST: 1
   REGN      QI  MW   MF                        NAME
   15780-56-0 563 212 C12H20O3           ACETOACETIC ACID, 1-
                                   ETHYLCYCLOHEXYL ESTER (8CI)
     730-46-1 603 254 C16H30O2           2(3H)-FURANONE, 5-DO
                                   DECYLDIHYDRO- (9CI)
                                     HEXADECANOIC ACID, 4-HYDROXY-
                                   , .GAMMA.-LACTONE (8CI)
                                     .GAMMA.-PALMITOLACTONE
     502-26-1 688 282 C18H34O2           2(3H)-FURANONE, DIHY
                                   DRO-5-TETRADECYL-(8CI9CI)
                                     .GAMMA.-STEAROLACTONE
                                     OCTADECANOIC ACID, 4-HYDROXY-
                                   , .GAMMA.-LACTONE
                                     4-OCTADECANOLIDE
    2420-38-4 728 314 C19H38O3           OCTADECANOIC ACID, 4
                                   -HYDROXY-, METHYL ESTER (8CI)

    52262-76-7 610 338 C22H42O2          7,9-DOCOSANEDIONE (9
                                   CI)
    26933-79-9 704 580 C37H72O4          HEXADECANOIC ACID, 1
                                   ,5-PENTANEDIYL ESTER (9CI)
                                     PALMITIC ACID, PENTAMETHYLENE
                                   ESTER (8CI)
                                     1,5-PENTANEDIOL DIPALMITATE

NEXT REQUEST:
```

FIG. 2. Typical NIH-type search performed using the MSSS.

the merged systems include worldwide access to the same data base, and continuous updating of the data base.

To attain the goal of a truly worldwide system it was decided to implement the MSSS on a commercial time-sharing computer system supported by a well-developed communications network. The Interactive Services Corporation time-sharing system was selected; this network is accessible by a local telephone call from many cities in the U.S., Canada,

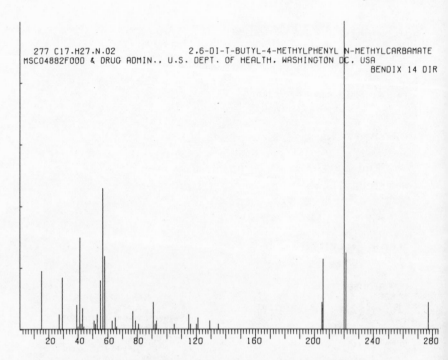

FIG. 3. Typical line diagram produced by the MSSS.

Mexico, Europe, Israel, Japan, Hong Kong, and Australia. A small subscription fee ($300) is paid by each user and this is used for maintenance, storage, and "update" costs for the entire NIH/EPA CIS including the MSSS. The time-sharing system fees ($36 per hour) are for computer and network connect-time only. A summary of current MSSS options is given in Fig. 4.

3.1. Current MSSS Use

By early 1978 there were more than 250 laboratories using the system, with an average of two new accounts being added each week. Over a hundred searches per day were performed; about a fifth of these from EPA laboratories. Both EPA and non-EPA use of the MSSS continues to grow and use of the MSSS has been incorporated into Agency policy by requiring that EPA contractors doing GC–MS contract analysis work use the MSSS. In fact, recent contracts for GC–MS analysis totaling over 2.0 million dollars will use the MSSS–Biemann option as the primary standard for identification, thus assuring the validity and comparison of results with those run by other contractors and in other laboratories. The heavy and growing use of the MSSS is clearly contributing to its improvement of data quality and quantity as well as its acceptance as a worldwide standard reference for mass spectrometry.

3.2. Future Development of the MSSS

Currently funded MSSS research and development projects include a vigorous effort to expand and improve the quality of the combined data base of mass spectra. The present data base consists of an expanded version of the original MSDC file, another collection acquired from John Wiley and Sons, and spectra collected by the EPA, and is available from the U.S. National Bureau of Standards (NBS) on magnetic tape in printed form. This amounts to over 31,000 unique spectra. The EPA and NIH have contracted to acquire new spectra of particular interest to each agency. In addition, contractors are evaluating the spectra in the present data base, removing erroneous and duplicate spectra, and developing guidelines for the establishment of a large, high-quality file. All participating agencies are working to collect existing files of spectra from spectrometrists throughout the world for inclusion in the data base. The EPA is establishing

MASS SPECTRAL SEARCH SYSTEM (MSSS)
CURRENT OPTIONS (10/77)

1. PEAK AND INTENSITY SEARCH
2. MOLECULAR WEIGHT SEARCH
3. CODE SEARCH (USING SSS)
4. MOLECULAR FORMULA SEARCH
   (a.) COMPLETE
   (b.) PARTIAL, STRIPPED
5. PEAK AND MOLECULAR WEIGHT SEARCH
6. PEAK AND MOLECULAR FORMULA SEARCH
7. MOLECULAR WEIGHT AND MOLECULAR FORMULA SEARCH
8. COMPLETE SPECTRUM (MANUAL OR MINICOMPUTER INTERFACED)
   (a.) BIEMANN
   (b.) PBM
9. DISSIMILARITY COMPARISON
10. SPECTRUM/SOURCE PRINT-OUT
11. SPECTRUM/SOURCE DISPLAY

12. SPECTRUM/SOURCE PLOTTING
13. SPECTRUM/SOURCE MICROFICHE
14. CRAB-COMMENTS AND COMPLAINTS
15. ENTERING NEW DATA
    (a.) MINI-COMPUTER INTERFACE
    (b.) DATA COLLECTION SHEETS
16. NEWS-NEWS OF THE MSSS
17. MSDC BULLETIN-LITERURE SEARCH
    (a.) AUTHOR SEARCH
    (b.) INDEX SEARCH
    (c.) SUBJECT SEARCH
18. SSS-SUBSTRUCTURE SEARCH OF CAS DATA
19. MOLECULAR FORMULA FROM ISOTOPE PATTERN
20. MOLECULAR WEIGHT FROM SPECTRAL DATA
21. MOLE FRACTION LABELING
22. ISOTOPE PATTERNS FROM MOLECULAR FORMULA
23. PROTON AFFINITY DATA

FIG. 4. Current and planned options for use with the MSSS.

collaborative efforts in data collection and software techniques with environmental groups in Europe and environmental and agricultural groups in Canada. Many contributions have been received from scientists around the world, making the MSSS a truly user-oriented and user-accepted system. A goal of 40,000 high-quality spectra of different compounds by 1980 has been set.

The original EPA-developed minicomputer-to-remote-computer direct transmission system was compatible with only one GC–MS system, which employed a Digital Equipment Corporation PDP-8 Processor. Work is in progress to develop minicomputer and remote computer programs for many of the minicomputers that are used on different GC–MS data systems. This effort is receiving some support from GC–MS manufacturers and data system houses such as Finnigan, INCOS, Hewlett-Packard, VG-Data Systems, Systems Industries, and Varian-MAT; these companies are participating in the development of direct transmission programs for their particular minicomputers.

It is emphasized that the data base and the software for accessing and searching it are separate and distinct. Therefore, a number of different and perhaps experimental software search systems may be operational simultaneously with the same data base. It is expected that in the near future new developments in software that use a mass spectral data base will be available. Indeed, a user may wish to develop specialized software and compare it to the existing operational software; this being encouraged by the system designers. Future software includes structural interrogating systems (e.g., searches for all spectra of β-chloroamines), the Self-Training Interpretive and Retrieval System (STIRS) developed by McLafferty and associates at Cornell University (see p. 182), software based on learning machine or pattern recognition techniques, software based of Wiswesser Line Notation (WLN) or Chemical Abstracts Service (CAS) Registry Numbers (REGN), and structure connection table files. Another item currently being implemented is to include the time and place of sampling for pollutants as input along with unknown spectra. This would permit retrievals by distribution of identified and even unidentified pollutants.

As virtually all GC–MS systems become computerized, another component of the MSSS is expected to evolve. This would be the capability of a GC–MS data system minicomputer to extract from the remote computer, by a dirct telephone connection, a subset of the large, continuously updated master data base. This 500–2000 spectra minidata base would be retained at the local computer site and minicomputer software would be used to search the small library locally. Specialized users who have large numbers of unknowns in one area of concern—pesticides, food additives, drugs—would have the benefit of decreased costs, yet would retain the advantages of a uniform, updated data base, and the backup of the large numbers of unknowns in one area of concern. This approach is far more economical and feasible than an attempt to develop a complete MSSS on a local spectrometer data system minicomputer. Support for a large data base requires costly peripherals such as large disks for each minicomputer; flexible search systems with many options require relatively large core memories for each computer. It is difficult to develop time-sharing minicomputer software to permit simultaneous data acquisition and data base searching. The maintenance of a large data base on a small system is costly and time consuming; thus it tends to become static.

With the further development of the MSSS, the effectiveness of computerized GC–MS should improve substantially at a small additional cost. The cost per identification should continue to decrease dramatically in the next few years.

4. Suggested Reading

MSSS User's Manual, available from the CIS project, Interactive Services Corp., Suite 500, 918 16th St. NW, Washington, D.C. 20006 (telephone number, 202-223-6503 or 800-424-9600).

S. R. HELLER, *Anal. Chem.* 47, 1972 (1951).

S. R. HELLER, G. W. A. MILNE, and R. J. FELDMANN, *Science* **195,** 253 (1977).

The MSSS data on tape can be leased from the U.S. NBS. For details please contact Dr. Lewis Gevantmann, NBS-05RD, A323/221, Gaithersburg, Maryland 20234. The MSSS in book form, *EPA/NIH Mass Spectral Data Base* by S. R. Heller and G. W. A. Milne, is available from the U.S. Gov't. Printing Office as NBS publication NSRDS-NBS 63, stock number 003-003-01987-9 ($65, add 25% for other than U.S. mailing).

10

Environmental Applications of Mass Spectrometry

Stephen R. Heller

1. Introduction

Qualitative and quantitative procedures to analyze organic pollutants in all media—air, water, solid waste—have received considerable attention in the past few years. Within the United States most of this attention has been a direct or indirect result of the formation and mission of the United States Environmental Protection Agency (EPA), established in 1971.

The enormous number of chemicals, their degradation products, and metabolites found in the environment is staggering. Probably over 100,000 different chemicals could be identified in air and water if the money, manpower, and technical facilities were available. How many of these are indeed potentially harmful, slightly toxic, or extremely toxic is unknown, and will remain unknown for a long time to come, if not forever. With the passage of the Toxic Substances Control Act, the United States will, for the first time, develop a public inventory of all chemicals (excluding drugs and chemicals covered by other legislation) manufactured, imported into, or exported from the United States. Even with such a list of chemicals, however, development of specific analytical techniques for each is virtually impossible. Hence the only practical solution to identifying pollutants will be to use more general, yet sensitive, procedures based on such properties as differences in solubility, chromatographic behavior, and spectral char-

Stephen R. Heller • U.S. Environmental Protection Agency, MIDSD, PM-218, Washington, D. C. 20460

acteristics. Many of the procedures in the area of wet chemistry are either slow, expensive, or lack the needed sensitivity. Hence the clear reason for the EPA expending considerable efforts and assets is toward finding and developing techniques which are fast, inexpensive, and sensitive. High on this list of methods is GC–MS. Other methods such as Fourier transform–gas chromatography–infrared (FT–GC–IR) and Fourier transform–nuclear magnetic resonance (FT–NMR) are now being developed as complementary tools in pollutant analysis.

GC-MS has many advantages in pollutant identification and analysis. During the 1960s the National Institutes of Health (NIH), National Science Foundation (NSF),and National Aeronautics and Space Administration (NASA) expended considerable sums of money in grants and contracts to improve GC–MS as an analytical technique. By adding a computer system (either totally dedicated or part dedicated and part centralized) the productivity of GC–MS can be made quite high. While a computer certainly does not reduce the need for highly skilled and trained manpower, it does provide great leverage for increasing a mass spectrometrist's productivity. The EPA, with a limited staff, has been able to analyze hundreds of thousands of samples and make many unambiguous identifications of specific organic pollutants found in environmental samples. Identification of pollutants in the micro- to nanogram level has become a simple routine experiment in most EPA laboratories. GC–MS has improved sensitivity by factors of 100–1,000,000, and productivity by a factor of 100, and has reduced identification time 10- to 100-fold.

The EPA has made a major commitment to GC–MS over the past 5 years by equipping its laboratories throughout the country with over two dozen GC–MS instruments having dedicated data systems and interfaced to a central host computer for standard qualitative identifications. Over 3 million dollars have been spent on instrumentation alone and about 0.5 million dollars on qualitative identifications. Figure 1 shows a typical EPA GC–MS system. Most of the GC–MS instruments in the EPA are Finnigan quadrupoles with DEC PDP-8 minicomputer data systems. There are also a few Varian, Hewlett-Packard, and LKB GC–MS instruments interfaced with Varian, Hewlett-Packard, and INCOS data systems.

The many needs for firm qualitative organic identifications include:

(i) the causes of taste or odor in drinking water,
(ii) the distribution of toxic compounds in surface or wastewater,
(iii) the accumulation of persistent compounds in wildlife tissue,
(iv) the cause of fish kills,
(v) the effectiveness of treatment facilities in removing classes or specific types of compounds,

(vi) the specific sources and degradation mechanisms of organic pollutants, and

(vii) the enforcement of effluent standards for toxic organic compounds.

Several examples are described in this chapter.

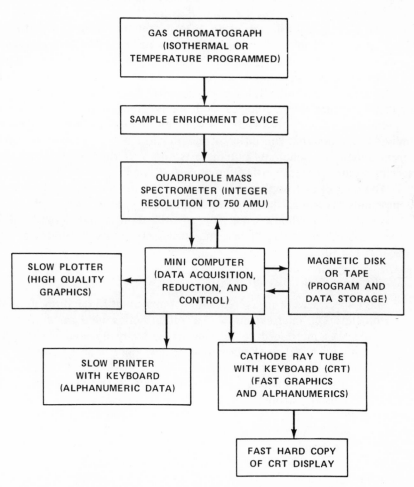

FIG. 1. Configuration of a typical GC–MS–computer system used by the EPA.

2. Drinking Water Analysis

The first example of GC–MS in the EPA is a detailed description of the Agency's analytical quality assurance (AQA) procedures applied to a study of qualitative analysis for trace organics of drinking water samples taken from five U. S. cities. The concentration, isolation, and identification procedures used by the EPA in this work are liquid–liquid extraction (GC–MS). However, some of the AQA techniques that are described also have applicability in other methods of trace organic analysis including: the entrainment of volatiles in an inert gas stream followed by trapping and GC–MS, and carbon or resin adsorption, extraction, and GC–MS (see also Chapter 5).

The data used to illustrate the AQA procedures were obtained from drinking water samples taken during January and February 1975 in Miami, Seattle, Philadelphia, Cincinnati, and Ottumwa, Iowa. The results of these analyses are a part of a larger survey of drinking water supplies that was conducted by several laboratories during early 1975. Some EPA facilities applied different methodologies of isolation and concentration of the organic contaminants, but GC–MS was always used for identification of individual pollutants. The different methodologies are effective with different classes of pollutants but there is some overlap between classes, serving as an excellent verification of certain results.

The overall philosophy of this survey was to analyze for all organic compounds present in the samples. Within this context, the emphasis of the survey was qualitative, i.e., the identification of individual organic compounds in the water. Precise measurement of concentration was not a goal of the survey.

2.1. Experimental Procedure

Mass spectra were measured with a Finnigan 1015 quadrupole mass spectrometer. The inlet system was a Varian Series 1400 gas chromatograph that was interfaced to the spectrometer by an all-glass jet-type enrichment device and an all-glass transfer line. Control of the quadrupole rod mass-set voltages, data acquisition, data reduction, and data output was accomplished with a Systems Industries data system that employed a Digital Equipment Corporation PDP-8E minicomputer and a 1.6-million-word Diablo disk drive. Data were displayed on a Tektronix 4010 cathode ray tube (CRT) display unit or a Houston plotter.

The GC column used in this study was a 2 m × 2-mm-i.d. coiled glass tube containing 1.5% OV-17 and 1.95% QF-1 on Supelcoport (80–100 mesh). The initial column temperature of 60° C was held for 1.5 min, then the temperature was programmed at 8° C min^{-1} to a final temperature of 220° C, which was held for 15 min. Conditions that were held constant throughout the analyses were as follows: helium carrier gas flow rate, about 30 ml min^{-1}; GC injection port, 190°C; interface and transfer line, 210°C; spectrometer manifold, 100° C; pressure in the MS, 10 Torr; electron energy, 70 eV; filament current, 500 mA; electron multiplier, 3000 V; mass range scanned from 33–450 amu at an intergration time of 8 msec amu^{-1}; and sensitivity, 10 AV^{-1}.

The method used applied to all organic compounds present that are extracted partially or completely into the methylene chloride–diethyl ether solvent. All compounds originally present in water at a concentration of approximately 10 ng liter^{-1} (0.01 ppb) or greater that elute from the GC column without decomposition within 35 min were observed. Very volatile compounds, e.g., chloroform and vinyl chloride, were not observed as they were either lost during extract concentration or masked during solvent elution from the GC. Compounds that were observed include the following: aliphatic hydrocarbons (C_{10} and larger), aromatic hydrocarbons (benzene derivatives, biphenyls, alkyl benzenes, polynuclears, etc.), pesticides (chlorinated, organophosphorus, some carbamates), phenols (all types), PCBs, plasticizers (phthalates, adipates, and sebacates), and various other types of compounds including sulfur compounds, amines, alcohols, aldehydes, ketones, and some carboxylic acids.

2.2. Interpretation of Results

Sampling information and results of the water analyses are given in Table 1. Table 2 is a summary of the application of AQA techniques to the analyses reported in Table 1.

Most of the concentration values in Table 1 are estimates that were based on conservative extraction efficiencies and average response factors. These estimates are probably accurate to within a factor of 10. In a few cases authentic samples were available and extraction efficiencies and response factors were determined. This permitted better estimates of concentration, and these results are probably accurate to better than ±50%. In no case was a precise concentration measurement attempted by development of a calibration curve with several standards and careful measurement of integrated instrument signals: high-precision measurements were beyond the scope of this survey.

The AQA that applies to qualitative organic trace analysis may be conveniently divided into four categories:

   (i) reagent and glassware control,
   (ii) instrumentation control,
   (iii) supporting experiments, and
   (iv) data evaluation.

TABLE 1. Results of Analysis Using Liquid-Liquid Extraction and GC-MS

Location of sampling	Date collected	Date received	Date extracted	Compounds identified	Approximate concentration[a] (ppb)
Miami, Fla. Miami-Dade Water and Sewer Authority	1/20/75	1/29/75	1/31/75	Bromoform	0.2
				Hexachloroethane	0.07
				Di-*n*-octyl adipate	20.0
Seattle, Wash. Seattle Water Dept.	1/27/75	2/5/75	2/10/75	Nicotine None	
Philadelphia, Pa. Philadelphia Water Dept.	2/3/76	2/12/76	2/18/75	1,2-*Bis*(2-chloroethoxy)-ethane	0.03
Cincinnati, Ohio Cincinnati Water Works	2/11/75	2/18/75	2/19/75	Dibromochloromethane	0.05
				Isophorone	0.02
				Trimethyl isocyanurate	0.02
Ottumwa, Iowa Ottumwa Water Works	2/17/75	2/26/75	2/28/75	Benzoic acid	15.0
				Phenylacetic acid	4.0

[a]Concentrations are estimated as accurate to within a factor of 10; with di-*n*-octyl adipate, nicotine, and benzoic acid, authentic samples were available and the concentrations of these are probably accurate to within 50%.

TABLE 2. Analytical Quality Assurance in the Identification of Organics

Compounds identified	In blank SIM trace	Spectrum matched	GC retention time	Extracted in fraction	Molecular ion observed	(M + 1) ion isotope[a] ratio Calcd.%	Found%	Spectrum checked for consistent major fragments
Bromoform	No	MSSS	Very short	Neutral	Yes	Br₃ pattern		Yes
Hexachloroethane	No	MSSS	Short	Neutral	No	Cl₅ pattern[b]		Yes
Di-n-octyl adipate	No	MSSS and standard	Matched standard	Acid same as standard	No	Not applicable		No
Nicotine	No	MSSS and standard	Matched standard	Base same as standard	Yes	10.8	12	Yes
1,2-*Bis*(2-chloro-ethoxy)ethane	No	MSSS	Not applicable	Neutral	No	Not applicable		Yes
Dibromochloro-methane	No	MSSS	Very short	Neutral	Yes	Br₂Cl pattern		Yes
Isophorone	No	MSSS	Not applicable	Neutral	Yes	9.7	9.1	Yes
Trimethyl isocyanurate	No	MSSS	Not applicable	Neutral	Yes	6.5	7.2	Yes
Benzoic acid	No	MSSS and standard	Matched standard	Acid	Yes	7.6	7.6	Yes
Phenylacetic acid	No	MSSS	Not applicable	Acid	Yes	8.6	7.3	Yes

[a] In non-halogen-containing compounds, the (M + 1) ion abundance is expressed a percentage of the molecular ion abundance in the calculated and experimental values.

[b] The Cl₅ pattern was observed in the M–Cl ion.

2.2.1. Reagent and Glassware Control

This is required to minimize the introduction of contamination from the materials used in the liquid–liquid extraction procedure. Effective glassware cleaning procedures have been developed. High-quality commercial reagents and solvents are available, but quality is still somewhat variable and unpredictable. In solvents that are used for extractions, impurities are amplified by about a factor of 2000 during extract concentration. Clearly, if background contaminants that are introduced from reagents or solvents seriously obscure compounds in the sample, purification of these materials is required.

2.2.2. Instrumentation Control

This is required to ensure that the total operating instrumentation system is calibrated and in proper working order. If a computerized GC–MS system is used to collect data, the computer data system must be included in the performance evaluation. The recommended instrumentation control procedure employs a standard reference compound and a set of reference criteria to evaluate the performance of the overall system. This evaluation should be performed each day the GC–MS system is used to acquire data from samples or reagent blanks. The records from the performance evaluations should be maintained with the sample and reagent blank records as permanent documentation supporting the validity of the data.

The reagent blank is a supporting experiment required for all samples. This is true even when contamination from glassware and reagents is well controlled. The reagent blank result is the documentation that proves that good control was exercised, and it defines precisely the level of background that was beyond control.

The reagent blank evaluation may be a straightforward comparison of corresponding peaks and mass spectra in the reagent blank and sample. A more rigorous procedure is required to make objective judgments in situations that are not obvious. An effective technique for comparing blanks and samples employs selective ion monitoring (SIM). The SIM profile provides an apparent increase in sensitivity by subtracting from the total ion current all the ion abundance data contributed by background, unresolved components, and other irrelevant ions. The SIM profile generator is a standard data reduction program on all modern computerized GC–MS systems (see Chapter 4). A fast graphics display device is required to facilitate reviewing a large number of SIM plots.

It is emphasized that it is not necessary to have even a tentative identification of a compound to apply this technique to reagent blank evaluation. To conduct an SIM comparison, the mass spectra of all GC peaks in the sample are examined. One or several ions that are prominent in a spectrum from each peak are selected, and the sample and reagent blank SIM profiles are generated on the CRT. Comparison of these permits, in most cases, straightforward judgments concerning the presence of compounds in the sample and the reagent blank.

If a corresponding peak is observed in the reagent blank and its concentration, as judged from the peak height, is approximately the same as or exceeds the sample concentration, the decision is clear and the compound must not be reported. A far more difficult judgment must be made when the concentration of a component in the sample exceeds its concentration in the reagent blank. The material in the sample could, of course, be a true sample component. Alternatively, it has been observed empirically that compounds in the blank sometimes merely appear to be at lower concentration than the same compounds in the corresponding sample. In view of the uncertainties, any compound that is observed in the sample should not be reported if it is part of an overall pattern of peaks that is repeated in the blank, although at a lower apparent concentration.

2.2.3. Supporting Experiments

Chemical ionization, field ionization, and high-resolution mass measurements are GC–MS techniques capable of generating very strong evidence in support of identifications. However, the production of this evidence is restricted because only a relatively few laboratories have developed capabilities with these techniques. High-resoltuion mass measurements are further limited by sample size, since some sacrifice in sensitivity is required to achieve sufficient resolution.

After a tentative identification is made, either by interpretation or empirical spectrum matching, several other types of supporting experiments become possible. The retention data from the GC–MS of a pure compound (standard) may be compared with analogous data from the sample component. Similarly, the mass spectra of the standard and sample component, obtained under identical conditions, may be compared. The standard may be dissolved in water at an appropriate concentration, extracted, and measured. Its recovery in the same fraction that the suspected component appeared in, and the observation of the same mass spectrum as the sample component, is strong confirmation of the correctness of the identification.

2.2.4. Data Evaluation

The evaluation of the data must weigh the available evidence in terms of its reliability and determine the cost and benefits to be gained by gathering additional information.

Clearly the most convincing evidence for an identification is obtained by examining pure standards that correspond to suspected sample components. However, the existence of this evidence is constrained by the availability of the pure standard and the additional cost and time required to examine it. Because it is not usually possible to predict which compounds will be found, some standards will not be available immediately. There are many very practical limitations imposed on the development and maintenance of a large library of pure authentic standards. Many compounds are obtainable from chemical supply houses, but procurement time is variable and may extend to weeks or months. Some compounds are not available from any supplier, frequently because they are metabolites or by-products of industrial processes rather than manufactured products. This same limitation of standard availability also precludes careful concentration measurements in many cases.

Because of the problem of standard availability, it is worthwhile to determine whether conditions could exist that would lead to reliable identifications without standards. One criterion for a reliable identification that might be used is a quantitative measure of the goodness of match between an experimental mass spectrum and a spectrum from the printed literature or a computer-readable data base. The Biemann similarity index (SI), calculated on a scale from option KB of the MSSS (see Chapter 9) has been used. Experience with this SI indicates that, in general, a value greater than about 0.4 corresponds to the existence of a reasonable match between two mass spectra.

A reasonable match often does imply an identification, but sometimes it does not; positional isomers and members of homologous series of compounds often give very similar mass spectra.

Another problem with identifications based on empirical spectrum matching is that differences in ion abundance measurements are sometimes observed when the mass spectrum of a compound is measured on two different spectrometers (see p. 21). Most of these differences are probably caused by nonuniform calibration procedures or a failure to use an ion abundance calibration procedure. Also, it is well known that different types of inlet systems often have effects on relative abundance measurements. With a GC or batch inlet system that is operated in the 100–250°C temperature range, temperature-dependent fragmentations are promoted with frequent reductions in the abundances of molecular and other higher mass ions. With a well-designed direct inlet system, these temperature effects may be largely precluded. As a result of these factors, it is quite common for two spectra of the same compound, measured with different inlet systems or spectrometers, to give a rather low SI. A low SI may also be caused by unresolved or partly resolved components which generate mass spectra containing extraneous ion abundance values.

It is concluded that the SI must be used with caution. A relatively high SI (>0.7) may be regarded as an indication of a reasonable match, but only as suggestive of the probability of an identification. A relatively low SI (<0.4) cannot be regarded as complete rejection.

2.3. Analytical Quality Assurance

These AQA concepts were applied to the analyses reported in Table 1. Table 2 summarizes the results of the tests. Bromoform, nicotine, dibromochloromethane, isophorone, trimethyl isocyanurate, benzoic acid, phenylacetic acid were not found by SIM in the corresponding reagent blanks and each gave spectra that were good matches by inspection with spectra in the MSSS data base. In each spectrum the molecular ion was observed and the M + 1 isotope accuracy or halogen isotope abundance distribution pattern was with the expected experimental error. In addition, each compound was extracted in the expected fraction, the more volatile components had the shorter retention times, and all the fragment ions in the mass spectra were reasonable and consistent with the assigned structure. On the basis of this evidence, these seven identifications were considered firm without recourse to authentic standards. Standards of two of the compounds were readily available, and these were used to supply additional supporting evidence.

Hexachloroethane gave a very good spectrum match by inspection and an expected short retention time; also, it was extracted in the appropriate fraction. The molecular ion was not observed, but ions were observed with halogen isotope distribution patterns that corresponded to the C_2Cl_5, C_2Cl_4, C_2Cl_3, C_2Cl_2, and CCl_3 ions. Therefore a consistent set of fragment ions was observed and these account for all the atoms of the proposed structure. The only other reasonable possibility for this peak was pentachloroethane, but the recorded spectrum of this compound in the MSSS data base did not exhibit a C_2HCl_5 molecular ion.

The compound di-*n*-octyl adipate was tentatively identified by the empirical matching procedure. However, no molecular ion was observed and the complexity of the fragmentation pattern precluded a rapid determination of its consistency with the proposed structure. This is a clear example of a spectrum that contains inadequate information to permit an accurate identification without a pure reference standard. The compound was found in the acid fraction. This aroused suspicion about its identity since dioctyl adipate might be expected to appear in the neutral fraction.

A sample of octyl decyl adipate containing di-n-decyl and di-n-octyl impurities was available in the laboratory. This was added to laboratory tap water and extracted according to the method. All the adipates were found in the acid fraction, as was the compound found in the Miami water sample. The retention time and mass spectrum of authentic dioctyl adipate was the same as that of the compound in the Miami water. This evidence strongly supports the identification of the authentic contaminant as di-n-octyl adipate. Its origin may be from vinyl plastic garden hose and similar materials that are in widespread use.

The compound 1,2-bis(2-chloroethoxy)ethane ($C_6H_{12}O_2Cl_2$) gave a spectrum that was in excellent agreement with the spectrum in the data base. Again, no molecular ion was observed and a reference standard was not available. However, the fragment ions at m/e 63 and 65 (C_2H_4Cl), 93 and 95 (C_3H_6OCl), and 107 and 109 (C_4H_8OCl) account for all the atoms in the compound and are consistent with the assigned structure. Two additional ions at m/e 137 and 139 correspond to the loss of a CH_2Cl group from the molecular ion. The overall evidence strongly supports the identification.

Seven other compounds were detected, but no identities are reported because inadequate evidence was available to permit reliable characterizations. Three of these appeared to be the same compound, and the available information about them is an excellent example of the application of the AQA concepts. Peaks were found at the same retention time near the detection limit of 10 ppt in the neutral fractions from Miami, Cincinnati, and Philadelphia. The three mass spectra were essentially the same except for variations in the abundance of the common background ion at m/e 43 (C_3H_2) and several other weak ions. This spectrum is a good match of the spectrum of chloropicrin. Since the molecular ion was not observed, however, interpretation of the match must be made with caution. The ions at m/e 117, 119, and 121 clearly indicate the presence of a CCl_3 group, and this is supported by the CCl_2 ions at m/e 82, 84, and 86 and the CCl ions at m/e 47 and 49. There is no ion that clearly points to an NO_2 group, and the spectrum therefore fails to account for all the atoms of the proposed structure. Under these circumstances an authentic standard was required in order to obtain additional information. Pure chloropicrin was shown to elute much earlier than the compounds in the three neutral fractions. Therefore, although this compound contains a CCl_3 group, it remains unidentified at this time. There are weak ions at m/e 103 and 145 in all three spectra; these suggest the saturated oxygenated hydrocarbon ions $C_5H_{11}O_2$ and $C_8H_{17}O_2$. This compound appears similar to the 1,2-bis(2-

chloroethoxy)ethane found in the Philadelphia neutral fraction; it has a longer chain and more chlorine atoms, and may be an intermediate in the formation of the ubiquitous chloroform. A chemical ionization mass spectrum may provide a valuable insight into the identity of this compound.

3. Study of a Landfill Leachate

The second example is a study of chemical elements and volatile organic compounds in a landfill leachate.

To determine a practical treatment for water containing pollutants leached from an abandoned landfill, Delaware's Department of Natural Resources requested analyses of several water samples.

In a residential and light-industrial area of Newcastle County, a resident complained about a persistent, unpleasant odor in his drinking water. Although the water was supplied by a private company, various governmental agencies became involved when the obnoxious odors were traced to contaminants leached from an abandoned landfill near the commercial wells.

For about 13 years, Newcastle County had used an abandoned sandpit as a receptacle for municipal and industrial garbage. In 1968 the landfill was closed after an area 5 km long and varying from 400 to 800 m in width was filled with refuse estimated to be 10–15 m deep. Within 1.5 km of the outer edge of the landfill is the Artesian Water Company's well field, from which 11–18 million liters of water are pumped daily.

Newcastle County hired a consulting firm to determine the pollution source and to propose a remedy to prevent further ground water contamination. The consulting engineers concluded that landfill leachate had polluted the aquifer. To prevent further contamination of the well field, the consulting firm proposed drilling recovery and blocking wells. Blocking wells would be located to remove water that would normally flow into the landfill area and become contaminated. Recovery wells would be drilled in or near the landfill area to remove the leachate-containing water which could then be treated and returned to the aquifer. Only treated water would then reach the private well field. To monitor aquifer pollution and to devise an effective treatment, state and county officials needed knowledge of the chemical composition of the leachate. Therefore Delaware's Department of Natural Resources requested through the EPA's Region III office in Philadelphia, Pennsylvania, that the Athens, Georgia Research Laboratory identify organic and inorganic pollutants in four water samples: water from a well inside the landfill, water from two recovery wells between the landfill and the well field, and finished water from the Artesian Well Company (Fig. 2).

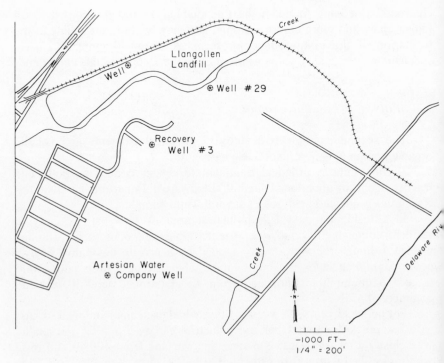

FIG. 2. Map showing landfill area and sampling sites for the leachate analysis.

Volatile organic components were identified and measured by GC and GC–MS. Each sample was extracted by a procedure designed to separate it into neutral, acidic and basic fractions. Preliminary examination by GC showed which fractions contained significant amounts of organic components, and mass spectral data were acquired for these fractions. Preliminary identifications of several compounds were obtained by computer matching of unknown mass spectra with standard mass spectra. When standard samples were available, these identifications were confirmed by comparing mass spectra obtained under the same conditions. Further confirmation was obtained by comparing GC retention times of standards and unknowns. Quantitative data were calculated from GC peak area measurements.

The relative contamination in each well was shown by comparison of the total amount of organic materials indicated by peak areas from the gas chromatograms. These data (Table 3) showed that the landfill well contained approximately 100 times more volatile organic matter than the recovery wells.

Thirty compounds were positively identified (using the MSSS) and quantitated in the landfill well sample (Table 4). Major components were short-chain acids and industrial chemicals. The most concentrated volatile organic component was 2,3-dibromo-1-propanol (23.8 mg liter^{-1}). Several compounds (various alcohols, hydrocarbons, and ethers) that were tentatively identified could not be positively identified, because no standards were available. One of these was 1-chloro-2,3-dibromopropane, a derivative of the major organic contaminant of the well.

Several compounds (caprolactam and nine short-chain acids) were found in both recovery well #3 and landfill well water (Table 4). Although the major component of the landfill well, 2,3-dibromo-1-propanol, was not positively identified in water from recovery well #3, gas-chromatographic evidence suggested that a trace amount of this compound was present.

Recovery well #29 was less contaminated than recovery well #3, but several compounds were identified: triethyl phosphate, mono- and dichlorobenzene, and eight short-chain acids (Table 4). Seven of the acids were also found in well #3.

The Artesian Well Company water extract did not contain volatile organic contaminants in quantities that could be detected by GC or GC–MS, and no organic compounds were identified. The neutral-fraction chromatogram showed only a few ill-defined humps with no sharp peaks; the acidic fraction chromatogram had no peaks. Compounds present in this finished water at concentrations of 0.001 mg liter^{-1} should have been detected, because the finished water extract represented a larger sample volume than the well water extracts.

Analysis by spark source mass spectrometry (SSMS) analysis produced quantilative data for the 21 elements detected in these four samples and supported the volatile organic component analyses. The landfill well sample was the most contaminated; well #29 was less contaminated than well #3. Several elements were present in considerably higher concentrations in the landfill well water than in other water samples, but the most

TABLE 3.   Volatile Organic Material in Water
Samples from Wells Shown in Fig. 2.

Sample	Concentration (ppm)[a]
Landfill well	89.3
Recovery well #3	1.5
Recovery well #29	0.4
Artesian Water Co. well	0.005

[a]Calculated from area of major portion of gas chromatogram.

TABLE 4. Volatile Organic Compounds Identified in Water Samples from Wells Shown in Fig. 2

Compound	Landfill well extract concentration[a] (ppm)	Recovery well #3 extract concentration (ppm)	Recovery well #29 extract concentration (ppm)	Detection thresholds in water[b] (ppm)	
				Taste	Odor
n-Butanol	5.4			0.5	2.5
Camphor		0.007		1.9	
Caprolactam	3.0	0.03[c]			
Chlorobenzene			0.003[c]		0.07–0.65
o-Cresol		0.03			
p-Cresol		0.16			3.5
Cyclohexanol	2.9	0.005[c]			
Diacetone alcohol	23.8				
2,3-Dibromo-1-propanol					
Dichlorobenzene			0.07		
Diethyl phthalate	1.2				270
2-Ethyl hexanol	0.2[c]				
Fenchone	0.2[c]				
1-Methoxy-2-propanol	0.3[c]				
Phenol	9.9	0.005[c]			0.02–0.03
1,4-Thioxane			0.01		
Triethyl phosphate	0.1				
p-Xylene					2.2

Acetic acid	4.7	0.3	0.19	20—40	2.4
Benzoic acid	2.4	0.02[c]		6.2–6.4	<7
Butyric acid	39	0.004	0.03		3
Heptanoic acid		0.03	0.005	2.5–50	3
Hexanoic acid	5.9		0.01		8.1
Isobutyric acid	1.3	0.07	0.005	1.6	0.7
Isovaleric acid	0.6	0.24	0.005		3
Octanoic acid	2.6			5.8	
Phenylacetic acid		0.07[c]			
Propionic acid	6.5	0.09	0.005		20
Toluic acid	0.1				
Valeric acid	0.5	0.03	0.005		3

[a] Because extraction efficiencies are unknown, reported concentrations are minimum values.
[b] From *Compilation of Odor and Taste Threshold Values Data* (W. H. Stahl, ed.), American Society for Testing and Materials, Philadelphia, Pa. 1973.
[c] Estimated value; not as accurate as unqualified concentrations.

significant differences were observed for concentrations of bromine and barium. The high concentration of bromine in the landfill well supported the identification of 2,3-dibromo-1-propanol as the major organic contaminant. Comparison of SSMS data from all four samples indicated that magnesium, iron, manganese, cobalt, and boron would be other useful elements to trace the movement of contaminants from the landfill toward the well field.

Analytical data were reported to the Delaware Department of Natural Resources. A series of recovery wells has been completed. Water from these wells is presently discharged into small creeks in the area, but the county is constructing aeration tanks that will permit effluent discharge into a tidal area beyond the freshwater zone.

4. A Study of the Houston Ship Channel

The last example of the use of GC–MS in EPA work is a study of organic pollutants associated with power plant cooling water in the Houston Ship Channel.

The Houston Ship Channel, which receives many industrial discharges, is the source of cooling water for a large power generating plant. Cooling water flows through a canal to the oil-fired power generating plant and is discharged into a nearby bay adjacent to important shrimp-producing estuaries (Fig. 3). Present discharges may be detrimental to receiving water quality, and even larger volumes of cooling water will be required for planned power plant expansion. The EPA needed data to evaluate biologic and hydrologic studies prepared by power plant officials and undertook the identification of organic contaminants by GC–MS and the comparison of samples by gas chromatographic fingerprinting to qualitatively establish relative degrees of pollution.

Samples were collected from nine states (Table 5) during a four day period, stored in 18-liter plastic containers on ice until received at a mobile laboratory, and kept frozen until extracted. A blank sample (10 liters of distilled water) was also stored frozen in an 18-liter plastic container for six days and extracted and analyzed by the same procedure used for the samples. Each sample was acidified and extracted with chloroform to obtain neutral and acidic materials; the water layer from this extraction was made basic and extracted again with chloroform to obtain basic materials. Each extract (3.2 liters total) was concentrated to about 3 ml, stored in a vial, and shipped on dry ice to the laboratory. A second extraction procedure was used to separate acidic and neutral organics. To facilitate GC analyses, acidic materials were converted to methyl esters or

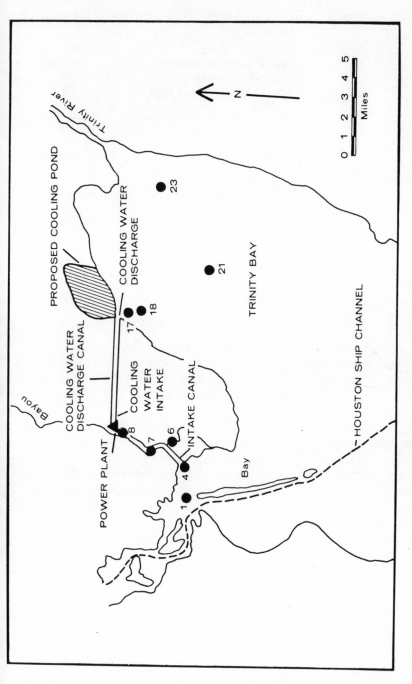

FIG. 3. Map showing the locations of the power plant, which draws cooling water from the Houston Ship Channel and discharges it into Trinity Bay, and sampling sites.

TABLE 5.   Collection Sites for Samples Indicated in Fig. 3

Sample number	Collection site
1	Dredged shipping channel water collected at center of channel near dredged intake canal
4	Water collected at mouth of dredged cooling system intake canal
6	Bayou water collected 15 m downstream (south) from intersection of bayou and cooling system intake canal
7	Bayou water collected 15 m upstream (north) from intersection of bayou and cooling system intake canal
8	Bayou water collected 15 m downstream of power plant cooling system intake
17	Bay water collected 800 m due south of cooling water canal discharge point
18	Bay water collected 1.6 km due south of cooling water canal discharge point
21	Bay water collected from center of bay approximately 8 km southeast of cooling water canal discharge point
23	Bay water collected near river mouth approximately 8 km northeast of site 21

ethers with diazomethane. The volume of each extract was reduced to 50 μl, and a 2-μl aliquot was injected into the gas chromatograph to be separated on a 50-ft Carbowax 20M-TPA SCOT column. For GC–MS analyses sample volumes were reduced as necessary for sensitivity requirements.

Few organic compounds were specifically identified. Comparison of mass spectra and GC retention times showed, however, that several compounds were present in the samples that were not present in the blank. Mass spectra of compounds unique to sample fractions were compared with mass spectra in data collections that contained spectra of only 17,000 compounds. The blank fractions contained several organic compounds. Some of these compounds identified from their mass spectra are plasticizers (phthalate and adipate esters) that were probably introduced during laboratory procedures. Others are due to solvent impurities and artifacts produced during methylation of the acid fractions.

In the neutral fraction of the cooling system influent (sample #8), triethyl phosphate and benzothiazole were identified; they were not found in neutral fractions of either the blank or the bay sample (#17) collected 800 m from the discharge point. Comparison of sample and standard GC peak heights indicated the concentration of each was approximately μg liter^{-1}. Azulene and 1- and 2-methylnaphthalene were identified in the bay sample neutral fraction. No spectral matches were found for other neutral water pollutants. Several acyclic aliphatics were probably alkanes or oxygenated alkanes but were not specifically identified. The neutral fraction of a bay water sample collected 1.6 km south of the cooling water canal discharge point (#18) contained triethyl phosphate (a compound also identified in the influent sample), two dichlorobiphenyl isomers, and two ethyl esters of long-chain saturated acids. Exact structures of the latter four compounds were not determined.

Although water pollutants in acidic fractions of the influent and bay water sample #17 were not specifically identified, the chromatograms show that the bay sample (#17) contained fewer organics in lower concentrations than the cooling system influent (#8).

The basic fraction of the influent (sample #8) contained few organic compounds not present in the blank; bay sample #17 contained more organic compounds than the influent sample. The basic pollutants were not identified.

Comparison of flame ionization gas chromatograms of neutral fraction of samples #1, #6, #7, #8, #18, and #23 indicated the relative organic content of waters from different areas (collection sites in Fig. 3).

This organic pollutant information was reported to the EPA Region VI Enforcement Division for inclusion with other analytical data (such as dissolved oxygen, total organic carbon, metals, total residue, total and Kjeldahl nitrogen, sulfates, salinity, pH, temperature, chlorides, chlorinated hydrocarbons, and various biological parameters), which were obtained from other EPA laboratories.

A report was prepared to support litigation. However, in January 1973 the issue was settled out of court when power plant officials agreed to monitor chemical and biological parameters of discharged cooling water and Trinity Bay water.

5. Summary

GC–MS plays a large and growing role in the EPA's analytical abilities to monitor the environment. While most of the Agency's work has been in the area of water studies, air and solid waste groups are also now beginning to use this very powerful technique. Vital to the use of GC–MS in the Mass Spectral Search System (MSSS) (Chapter 9), which was and still is a system in which EPA provides substantially for its development and use.

6. Exercises

1. Why is there a need to use solvents of extremely high purity in trace analyses of environmental samples?

2. Some solvents are deliberately adulterated. Investigate this practice.

3. Phthalate ester plasticizers are ubiquitous contaminants. Where might they be found in the laboratory?

4. A seawater sample is shown by GC with a flame ionization detector to contain a series of n-alkanes, presumably derived from petroleum. During GC–MS, an additional component is detected, coeluting with n-octadecane. The major ions in the spectrum of this compound are at m/e 32, 64, 96, 128, 160, 192, 224, and 256. What is it?

5. What precautions should you take before believing a computer "identification" of a compound during GC–MS?

7. Suggested Reading

A. L. ALFORD, Environmental Applications of Mass Spectrometry, *Biomed. Mass. Spectrosc.* **2**, 229 (1975).

A. L. ALFORD, EPA Research Report No. 660/4-75-004, June 1975.

J. W. EICHELBERGER and W. M. MIDDLETON, EPA Research Report No. 600/4-75-007, September 1975.

A. L. ALFORD, EPA Research Report No. 600/2-73-103, September 1973.

Applications of Mass Spectrometry in the Pharmaceutical Industry

Anthony G. Zacchei and William J. A. VandenHeuvel

1. Introduction

Applications of mass spectrometry performed in the Merck research laboratories and which illustrate the use of techniques discussed in previous chapters are described here. The data presented were obtained by using the following instruments (Chapter 2):

(i) LKB 9000S low-resolution mass spectrometer equipped with a gas chromatographic inlet and a Systems Industries 150 data acquisition system,

(ii) LKB 9000 instrument with Varian 620i data system, and

(iii) AEI MS902 high-resolution mass spectrometer equipped with a gas chromatographic inlet and a modified AEI DS30 data system.

2. Drug Metabolism

As applied to drug metabolism studies, mass spectrometry centers on (i) metabolite identification (structure elucidation) and (ii) quantitative

Anthony G. Zacchei • Merck Institute for Therapeutic Research, West Point, Pennsylvania 19486. William J. A. VandenHeuvel • Merck, Sharp and Dohme Research Laboratories, Rahway, New Jersey 07065

determination of the therapeutic agent and its metabolites. The most common use has been in structural determination; recently, however, the major emphasis appears to be moving towards the quantitative aspects of GC–MS. We will illustrate both aspects of mass spectrometry employing a variety of approaches in order to unravel some of the problems encountered in drug metabolism.

2.1. Gas Chromatography–Mass Spectrometry

Since the development of modern drugs tends to lean towards more and more potent agents, smaller doses are required to produce the desired pharmacological effect. This trend has been accompanied by increasing problems in drug metabolism studies; that is, minute amounts of drugs and their metabolites, usually in the microgram–nanogram range, have to be detected and characterized in biological fluids in the presence of a variety of interfering endogenous compounds. The superb separating power of gas chromatography combined with the structure-elucidating capability of the mass spectrometer, and the prudent choice of derivatization permit the identification and quantitation of microgram quantities of drug metabolites.

2.1.1. Metabolism of MK-196

Typical TIC profiles obtained from a pH 1 benzene extract of chimpanzee urine following the administration of the polyvalent uricosuric-saluretic agent, (6,7-dichloro-2-methyl-l-oxo-2-phenyl-5-indanyloxy)acetic acid (MK-196) are presented in Fig. 1. The lower curve was obtained following esterification of the extract with ethereal diazomethane; the upper curve was obtained on the esterified and trimethylsilylated extract. Upon comparison of the two TIC profiles, a change in retention time is noted for the first component, suggesting the presence of a derivatized (TMS) functional group; no change was noted in the retention time of the second component. The third peak increased considerably in area and a minor fourth component was detected after trimethylsilylation. The mass spectra (Fig. 2), which were obtained on each of the components (as the methyl esters), indicated that the compounds were drug-related since the characteristic two-chlorine isotopic patterns were observed. In each instance, the metabolite was tentatively assigned a structure (Fig. 2) based upon the results obtained from MS and GC–MS analyses before and after derivatization with a variety of agents. TLC separation techniques on the extracted acidic metabolites were utilized to isolate sufficient quantities of the metabolites for unequivocal assignment of the position of the substituents on the 2-phenyl moiety and for the configuration of the hydroxyl group at the C-1 position. Synthetic reference compounds were subsequently prepared to provide direct comparisons (GC, GC–MS, MS, TLC, and NMR) with the metabolites. In all cases the metabolites were identical to the authentic material.

The first component which eluted from the GC column (Fig. 2a) exhibited a shorter retention time than the parent drug and the molecular ion showed a net increase of 2 amu over the parent drug to 380 (as the methyl ester). The base peak in the spectrum occurred at m/e 105 and results from cleavage of the bonds alpha to C-2 with a concomitant hydrogen transfer. The most significant fragment ion resulted from a dehydration to produce the $[M-18]^+$ ion at m/e 362. Following trimethylsilylation of the esterified extract, the retention time of this metabolite decreased and the molecular weight increased by 72 mass units to 452, confirming the presence of the hydroxyl group. Following synthesis of the α (*cis* with respect to the 2-phenyl substituent) and β isomers and subsequent GC analysis, the metabolite was identified as the 1α-hydroxy analog of MK-196 since it exhibited the same retention time (1.5 min) as the authentic material. The β isomer exhibited a retention time of 1.1 min under identical GC conditions. NMR analysis confirmed that the metabolite was indeed the 1α-isomer when a direct comparison was made with the metabolite and the two synthetic reference compounds.

The second component which eluted from the column was characterized as unchanged drug (Fig. 2b) based upon direct comparison of GC and GC–MS properties. No change in retention time was noted upon trimethylsilylation. Little fragmentation was obtained, the molecular ion being

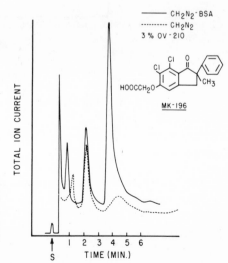

FIG. 1. TIC profiles of chimpanzee urine. The lower (broken) curve was obtained following methylation of the acidic extract; the upper (continuous) curve was obtained after methylation and trimethylsilylation.

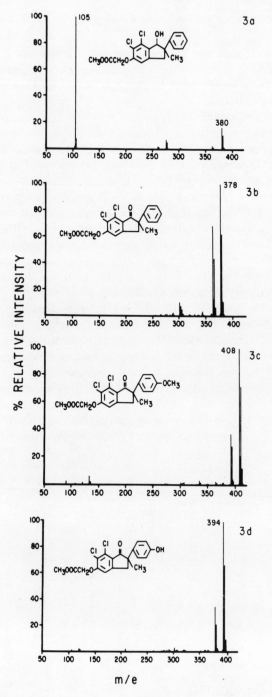

FIG. 2. Mass spectra of methyl esters of chimpanzee metabolites of MK-196.

the base peak. The most significant fragment ions are $[M-15]^+$ (loss of methyl radical) and $[M-77]^+$ (loss of 2-phenyl moiety).

GC–MS analysis on the next metabolite (peak 3) following derivatization with diazomethane, and then BSA indicated no change in retention time after trimethylsilylation. Mass spectral analysis (Fig. 2c) indicated a molecular ion of m/e 408 (base peak), a net increase of 30 amu over the parent drug, and minimal fragmentation. The metabolite was identified as the methoxy analog of MK-196. NMR analysis established unequivocally that the methoxy group was *para* substituted on the phenyl ring.

The major chimpanzee metabolite, which also eluted in peak 3, was identified as the *para* hydroxy analog of MK-196, namely, [6,7-dichloro-2-(4-hydroxyphenyl)-2-methyl-1-oxo-5-indanyloxy]acetic acid. Mass spectral analysis of the metabolite as the methyl ester (Fig. 2d) indicated a molecular ion (base peak) at m/e 394 and minimal fragmentation. Only four fragment ions with relative intensities greater than 3% were observed, namely, $[M-15]^+$ ($m^* = 364.6$, $394 \rightarrow 379$), $[M-59]^+$, $[M-73]^+$, and $[M-93]^+$ corresponding to the loss of CH_3, CH_3OOC, CH_3OOCCH_2, and C_6H_5O radicals, respectively. Following trimethylsilylation (which produced a large TIC peak), the molecular weight increased to 466, a net gain of 72 amu corresponding to the addition of one TMS group. NMR analysis confirmed the position of the hydroxyl group.

As stated previously, TLC separation techniques were used on the extracted acidic metabolites to isolate sufficient quantities of the metabolites for unequivocal characterization of the position of the substituents on the phenyl moiety and the configuration of the hydroxyl group at C-1. During this isolation procedure, an additional minor metabolite was isolated. GC and GC–MS data indicated that the 2-phenyl moiety was disubstituted, containing a free hydroxyl group and a methoxyl group. The mass spectrum of the metabolite as the methyl ester (Fig. 3, top) established the molecular weight as 424. Major fragment ions were observed at m/e 409, corresponding to a loss of a methyl radical (consistent with this series of compounds), m/e 365 and 351 (loss of 59 and 73 from side chain, respectively), and m/e 301 (loss of entire 2-phenyl substituent). Trimethylsilylation of the methyl ester derivative resulted in a decrease in retention time from 6.3 to 4.9 min indicating a reactive site. Upon derivatization with BSA, the molecular weight increased by 72 to 496 (Fig. 3, bottom). The base peak in the spectrum is again the molecular ion. The data are consistent with [6,7-dichloro-2-(x-hydroxy-y-methoxyphenyl)-2-methyl-1-oxo-5-indanyloxy]acetic acid. The NMR results confirmed that the metabolite was disubstituted on the phenyl group and also indicated a 1,2,4-trisubstituted aromatic system. The exact positions are unknown since a reference sample was not available.

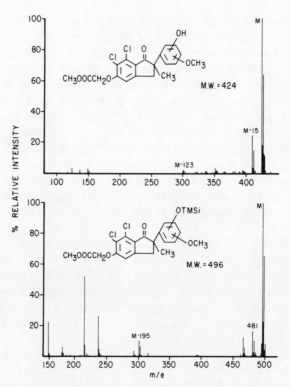

FIG. 3. Mass spectra of the disubstituted MK-196 metabolite. The upper part presents the spectrum of the methylated metabolite; the lower part illustrates the spectrum obtained following methylation and trimethylsilylation of the metabolite.

2.1.2. Metabolism of MK-473

Similar studies were performed on a related indanone, MK-473, where the 2-phenyl moiety is replaced with a cyclopentyl group. A gas chromatogram of an acidic urine extract from human urine following esterification with diazomethane is illustrated in Fig. 4. GC–MS analysis of the esterified extract established that the first peak was unchanged drug. The second peak (I.S.) was the internal standard used for drug quantitation by GC analysis. Peaks A, B, C, and D were characterized as drug metabolites based upon the observed mass spectral properties; all contained the expected dichloro isotope pattern. The spectra obtained for metabolites A, B, and C were similar (Table 1). The data presented are results of GC–MS analyses of the urine extract following derivatization with (1) diazomethane, (2) BSA, and (3) diazomethane and BSA. Notice the similarity in fragmentation for metabolites A, B, and C following derivatization. These metabolites all exhibited a decrease in retention time following trimethylsilylation of the esters, confirming the initial mass spectral interpretation of the presence of hydroxyl groups on the cyclopentyl moiety. Metabolite A was tentatively assigned as a tertiary alcohol based upon chromatographic properties.

Typical spectra of the metabolites are presented in Fig. 5. The spectra are of metabolite B after derivatization with diazomethane (top spectrum) and after trimethylsilylation of the methyl ester (bottom spectrum). A weak molecular ion was observed for the methyl ester derivative at *m/e* 386, whereas, following BSA treatment, the molecular ion increased 72

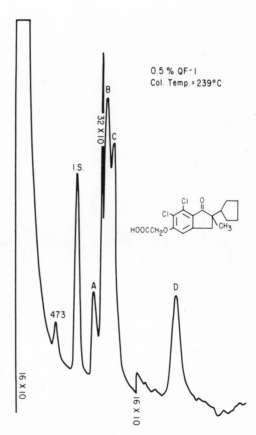

FIG. 4. A gas chromatogram of an acidic extract of human urine after ingestion of MK-473, obtained following methylation of the acidic extract.

TABLE I.   Summary of the Characteristic m/e Values for the Isolated Human Urinary Metabolites of MK-473

Characteristic m/e values

Fragmentation	Metabolite A			Metabolite B			Metabolite C		
	1[a]	2[b]	3[c]	1[a]	2[b]	3[c]	1[a]	2[b]	3[c]
M^+					516	458		516	458
M-15 ($-CH_3$)		501	443		501	443		501	443
M-18 ($-H_2O$)						423 (loss of Cl)			423 (loss of Cl)
M-15-18 ($-CH_3, -H_2O$)									
M-59 ($-CH_3OCO$) or $M^+-C_3H_7O$						399			399
M-90 [$-(CH_3)_3SiOH$]	368		368	368	426	368	368	426	368
M-90-15 [$-CH_3, -(CH_3)_3SiOH$]	353		353	353	411	353	353	411	353, 339
M-90-41 [$-(CH_3)_3SiOH, -C_3H_5$]	327		327	327	385	327	327	385	327
Base peak[d]	302[d]	360[d]	302[d]	302[d], 301[e]	360[d], 359[e]	302[d], 301[e]	302[d]	360[d]	302[d]
Base peak - 15 (loss of CH_3)	287	345	287	287	345	287	287	345	287
Base peak - 35 (loss of Cl)	267	325	267	267	325	267	267	325	267, 251
Base peak - 73 ($-CH_3OCOCH_2$)	229	271	229	229	271	229	229	271	229
					229, 155	155		229, 155	155

[a] Isolated acid metabolites treated with diazomethane (methyl esters).
[b] Isolated acid metabolites treated with BSA (silyl esters and ethers).
[c] Isolated acid metabolites treated with diazomethane then BSA (methyl esters and silyl ethers).
[d] Base peak (resulting from McLafferty rearrangement).
[e] Intensity almost equal to the base peak.

amu to *m/e* 458. The base peak in each spectrum occurred at *m/e* 302, resulting from a McLafferty rearrangement, to produce the following ion:

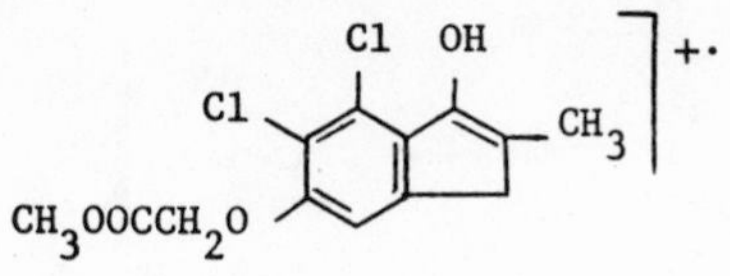

In contrast to the spectra obtained on MK-196 (2-phenyl substituent) and its metabolites where the molecular ion was the base peak, the base peak in the spectra of MK-473 (2-cyclopentyl substituent) and its metabolites always occurred at *m/e* 302 (McLafferty rearrangement ion) regardless of the position of the hydroxyl group on the 2-cyclopentyl moiety. As a consequence, this ion is diagnostic of changes that would occur in the indanone ring system. The positions of the hydroxyl groups on the cyclopentyl moiety were established by NMR analysis on isolated samples. The metabolites were identified as the following compounds:

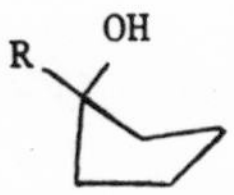

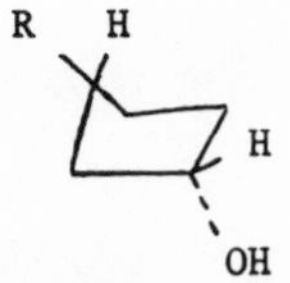

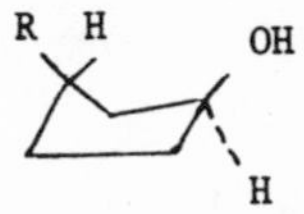

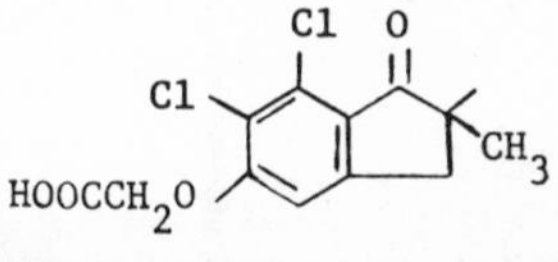

Metabolite D was characterized as the 3-keto analog of metabolite C upon mass spectral analysis, derivatization with 2,4-dinitrophenylhydrazine and hydroxylamine, and NMR analysis.

2.2. Gas Chromatography–High-Resolution Mass Spectrometry

The use of high-resolution MS in GC analysis (GC–HRMS) has been extremely diagnostic in the identification of a number of metabolites. The GC system separates the various components of a mixture and the HRMS permits elemental composition determinations for the ions as a result of the accurate mass values obtained for the ions. These calculations are usually undertaken by computer. The final output from the GC–HRMS

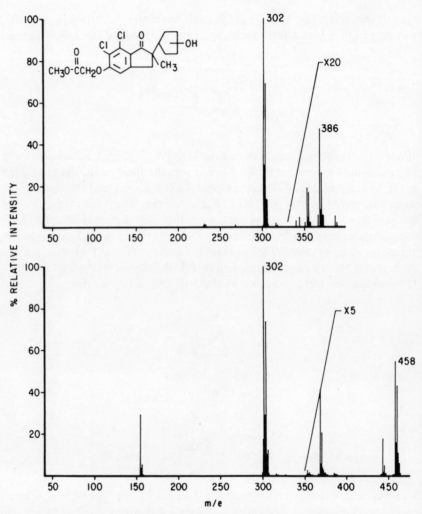

FIG. 5. Mass spectra of the major human MK-473 metabolite (peak B in Fig. 4). Upper part illustrates spectrum obtained following methylation; the lower part presents the spectrum obtained after methylaton and subsequent trimethylsilylation.

run is a list of the peaks for each ion, their intensities, accurate mass values, and elemental compositions for each GC peak. Appropriate data manipulation can then be performed.

2.2.1. Metabolism of MK-251

During our studies on the metabolic transformation of the antiarrhythmic agent, MK-251, in the monkey and baboon,

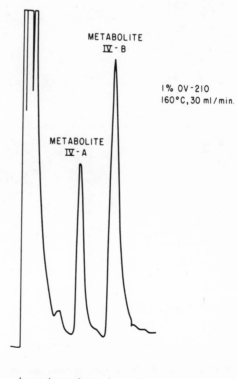

MK-251

a particular radioactive fraction was isolated which appeared to be one component based upon TLC analysis in three solvent systems and GC analysis on a 1% QF-1 column. Initial low-resolution mass spectral behavior, which appeared to be temperature dependent, indicated that the nucleus was intact and that the sample might be a mixture since unrealistic fragment losses were observed. IR analysis gave further support for a mixture by the presence of both intact methyl group absorption bands and carbonyl absorption bands. Subsequent GC analysis on different columns confirmed the presence of a mixture, and gas-liquid radiochromatography (GLRC) established that both components were drug-related (contained ^{14}C activity). The TIC profile of these metabolites following GC–MS on a 1% OV-210 column is presented in Fig. 6. The low-resolution spectra

METABOLITE
IV-B

1% OV-210
160°C, 30 ml/min.

METABOLITE
IV-A

FIG. 6. GC–MS profile of monkey and baboon metabolites of MK-251.

0   1   2   3   4

TIME (MIN.)

obtained for each component and for authentic MK-251 are presented in
Fig. 7. The upper spectrum is that of MK-251. As indicated, no molecular
ion (MW = 311) was noted; however, an intense $[M-15]^+$ ion at m/e 296
(base peak) was observed. Likewise, no apparent molecular ion was
observed in the spectra of the metabolites, which suggested that labile

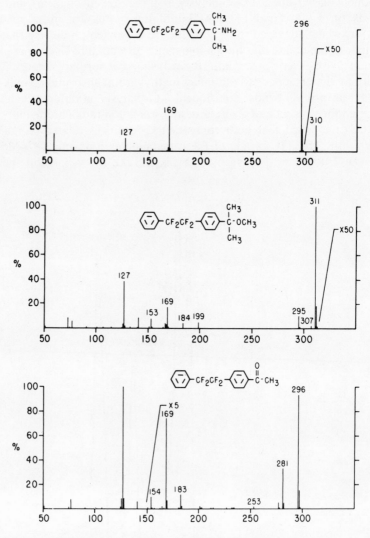

FIG. 7. Low-resolution spectra of MK-251 and the metabolites shown in Fig. 6. The upper
part shows the mass spectrum of MK-251, the middle part presents the spectrum of methyl
ether metabolite of MK-251, and the lower part illustrates the spectrum of the acetophenone
metabolite.

functional groups might be present. Also, no changes were observed in retention times of the metabolites following treatment of the metabolite mixture with a number of derivatizing reagents, suggesting no reactive sites. Consequently, GC−HRMS was performed to gain insight into the empirical formula of the observed fragment ions, specifically to determine if nitrogen was present. Tables 2 and 3 present a comparison of the high-resolution data obtained with the postulated metabolites (see structures, Fig. 7) and authentic reference material which was subsequently synthesized. The data presented are mean values from 5 or 6 analyses on each sample. The spectra were obtained at 34 sec/decade. In no instance was nitrogen detected in any of the fragment ions. In the case of metabolite IV-A, the molecular weight was assumed to be 326, since the ion at m/e 307 shows the loss of F from the molecule and the presence of oxygen. Furthermore, the ions of m/e 311 and 295 must have a common origin, inasmuch as the loss of 16 amu is rare. The mass spectral evidence of the metabolite and the lack of derivatization with a variety of agents are consistent with structure for metabolite IV-A as α,α-dimethyl-4-($\alpha,\alpha,\beta,\beta$-tetrafluorophenethyl)benzyl methyl ether. Using similar reasoning, metabolite IV-B was held to be 4-($\alpha,\alpha,\beta,\beta$-tetrafluorophenethyl)acetophenone. An excellent correlation was obtained upon direct comparison with the

TABLE 2. A Comparison of the GC-High-Resolution Data for Metabolite IV-A of MK-251 and Synthetic Material[a]

Empirical formula	Calculated m/e	Measured m/e Synthetic	Measured m/e Metabolite	Ion identity
$C_{18}H_{18}OF_4$	326.1294	326.1250	—	$M^{+\cdot}$
$C_{18}H_{17}OF_4$	325.1215	325.1254	325.1257	$M-H$
$C_{17}H_{15}OF_4$	311.1059	311.1047	311.1048	$M-CH_3$
$C_{18}H_{18}OF_3$	307.1310	307.1296	307.1319	$M-F$
$C_{17}H_{15}F_4$	295.1110	295.1112	295.1109	$M-OCH_3$
$C_{17}H_{14}F_4$	294.1032	294.1019	294.1027	
$C_{11}H_{13}OF_2$	199.0934	199.0924	199.0931	$M-PhCF_2$
$C_{10}H_{10}OF_2$	184.0700	184.0702	184.0695	$199-CH_3$
$C_{10}H_{11}F_2$	169.0829	169.0827	169.0829	
$C_9H_7OF_2$	169.0465	169.0464	169.0463	
$C_{10}H_{10}F_2$	168.0751	168.0738	168.0737	
$C_{10}H_9F_2$	167.0627	167.0656	167.0674	$199-HOCH_3$
$C_9H_7F_2$	153.0516	153.0505	153.0512	
$C_8H_7F_2$	141.0516	141.0522	141.0516	
$C_7H_5F_2$	127.0359	127.0366	127.0357	$PhCF_2^+$
$C_7H_4F_2$	126.0281	126.0289	126.0282	$127-H$
C_6H_5	77.0391	77.0403	77.0388	Ph^+

[a] See Fig. 7 (middle).

TABLE 3.   A Comparison of the GC-High-Resolution Data
For Metabolite IV-B of MK-251 and Synthetic Material[a]

Empirical formula	Calculated m/e	Measured m/e Synthetic	Measured m/e Metabolite	Ion identity
$C_{16}H_{12}OF_4$	296.0824	296.0827	296.0828	$M^{+ \cdot}$
$C_{15}H_9OF_4$	281.0589	281.0585	281.0580	$M-CH_3$
$C_{16}H_{12}OF_3$	277.0840	277.0843	277.0829	$M-F$
$C_{16}H_{12}OF_2$	258.0856	258.0862	258.0859	$277-F$
$C_{14}H_9F_4$	253.0640	253.0674	253.0651	$M-COCH_3$
$C_{15}H_9OF_2$	243.0621	243.0611	243.0621	$258-CH_3$
$C_{14}H_8F_2$	214.0594	214.0591	214.0579	
$C_{10}H_9OF_2$	183.0621	183.0609	183.0608	
$C_9H_7OF_2$	169.0465	169.0462	169.0458	$M-PhCF_2$
$C_8H_4OF_2$	154.0230	154.0229	154.0228	$169-CH_3$
$C_8H_7F_2$	141.0516	141.0515	141.0515	
$C_7H_5F_2$	127.0359	127.0353	127.0351	$PhCF_2{}^+$
$C_7H_4F_2$	126.0281	126.0279	126.0285	
C_6H_5	77.0391	77.0379	77.0377	Ph^+
C_2H_3O	43.0184	43.0198	—	

[a]See Fig. 7 (bottom).

spectral characteristics of authentic material. A rationalization for the fragmentation pattern of this metabolite is illustrated in Fig. 8.

In a similar manner, GC–HRMS was used to characterize the nitro metabolite of MK-251:

$$\langle\!\!\bigcirc\!\!\rangle - CF_2CF_2 - \langle\!\!\bigcirc\!\!\rangle - C(CH_3)_2NO_2$$

The low-resolution spectrum (Fig. 9) established that the nucleus was intact; however, no apparent molecular ion was detected. Attempts to derivatize the sample were unsuccessful. The spectrum obtained after probe analysis was identical to that following GC–MS analysis, indicating that no degradation occurred upon GC analysis. The IR spectrum, when compared to spectra of related reference compounds, indicated that the nucleus was intact; however, a strong absorption band at 1540 cm^{-1} was observed, suggesting the presence of a nitro group. The metabolite, upon chemical reduction, gave rise to the parent drug. GC–HRMS performed on the sample established the empirical formula of the major fragment ions. An extremely weak molecule was again observed, so no accurate mass measurements were possible. The weak ion at m/e 322 [M−F]$^+$ was diagnostic and established the presence of nitrogen. The averaged data

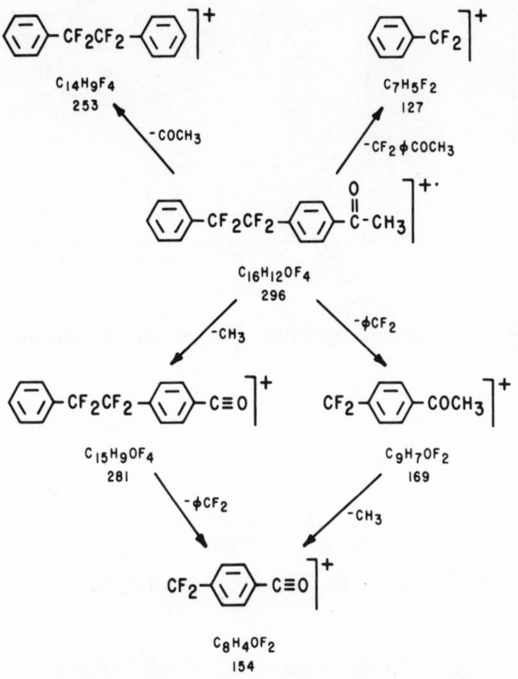

FIG. 8. Mass spectral fragmentation of 4-($\alpha,\alpha,\beta,\beta$-tetrafluorophenethyl)acetophenone.

from the GG–HRMS analyses of the metabolite and synthetic material are presented in Table 4 for comparison.

2.3. Direct-Probe Mass Spectrometry

Mass spectral studies using the direct insertion probe technique have also been of value in drug metabolism studies where structural information was desired.

2.3.1. Metabolism of MK-647

In the process of drug elimination many compounds are excreted as glucuronide conjugates. In the case of MK-647 [5-(2',4'-difluorophenyl)salicylcic acid],

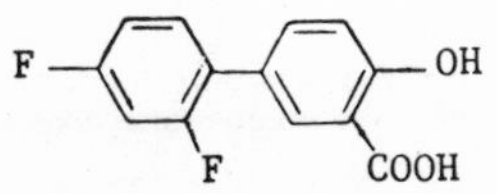

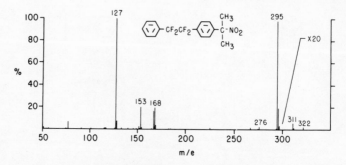

FIG. 9. Low-resolution spectrum of the nitro metabolite of MK-251.

an analgesic and anti-inflammatory agent, two radioactive fractions were isolated from human urine. The first fraction, which accounted for 20% of the urinary radioactivity, was extracted into ether from an acidic pH. The metabolite was identified as the glucuronide ester conjugate of MK-647 following probe MS analysis of the TMS derivative. The major fragment ions detected are illustrated in Fig. 10. A molecular ion was observed at

TABLE 4.   A Comparison of the GC-High-Resolution Data
for a Metabolite of MK-251 and Synthetic Material[a]

Empirical[b] formula	Calculated m/e	Measured m/e Synthetic	Measured m/e Metabolite	Ion identity
$C_{17}H_{15}OF_4$	311.1059	311.1070	311.1075	M–NO
$C_{17}H_{15}F_4$	209.1110	295.1109	295.1119	M–NO$_2$
$C_{17}H_{14}F_4$	294.1032	294.1031	294.1041	M–HNO$_2$
$C_{17}H_{15}F_3$	276.1126	276.1114	276.1107	295–F
$C_{17}H_{14}F_3$	275.1048	275.1042 ·	275.1044	
$C_{17}H_{14}F_2$	256.1064	256.1063	256.1063	275–F
$C_{14}H_9F_4$	215.0672	215.0674	215.0673	
$C_{14}H_{11}F_2$	169.0829	169.0820	169.0821	
$C_9H_7OF_2$	169.0465	169.0465	169.0463	
$C_{10}H_{10}F_2$	168.0751	168.0725	168.0724	295–PhCF$_2$
$C_{10}H_9F_2$	167.0672	167.0671	167.0669	294–PhCF$_2$
$C_9H_7F_2$	153.0516	153.0518	153.0514	
$C_9H_5F_2$	151.0359	151.0361	151.0355	
$C_7H_5F_2$	127.0359	127.0358	127.0358	PhCF$_2{}^+$
$C_7H_4F_2$	126.0281	126.0278	126.0272	
C_6H_5	77.0391	77.0381	77.0390	Ph$^+$

[a] See Fig. 9.
[b] No molecular ion was observed; however, a peak at m/e 322 corresponding to M–F (322.1061) was observed upon probe analysis of both synthetic and metabolite.

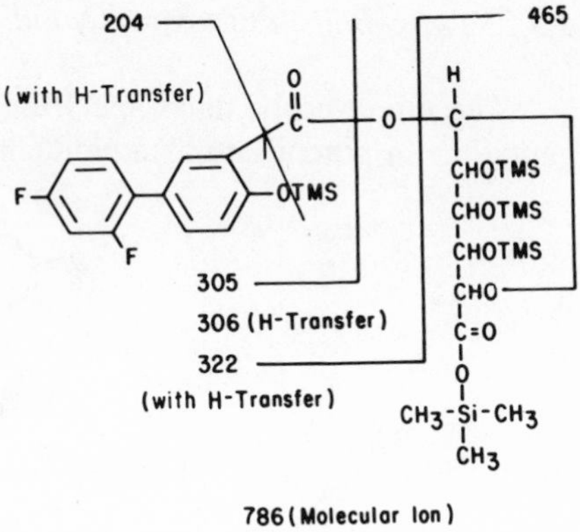

Fig. 10. Mass spectral fragmentation of the TMS derivative of the glucuronide ester of MK-647.

m/e 786 along with the characteristic $[M-15]^+$ ion. The second major fraction (64% of urinary radioactivity) was isolated from a 90-tube countercurrent distribution study. Following permethylation with dimethyl sulfoxide and methyl iodide, the fragmentation pattern (Fig. 11) obtained established that the metabolite was the phenolic glucuronide. The most significant ions were *m/e* 496 ($M^{+\cdot}$), 465 $[M-OCH_3]^+$, 437 $[M-COOCH_3]^+$, 264 (representing hydrogen transfer to the aglycone moiety following glycoside base cleavage), and 232 (resulting from base peak of aglycone and also from the permethylated glucuronide moiety minus one proton; this ion appears in conjunction with the protonated aglycone, *m/e* 264). The *m/e* 232 ion is generally observed when the glucuronide moiety is conjugated to phenolic hydroxyl groups along with ions at *m/e* 201, 169, 141, 116, 101, and 75. Ester glucuronides produce an *m/e* 233 ion rather than the *m/e* 232 ion.

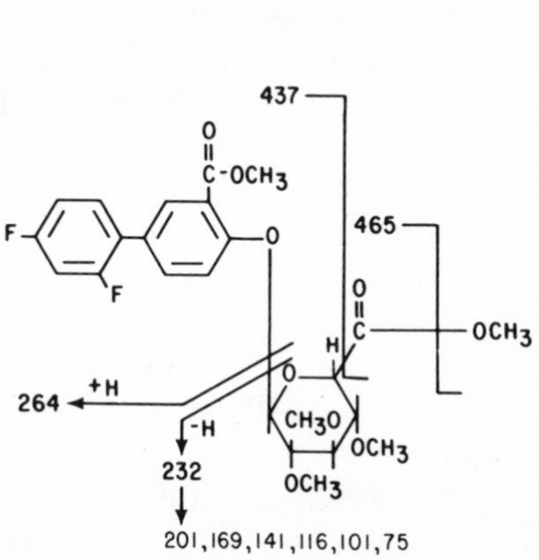

FIG. 11. Mass spectral fragmentation of the permethylated glucuronide ether of MK-647.

2.3.2. Metabolism of Cyproheptadine

The direct-probe mass spectrum of a crystalline metabolite of cyproheptadine, a potent antihistaminic and antiserotonergic agent.

was indistinguishable from that of cyproheptadine; however, the metabolite required a significantly greater vaporization temperature than the parent compound. The trimethylsilylated metabolite gave the mass spectrum of cyproheptadine, plus a number of other ions (e.g., m/e 204, 217, 305), but not all of those which are characteristic of TMS derivatives of glucuronides and carbohydrates. The apparent molecular ion of the trimethylsilylated species shifted from m/e 392 (TMS) to 419 (TMS-d_9); no ions of higher mass were observed.

GC of the metabolite resulted in the elution of one radioactive component, identical to cyproheptadine in retention time and mass spectrum. This metabolite, subjected to trimethylsilylation prior to GC, yielded cyproheptadine and several nonradioactive substances, one of which seemed to be the TMS derivative of a sugarlike compound. The molecular ions are found at $m/.$ 392 (TMS) and 419 (TMS-d_9); the molecular weight of the parent compound is 176 since the addition of three TMS groups is indicated. Many of the fragment ions observed with the TMS derivatives of carbohydrates and glucuronides are present in the spectrum of the TMS derivative of the metabolite-related compound, and the deuterium shifts confirm the structures.

Glucuronic acid, molecular weight 194, forms a penta-TMS derivative. The molecular weight of the metabolite-derived compound is 176, 18 amu less than that of glucuronic acid. Lactonization of a uronic acid reduces the molecular weight by 18 and the number of derivatizable groups to three. Glucofuranurono(6→3)lactone forms a tri-TMS derivative with the same molecular weight (392) as the TMS derivative of the metabolite; these two compounds possess similar retention times and mass spectra. Significant differences do exist between the spectra; $M^{+\cdot}$ is more intense than $[M-15]^+$ with the tri-TMS-metabolite, whereas the reverse is true with the TMS derivative of the reference lactone.

The NMR spectrum of the metabolite demonstrated that the endocyclic and exocyclic double bonds of the parent drug were unaltered.

Location of a substituent group on the piperidyl nitrogen atom was inferred from a 0.5 ppm downfield shift of the $-NCH_3$ signal relative to that in cyproheptadine. A displacement of similar magnitude was produced by *N*-methylation (i.e., quarternization) of the parent drug. The NMR spectrum of the metabolite also showed signals which suggested the presence of a carbohydrate moiety. The IR spectrum showed bands which were consistent with a glucuronide moiety. An absorption peak at 1715 cm^{-1} (COOH) was observed upon acidification of the sample.

It has been reported that trimethylsilylation of acyl glucuronides followed by GC results in elution of the tri-TMS ether of glucopyranurono(6→1)-lactone. Trimethylsilylation of an authentic acyl glucuronide, followed by GC, yielded a compound with GC and MS properties identical to those of the tri-TMS compound derived from the metabolite.

Since it has been reported that the β-D-galactopyranosylpyridinium cation undergoes β-galactosidase-catalyzed hydrolysis to yield galactose and pyridine, the structure illustrated below can be suggested by analogy for the metabolite:

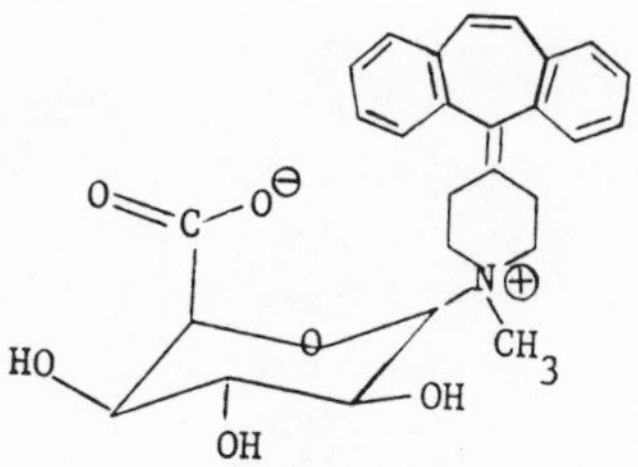

Injection of such a compound in a GC column at elevated temperatures should result in its thermal conversion to cyproheptadine and glycopyr-anurono(6→1)-lactone (which probably would not be eluted from the column). When' injected into the column after trimethylsilylation, the lactone, an isomer of glucofuranurono(6→3)-lactone, would form the tri-TMS derivative shown below and be eluted:

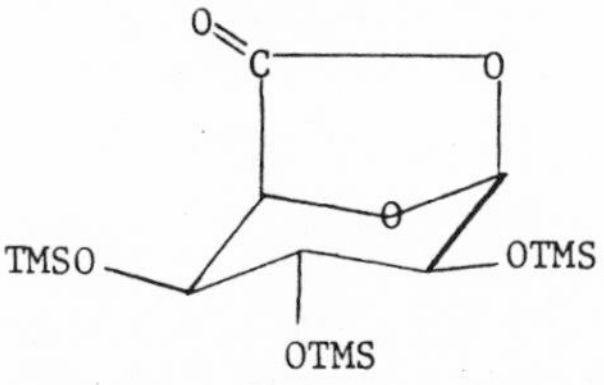

Benzene extraction of the hydrolyzed (β-glucuronidase) metabolite yielded a single radioactive compound which was found to be identical to cyproheptadine. GC–MS analysis of the trimethylsilylated hydrolysate indicated that a uronic acid is liberated from the metabolite by the glucuronidase treatment. Based on the available data, the metabolite is a previously unreported type of conjugate, and it is possible that other tertiary amines may form such quaternary, glucuronide-like conjugates.

This same compound, cyproheptadine, is extensively metabolized in the cat and dog. Figure 12 shows some of the metabolites identified in the urine of rat, cat and/or dog by using probe techniques before and after derivatization. A typical mass spectrum is presented in Fig. 13 for one of the metabolites, namely, 10,11-epoxy-desmethylcyproheptadine, isolated from rat urine. A molecular ion was detected at m/e 289 corresponding to the addition of an oxygen atom and the loss of a CH_2 group from the parent drug. The most characteristic features of this fragmentation pattern are the loss of 29 (CHO) and 17 (OH) from the molecular ion. The former observation (loss of 29) has been observed for a number of compounds in this series (*vide infra*). NMR analysis confirmed the presence of an epoxide

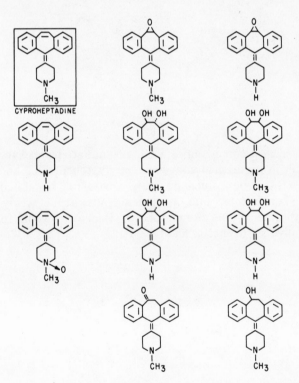

FIG. 12. Urinary metabolites of cyproheptadine identified in dog, rat, and cat.

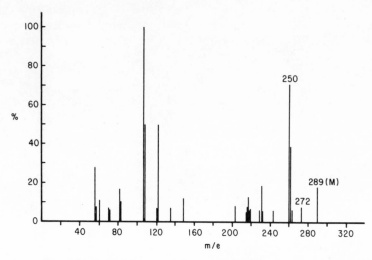

FIG. 13. Mass spectrum of 10,11-epoxy-cyproheptadine isolated from dog urine.

function with a peak at δ = 4.32 ppm and the absence of the characteristic 10,11 vinylic proton signal near δ = 7 ppm. The fact that the protons on the piperidyl ring appeared as two resolved complex multiplets of comparable area strongly infers that the methyl group on the piperidine ring was absent.

The epoxide metabolites were also detected following administration of the tricyclic antidepressant protriptyline at a dose of 25 μg g^{-1} to rats. The mass spectral data obtained following probe analysis of several isolated metabolites are present in Figs. 14 and 15. The mass spectrum of protriptyline is presented in the upper portion of Fig. 14 for comparison with that of the 10,11-epoxide metabolite. As expected, minimal fragmentation was observed. The ions at m/e 70 (base peak) and m/e 44 are indicative of the intact side chain. The ion at m/e 191 results from homolytic cleavage of the C–C bond adjacent to the tricyclic ring system. A low-intensity molecular ion was detected at m/e 263. The lower spectrum (Fig. 14) is that of the 10,11-epoxide metabolite (I). A net gain of 16 amu (M$^+$ = 279) was observed over the parent drug, suggesting the addition of an oxygen atom to form this metabolite. The ions at m/e 44 and 70 (cleavages a or b in Fig. 16) demonstrate the integrity of the side chain. The characteristic loss of 29 amu as CHO from the molecular ion (rearrangement c) forming the ion of m/e 250 is indicative of the epoxide at the bridgehead carbon atoms. This loss of 29 amu was noted earlier for the cyproheptadine metabolite. Additional structural confirmation was obtained by the lack of absorption at 290 nm in the UV spectrum, where

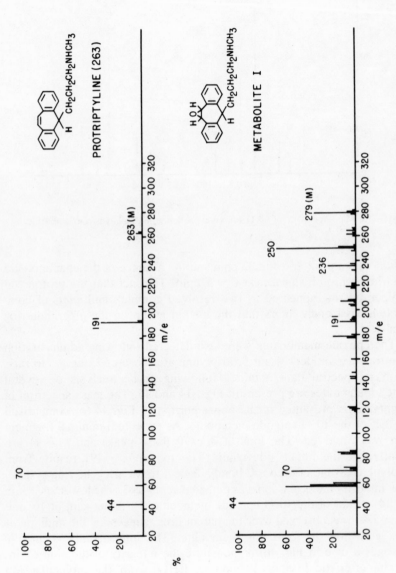

FIG. 14. Mass spectra of protriptyline and its 10,11-epoxide metabolite.

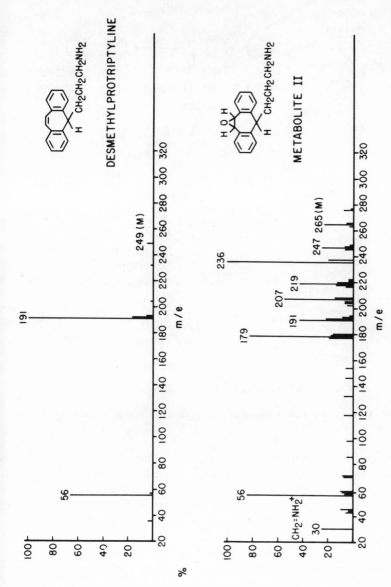

FIG. 15. Mass spectra of desmethylprotriptyline and the corresponding 10,11-epoxide metabolite.

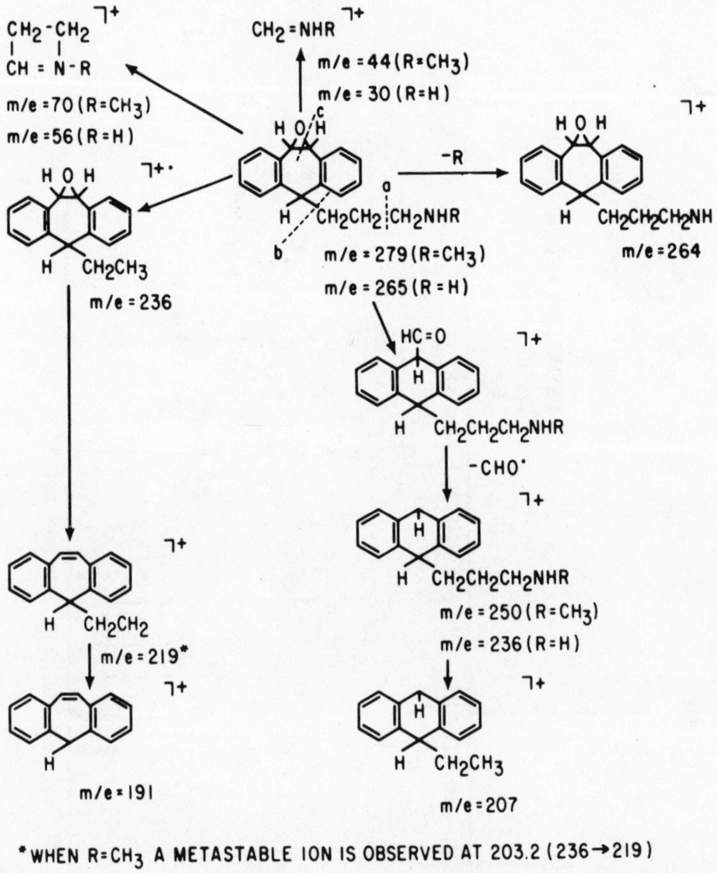

CH_2-CH_2 ⌉+
$CH = N-R$
m/e =70 (R=CH$_3$)
m/e =56 (R=H)

H O H ⌉+·
H CH$_2$CH$_3$
m/e = 236

CH$_2$=NHR ⌉+
↑ m/e =44 (R=CH$_3$)
m/e = 30 (R=H)
H O H

H CH$_2$CH$_2$ CH$_2$NHR
b   m/e = 279 (R=CH$_3$)
m/e = 265 (R=H)

-R →

H O H ⌉+
H CH$_2$CH$_2$CH$_2$NH
m/e = 264

HC=O ⌉+
H CH$_2$CH$_2$CH$_2$NHR
-CHO·
⌉+
H CH$_2$CH$_2$CH$_2$NHR
m/e = 250 (R=CH$_3$)
m/e = 236 (R=H)
⌉+
H CH$_2$CH$_3$
m/e = 207

⌉+
H CH$_2$CH$_2$
m/e = 219*
⌉+
H
m/e = 191

*WHEN R=CH$_3$ A METASTABLE ION IS OBSERVED AT 203.2 (236→219)

FIG. 16. Mass spectral fragmentation of protriptyline metabolites isolated from rat urine.

a characteristic maximum for protriptyline occurs. This result indicated that the metabolite did not possess the olefinic bond at the 10,11-position and is consistent with the epoxide structure. Subsequent NMR analysis and color tests confirmed the initial conclusions. Figure 15 presents the mass spectrum of authentic desmethylprotriptyline (upper) and that of the corresponding 10,11-epoxide metabolite (II). As seen previously for this series of compounds, minimal fragmentation occurs; a molecular ion was detected at m/e 249, the base peak at m/e 191 results from homolytic cleavage of the C–C bond adjacent to the ring system, and the ion at m/e 56 results from rearrangement of the side chain in a manner analogous to that observed for protriptyline. The lower spectrum was rationalized as arising from the 10,11-epoxide of desmethylprotriptyline. The molecular ion at m/e 265 shows a net gain of 16 amu over that of the corresponding reference material. However, more significant are the losses of 29 amu (CHO·) and 18 amu from the molecular ion to form the ions at m/e 236 (base peak, rearrangement c, Fig. 16) and m/e 247, respectively. The rationale for interpretation of the mass spectra of these two metabolites (I and II) is outlined in Fig. 16. Treatment of both epoxide metabolites with acid resulted in the formation of two radioactive fractions which were identified as the *cis* and *trans* isomers of the corresponding 10,11-dihydrodiols.

2.3.3. Metabolism of Cis-5-fluoro-2-methyl-1-[p-(methylsulfinyl)-benzylidenyl]indene-3-acetic acid

Probe analyses using high-resolution techniques have been extremely informative in metabolite structural assignment. During the course of studying the metabolism of the new anti-inflammatory agent, *cis*-5-fluoro-2-methyl-1-[*p*-(methylsulfinyl)benzylidenyl]indene-3-acetic acid,

a metabolite (IV) was isolated which exhibited an apparent molecular ion at m/e 372, suggesting that IV was formed by the addition of one oxygen atom (356 + 16) to the parent compound. A major peak at m/e 221 ($C_{12}H_{10}O_3F$) was inferred as representing the α-hydroxymethylindene (Fig. 17) since major fragment ions resulting from its apparent lactonization (m/e 203, $C_{12}H_8O_2F$) and subsequent decarboxylation (m/e 159, $C_{11}H_8F$) were also noted. This evidence strongly suggested oxidation of the methyl moiety since lactone formation was not observed in the fragmentation pattern of the parent drug. Lactone formation from α-hydroxy acids with a loss of CO_2 is known to occur. The ion at m/e 169 ($C_8H_9O_2S$) could only be accounted for if a second oxygen atom had been added to the drug. The structure illustrated for this ion was preferred since the ion at m/e 139 (C_7H_7OS) excluded introduction of oxygen to the aromatic nucleus. Treatment of metabolite IV with BSA established the addition of two TMS groups to the molecule ($m/e = 534$). The major fragment ions at m/e 365, 169, and 117 indicated that the benzyl alcohol group was not derivatized

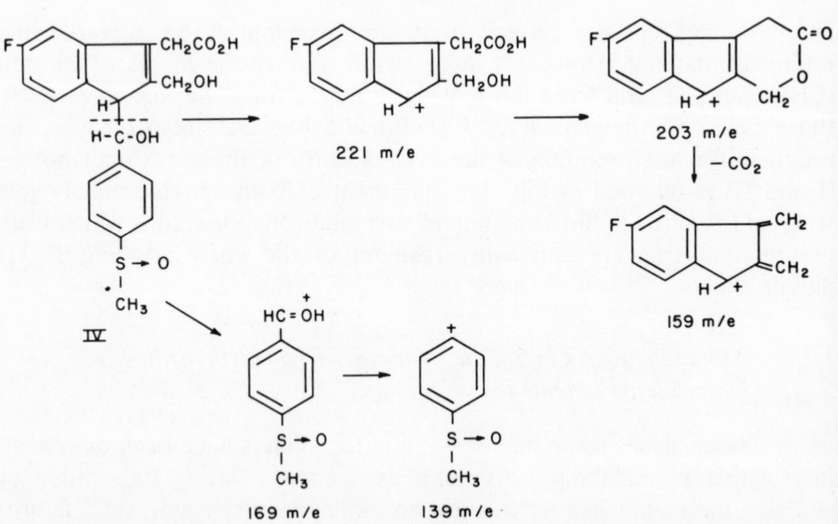

FIG. 17. Mass spectral fragmentation of a metabolite of *cis*-5-fluoro-2-methyl-1-[*p*-(methylsulfinyl)benzylidenyl]indene-3-acetic acid.

under the gentle reaction conditions (5 min, 25°C). Lactone formation of the TMS derivative was almost completely inhibited. The mass spectral evidence for this metabolite is presented in Fig. 17. NMR and UV analyses supported the key structural features of this metabolite.

2.4. Derivatization

Derivatization is a key component in GC–MS analysis (see Chapter 3). The preparation of derivatives is standard practice in gas chromatography in order to (1) increase volatility of polar compounds, (2) improve chromatographic performance *via* reduction of adsorption resulting from interaction of free hydroxyl or amino groups with reactive sites, and (3) introduce specific groups for use with selective detectors (electron capture, nitrogen, radioactivity, etc.). However, in the preparation of a derivative for quantitative applications, one must be certain that (1) the derivative is formed quantitatively even in the presence of often massive amounts of extraneous material, (2) no loss occurs as a function of time due to the lack of stability of the derivative or unexpected side reactions, and (3) no abnormal functional group alterations occur during the reaction. The last point is extremely critical when one is using this technique for structure elucidation. Consequently, derivative formation is employed in GC–MS analyses to provide additional structural information when the mass spectrum of an unknown cannot fully be interpreted (e.g., no molecular ion, labile functional groups) and to increase sensitivity and specificity when the mass spectrometer is used as a selective gas chromatographic detector.

The most frequently employed derivatization approaches are methylation, trimethylsilylation, O-methyloxime formation, acylation, or any combination of these methods. The use of deuteriated reagents has facilitated spectral interpretation as a result of shifts in amu values for various fragment ions (*vide infra*). BSA-d_{18} is a commonly used reagent to induce such fragment ion shifts which are used to determine the number of reactive sites and to aid in ion assignment. Hydroxyl, amino, phenolic, carboxylic, and thiolic functional groups all react with the trimethylsilylating reagents. The use of selected derivatives, especially those containing stable isotopes, has permitted the quantitative analysis of drugs and their metabolites using SIM and repetitive-scanning techniques.

2.5. Repetitive Scanning and Selective Ion Monitoring

Having discussed briefly the structural identification aspects of mass spectrometry in its application to drug metabolism studies, a short discussion of the quantitative aspects of mass spectrometry will now be presented. A thorough knowledge of mass spectral fragmentation patterns can be extremely valuable in drug metabolism studies, particularly when the mass spectrometer can be used as a highly specific and sensitive gas chromatographic detector. When the mass spectrometer is operated in a repetitive-scanning mode, an approximation of the amount of individual components is possible upon display of the ions specific to the compounds being examined. Figure 18 shows the results obtained following repetitive

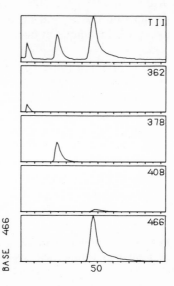

FIG. 18. Repetitive scanning data for a trimethylsilylated extract of MK-196 chimpanzee urine.

scanning in the 150–500 mass range for the trimethylsilylated extract of chimpanzee urine after administration of MK-196 (6,7-dichloro-2-methyl-1-oxo-2-phenyl-5-indanyloxy)acetic acid. The plots illustrate specific m/e values as a function of scan number. As indicated, the top plot presents the TIC profile obtained from the GC–MS run and is similar to the tracing previously obtained (Fig. 1). The next four plots show the relative intensities of ions associated with the metabolites previously discussed (as the TMS derivatives). The ion at m/e 362 resulted from the 1-α-hydroxy metabolite (M−90 fragment), the m/e 378 ion is the molecular ion of the parent compound, the m/e 408 ion is the molecular ion of the p-methoxy metabolite, and the m/e 466 ion represents the molecular ion of the p-hydroxy metabolite. Using data manipulation techniques, integration of the areas associated with each of the ions gives an estimate of the quantity of each drug-related component present. The values obtained by using this approach were comparable to those values determined by gas chromatography. The repetitive-scanning technique can also be used to deconvolute unresolved gas chromatographic peaks (thereby providing minimal interference from unrelated peaks) and for the qualitative identification of trace components. By judicious choice of ion, added specificity can be achieved when the ions characteristic of the compound are examined as a function of scan or retention time. This is an easy task since all ions and their intensities are stored and can be recalled at will. All ions pertinent to the compound should maximize at the same time; however, their intensity values would differ depending upon the associated fragmentation processes.

SIM is frequently employed (1) to identify various metabolites present in a biological extract, (2) to quantitate drug and metabolite levels following therapeutic doses, and (3) to establish isotope ratios. Selected examples using this technique at our laboratories are presented.

2.5.1. Metabolism of Carbidopa

Carbidopa (shown below)

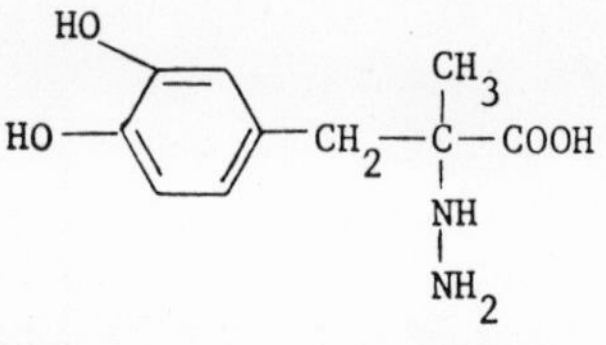

is a potent inhibitor of extracerebral aromatic amino acid decarboxylase and enhances the efficiency of L-dopa in increasing brain dopamine levels.

Therefore, carbidopa is used in conjunction with L-dopa for the treatment of Parkinson's disease. At one point in our urinary metabolism study of carbidopa all three known trimethylsilylated metabolites (I, II, and III)

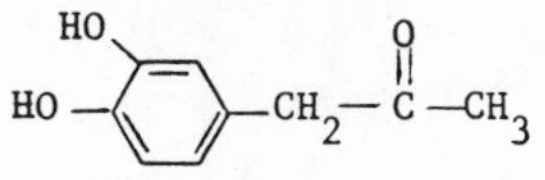

I; M = 310 (di-TMS)

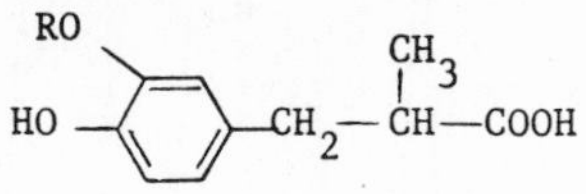

II; R = CH_3, M = 354 (di-TMS)

III; R = H, M = 412 (tri-TMS)

were found to possess one of two ring systems which were characterized by ions of m/e 179 and 209 or 267:

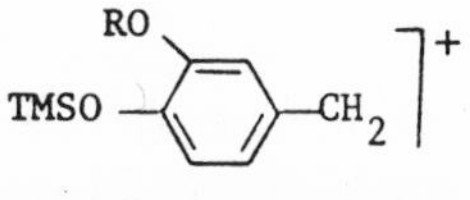

R = CH_3; m/e 209 (218)

R = TMS; m/e 267 (285)

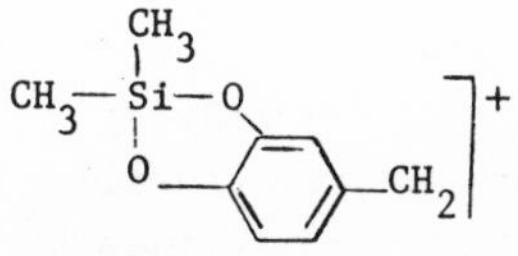

m/e 179 (185)

The values in parentheses are those obtained when deuteriated trimethylsilylating reagent (BSA-d_{18}) is used.

Several metabolite fractions from various animals were trimethylsilylated and then subjected to temperature-programmed GC–MS with repetitive scanning (m/e 10–550 every 10 sec) for 15 min. All mass spectral data were taken into a computer and a program was employed to plot the intensities of the ions m/e 179, 209, and 267 vs. scan number (or time) to yield reconstructed mass chromatograms. It was thought that this approach would offer a facile means for recognizing the presence of carbidopa metabolites in crude isolates and extracts. The intensities of the appropriate pairs of ions (e.g., 179 and 209, 179 and 267) were found to exhibit maxima at the retention times of the known metabolites. However, many normal urinary components yield these ions, and much more facile and definitive recognition of known metabolites in extracts was achieved by including the molecular ions in the multi-ion mass chromatograms.

Reconstructed mass chromatograms (m/e 179, 267, 412) resulting from analysis of a trimethylsilylated ether extract from urine of a monkey dosed with ^{14}C-carbidopa are presented in Fig. 19. Note that many components yield the ions of m/e 179 and 267. Both of these ions exhibit maxima at scan numbers ~ 420 and ~ 456, the retention positions for metabolites I

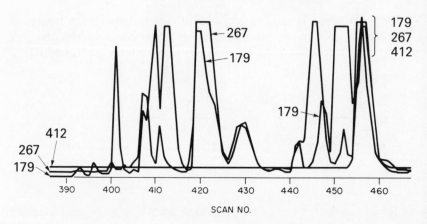

FIG. 19. Reconstructed mass chromatograms (m/e 179, 267, 412) resulting from SIM of trimethylsilylated urine extract from a monkey dosed with ¹⁴C-carbidopa.

and III (as TMS derivatives). The ion of m/e 412 (M⁺˙ of III) also exhibits its maximum at scan number ∼ 456 and the presence of this metabolite (III) in the extract is demonstrated. Reconstructed mass chromatograms from this extract for the ions of m/e 310, 354, and 412 (molecular ions of the three trimethylsilylated metabolites) are seen in Fig. 20. These three ions yield maxima at the retention positions (scan numbers) of the three metabolites (I, ∼ 420; II, ∼ 448; III, ∼ 456), indicating the selectivity resulting from searching for the molecular ions.

A TLC isolate from the above monkey urine ether extract known to contain a metabolite (because of the presence of radioactivity) was

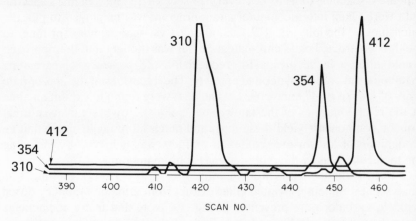

FIG. 20. Reconstructed mass chromatograms (m/e 310, 354, 412) obtained from the analysis described for Fig. 19.

subjected to SIM. Components yielding ions of m/e 179 were observed, but no corresponding 209 or 267 ions were noted. Use of gas–liquid radiochromatography demonstrated the presence of a volatile radioactive component, and hence a metabolite of carbidopa, in the isolate. A mass spectrum was obtained via combined GC–MS (Fig. 21). The molecular ion, m/e 324, shifted to 342 for the TMS-d_9 derivative, and the ion of m/e 179 shifted to m/e 188, not 185. The structure shown below was proposed for the ion of m/e 179 (188),

$$\left[\text{TMSO}\!\!-\!\!\langle\text{ring}\rangle\!\!-\!\!CH_2 \right]^+$$

and the metabolite was tentatively assigned the following structure:

$$\text{HO}\!\!-\!\!\langle\text{ring}\rangle\!\!-\!\!CH_2\!\!-\!\!\underset{\underset{H}{|}}{\overset{\overset{CH_3}{|}}{C}}\!\!-\!\!COOH$$

Since the ion of m/e 179 was of relatively low intensity, it was presumed that the trimethylsilyloxy group was in the *meta* rather than *para* position, as the latter would be expected to strongly stabilize the benzylic system. This was confirmed by comparison with the GC–MS behavior of two model compounds, and the metabolite structure was shown to be 2-methyl-3'-hydroxyphenylpropionic acid.

Mass chromatograms for the molecular ions of the trimethylsilylated metabolites are shown in Fig. 22. An intense signal for m/e 324 is observed only at scan number $\sim$ 423, for which m/e 179 shows a maximum (shoulder

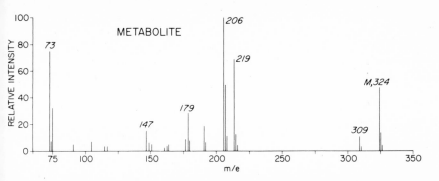

FIG. 21. Mass spectrum of trimethylsilylated monkey urinary metabolite of carbidopa.

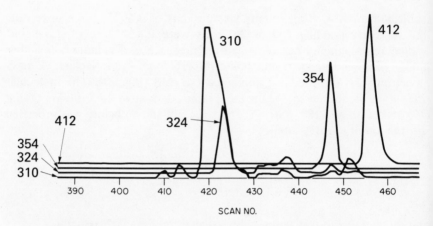

FIG. 22. Reconstructed mass chromatograms (*m/e* 310, 324, 354, 412) obtained from the analysis described for Fig. 19.

on the more intense peak, Fig. 19); these signals appear at the retention position of the authentic monohydroxy metabolite.

2.5.2. Determination of Specific Activity

Multiple-ion detection for establishing isotope ratios can be applied, in some cases, to the determination of specific activities of ^{14}C-labeled compounds. Figure 23 shows the molecular ion region of the mass spectrum of methyl thienylacetate. Because of the ^{34}S isotope there is an $[M + 2]^{+·}$ ion (*m/e* 158) of ~ 5%. The mass fragmentogram obtained by

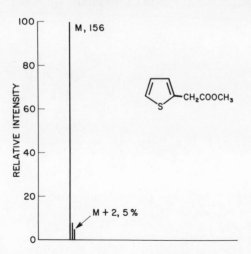

FIG. 23. Molecular ion region of mass spectrum of methyl thienylacetate.

monitoring m/e 156 and m/e 158 for a ^{14}C-labeled methyl thienylacetate are shown in Fig. 24. The intensity ratio for m/e 158/156 is clearly greater than that for the unlabeled species. The ^{14}C excess can be calculated as 24.5% (see below), leading to a specific activity of 15.7 mCi/mM, which compares favorably to the value of 16.2 mCi/mM determined by liquid scintillation counting of a known mass of the compound:

$$
\begin{array}{lcc}
 & ^{12}\text{C} & ^{14}\text{C} \\
\text{M} & 100 & 100 \\
\text{M} + 2 & 5 & 38 \\
33/133 \times 100 & = & 24.5\% \\
0.245 \times 64.0 \text{ mCi/mM} & = & 15.7 \text{ mCi/mM}
\end{array}
$$

2.5.3. Quantitative Determinations

Quantitative determinations of drug levels using selective ion monitoring techniques are widely employed. Figure 25 shows the mass spectra for MK-251, the novel antiarrhythmic agent previously discussed, and for a related compound which was used as an internal standard for the quantitative analysis of the drug. The spectra are of the trifluoroacetyl derivatives. As indicated, minimal fragmentation occurs. Low-intensity molecular ions were detected at m/e 407 and m/e 379. Consequently, the ions at m/e 280 and m/e 252, which are characteristic of the compounds and represent a major portion of the total ion current, were utilized in the development of a GC–MS method for the analysis of unchanged drug.

Typical results obtained from a GC–MS analysis are illustrated in

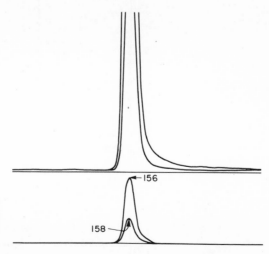

FIG. 24. Mass fragmentogram resulting from monitoring the ions m/e 156 and 158 for ^{14}C-labeled methyl thienylacetate.

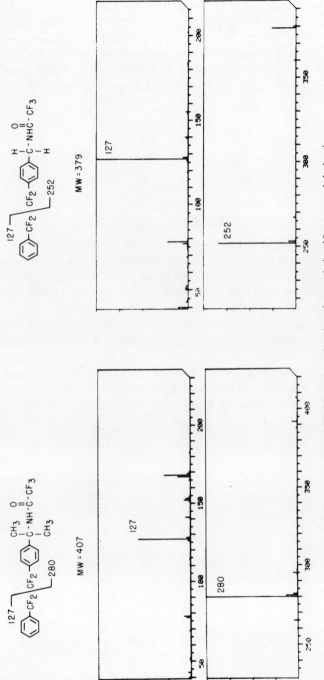

FIG. 25. Mass spectra of MK-251 and a related analog as their trifluoroacetyl derivatives.

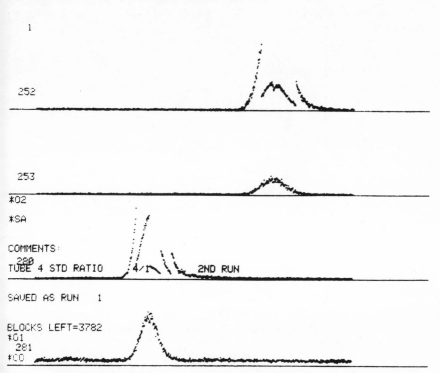

FIG. 26. Typical GC–MS analysis under selective ion monitoring conditions. Instrument set to monitor the ions at *m/e* 252, 253, 280, and 281.

Fig. 26. The ions at *m/e* 252, 253, 280, and 281 were monitored by computer control of the accelerating voltage as a function of time for a standard mixture of MK-251 and the internal standard (I.S.) carried through the extraction and derivatization procedure. A standard recovery curve (280/252 area ratio vs. weight ratio of MK-251/I.S.) for MK-251 from plasma is presented in Fig. 27. The data were obtained following single analyses of duplicate samples. As indicated, a linear relationship was observed in the 25–400 ng range. A 100-ng sample of internal standard and various amounts of MK-251 (25–400 ng) were each added to 2-ml aliquots of control plasma. The samples were subsequently carried through the extraction, derivatization, washing, and GC–MS steps to give the results. A comparison of MK-251 human plasma levels was made using this approach with those levels obtained using an established electron-capture–GC method. The values obtained are presented in Table 5. As evident, a good comparison was obtained. In our experience, the sensitivity attained with selective ion monitoring is comparable to that obtained using electron capture detection methods.

It is preferable to add the internal standard to the biological specimen

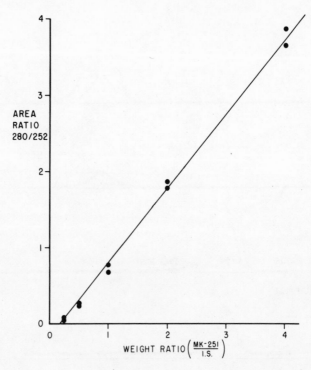

FIG. 27. Standard recovery curve of MK-251 from plasma using selective ion monitoring technique. Ions at *m/e* 252 and 280 were monitored.

before extraction and derivatization. The internal standard should have the same or very similar properties to the compound being studied. Using this approach, any losses due to extraction, derivatization, or volatization are compensated for by using the ratio method. In the aforementioned example we employed a related analog to MK-251; an ideal standard would be the compound itself modified to contain an increased mass by the introduction of stable isotopes (*vide infra*). Commonly employed isotopes are those of hydrogen (D), carbon (^{13}C), nitrogen (^{15}N), and oxygen (^{18}O). This procedure (stable-isotope dilution) ensures minimal distinction between drug and internal standard throughout the entire recovery sequence. Final differentiation is attained by the ratio method of the specific ions in a manner analogous to the situation where a related compound is used. It is imperative that the stable isotope be in a position where no exchange is possible during sample workup. Stable-isotope labeling would, because of a carrier effect, also tend to decrease sample losses on GC analysis due to adsorptive processes, especially at low sample levels.

2.6. Stable Isotopes

The use of stable isotopes in research has been expanding. They have been utilized to study biosynthetic and catabolic reactions *in vivo* and *in vitro*, to study enzyme kinetics, and to quantitate drug levels (see above). The stable isotopes of nitrogen and oxygen have been extremely important in examining the disposition of these atoms since there are no long-lived radioactive nuclides for these elements. The impetus for studies involving stable isotopes in medical research is provided by the increasing concern (safety, toxicity) over the use of radioactive tracers in clinical studies (especially pregnant women and children) and the increased availability of the isotopes.

The use of stable-isotope labeling in fragment ion structure assignment is illustrated in the following example. An acidic canine urinary metabolite of timolol-^{14}C, 3-(N-morpholino)-4-(3-t-butylamino-2-hydroxypropyl)-1, 2,5-thiadiazole, which was isolated by ion exchange and TLC, was subjected to direct-probe mass spectrometry. The presence of a large amount of extraneous material resulted in complex spectra, and yielded little structural information concerning the metabolite. An aliquot containing 2000 dpm of the sample was analyzed by temperature-programmed

TABLE 5. Comparison of MK-251 Levels in Human Plasma Using an Electron Capture Method or GC-MS Method, Subjects Dosed with 250 mg of MK-251

Time (day–hour)	Subject A		Subject B	
	ECD	GC–MS	ECD	GC–MS
1–1	25	20	37	30
1–2	80	91	80	90
1–4	132	139	146	155
1–6	154	168	166	178
1–8	142	150	134	145
1–10	102	125	130	137
1–12	106	108	114	120
2–0	86	95	78	85
3–0	62	60	61	60
4–0	39	36	39	35
5–0	19	15	24	22
6–0	16	10	17	12
7–0	13	—	9	—

ng ml^{-1}

GLRC; although a number of peaks were observed by flame ionization detection, no radioactive components were detected (300 dpm/component, lower detection limit). When analysis was carried out following exposure of the sample to trimethylsilylation (BSA) conditions, a single radioactive peak was readily observed. The derivatized sample, which was then analyzed by combined GC–MS, produced a mass spectrum for the drug-related compound which indicated a molecular ion of m/e 419. In order to determine the molecular weight of the underivatized metabolite, a second portion of the metabolite fraction was derivatized, this time with BSA-d_{18} to form the TMS-d_9 derivative; the resulting radioactive component exhibited a molecular ion at m/e 437, an increase of 18 mass units. The metabolite thus possessed two reactive functional groups, and a molecular weight of 275 was indicated [419 − (2 × 72) = 275]. The m/e values of fragment ions of particular interest are presented in Table 6.

Fragment A bears neither reactive functional group, whereas B carries both; together they comprise the entire molecule (186 + 233 = 419). Fragment C is fragment A plus a TMS group. These data require the derivatized metabolite structure shown below:

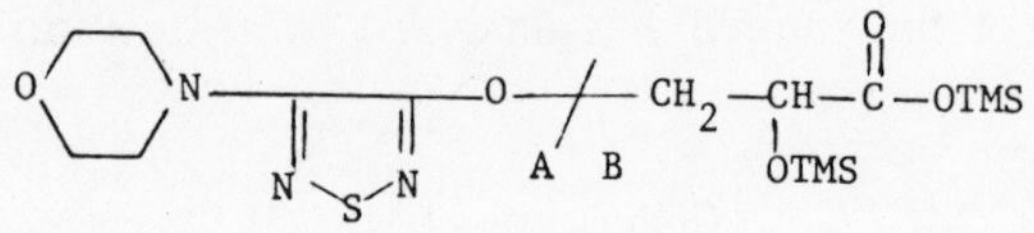

$$\text{O}\underset{\text{N} \diagdown \text{S} \diagup \text{N}}{\bigg\langle \bigg\rangle}\text{N} \underset{A}{\text{——O}} \diagup \underset{B}{\text{CH}_2} \text{—CH—}\overset{\displaystyle\text{O}}{\overset{\|}{\text{C}}}\text{—OTMS}$$

Scission of the side-chain ethereal O–C bond would produce fragments A and B, and transfer of a TMS group from the side chain onto the oxygen atom of fragment A would result in production of fragment C. This α-hydroxy acid was synthesized and found to be identical to the metabolite.

The extent of formation of polar metabolites resulting from the incubation of cambendazole (CBZ), isopropyl-2-(4-thiazolyl)-5-benzimid-azole-carbamate (shown below)

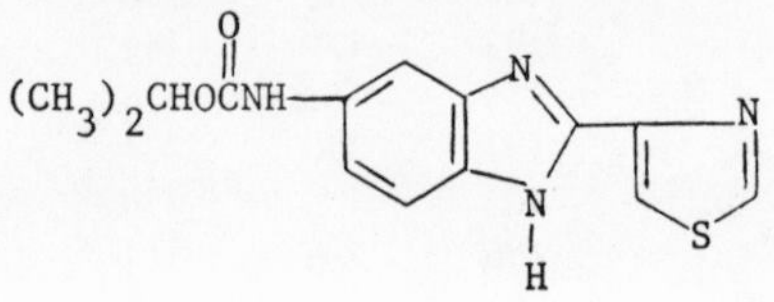

$$(\text{CH}_3)_2\text{CHOCNH——}$$

CBZ

with liver microsomes from rats or hamsters is increased severalfold by incorporation of glutathione (GSH) into the reaction mixture. The major polar metabolite, following GSH incorporation, was isolated and mass

TABLE 6. Representative m/e
Values for Fragment Ions Found Using
Stable-Isotope Labeling

Fragment	TMS	TMS-d_9
A	186	186
B	233	251
C	259	268

spectrometric techniques were particularly helpful in establishing the structure of the major metabolite.

A direct-probe temperature of ~ 250°C was required to obtain a mass spectrum from the metabolite. The only intense ion of high mass was found at *m/e* 274. No signal was observed in the mass spectrum at [M−42]$^{+\cdot}$, suggesting that if this ion of *m/e* 274 represents the molecular ion of a cambendazole-related compound, the isopropoxycarbonylamino side chain is altered (spectra of CBZ and its isopropoxycarbonylamino group-containing metabolites show strong [M−42]$^{+\cdot}$ signals resulting from the loss of C_3H_6 from the side chain). The *m/e* value 274 is 32 mass units greater than the molecular ion of the isocyanate from CBZ which arises via thermal elimination of isopropanol from the side chain. The mass spectrum of the isocyanate contains a number of singly and doubly charged ions for which the analogous ions are found in the metabolite. The *m/e* differences from the analogous ions reflect a difference in molecular weight of 32 mass units between the known isocyanate of CBZ and the volatilized compound from CBZ–GSH, suggesting that the latter is the analogous isocyanate possessing an additional sulfur atom. This is also supported by the relative intensities of the ^{34}S isotope signals.

This study was followed by one in which the CBZ was labeled with ^{13}C (thiazole ring, C_2, 20% excess). The major GSH reaction product resulting from use of this CBZ was examined by direct-probe MS. The resulting spectrum was identical to that from the unlabeled CBZ, except that it indicated ~ 20% excess ^{13}C, characterizing the volatilized species as being drug-related.

When an aliquot of the metabolite was exposed to trimethylsilylation conditions and then subjected to direct-probe MS, the presence of a species with ions *m/e* 346 and 331 (both displaying the enriched ^{13}C isotope) was noted. For comparison purposes, the mono-TMS derivative of the isocyanate from CBZ possesses signals at *m/e* 314 (M) and 299 (M−15), 32 mass units less.

On-column methylation of the metabolite using trimethylanilinium hydroxide in methanol led to the production of a volatile radioactive component as demonstrated by temperature-programmed GLRC. The sample was then subjected to combined GC-MS. The eluted compound from the isolate possessed an apparent molecular ion of *m/e* 376, with an intense fragment ion at *m/e* 287, M-89. Both of these ions exhibited the enhanced ^{13}C isotope peak (see Fig. 28), confirming that they arose from a drug-related compound. Under these on-column derivatization conditions, CBZ forms a dimethyl product with a molecular ion of *m/e* 330 and an intense [M-42]$^{+\cdot}$ ion. One can propose for the compound of molecular weight 376 a structure involving a trimethylated cambendazole molecule containing one additional sulfur atom. The fragment, [M-89]$^{+}$, would thus arise from loss of 42 and 47 (SCH$_3$). 4-Methylmercapto-*N,N'*-dimethyl-cambendazole was synthesized and found to exhibit GC and GC-MS behavior indistinguishable (except for the ^{13}C content) from that of the compound resulting from on-column methylation of the isolate. These and other data obtained by additional techniques suggest that the metabolite is a GSH conjugate of CBZ with attachment at the 4-position of the benzene ring.

A urinary metabolite of CBZ in the rat has been identified as the carboxamide:

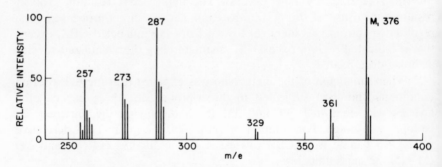

Since the corresponding carboxylic acid is also a known metabolite, it was of interest to ascertain whether the carboxamide is formed directly from the parent drug or from the acid via amidation. Use of CBZ labeled with

FIG. 28. Partial mass spectrum of radioactive product resulting from on-column methylation and GC-MS analysis of CBZ *in vitro* metabolite.

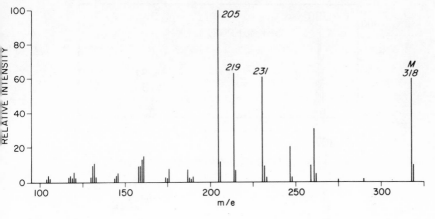

FIG. 29. Mass spectrum of the tetramethyl derivative of a metabolite of CBZ.

[15]N in the thiazole ring should indicate the source of the nitrogen atom. Retention of the label in the metabolite would require that the thizole ring was the source of the carboxamide nitrogen atom, whereas loss of label would require that it came from the body pool. ^{15}N-labeled (and ^{14}C-labeled) drug was administered to a rat (the cost of the labeled drug precluded its use with a large animal), and a crude TLC isolate containing ~ 1 μg of urinary metabolite was obtained. In order to carry out selective ion monitoring to obtain an isotope ratio on such a sample, the carboxamide was converted by on-column methylation using trimethylanilinium hydroxide in methanol to its tetramethyl derivative:

The mass spectrum of this compound is presented in Fig. 29; note the intense molecular ion at *m/e* 318. 4,6,7,-Trideuteriocarboxamide was used as a carrier in an additional TLC purification prior to GC–MS. This $^2H_3$ analog should also serve as a carrier during GC–MS, and would not interfere with measurement of the *m/e* 318/319 intensity ratio of the isolated metabolite. The sample was dissolved in the methylating reagent and examined by GC–MS monitoring the ions 318 and 319 (M and M + 1 of the carboxamide). The isotope ratio for the metabolite was the same as

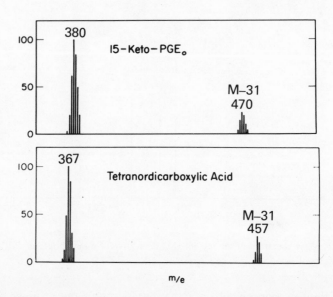

FIG. 30. Partial mass spectra of ^{2}H,^{3}H-labeled 15-keto-PGE$_0$ (top) and 7α-hydroxy-5,11-diketotetranorprostane-1,16-dioic acid (bottom) as their methylester-O-methyloxime-TMS ethers.

that for the ^{15}N-labeled drug, demonstrating complete retention of the ^{15}N label.

Considerable interest exists in ascertaining the effect drugs may have upon endogenous compounds. Several groups of workers (Vanderbilt University and Karolinska Institutet) have reported on the effect of anti-inflammatory agents upon urinary excretion levels of 7α-hydroxy-5,11-diketotetranorprostane-1,16-dioic acid, the major human urinary metabolite of prostaglandins E$_1$ and E$_2$. A GC–MS isotope dilution assay, using a deuteriated internal standard, has been employed for the measurements.

Preparation of a suitably labeled internal standard is often a problem in quantitative mass spectrometric work. We undertook to prepare the labeled (^{2}H,^{3}H) PG metabolite by dosing a rabbit with ^{2}H,^{3}H-15-keto-13,14-dihydroprostaglandin E$_1$,* a suitable precursor for the desired compound. A partial mass spectrum of the methyl ester-di-O-methyloxime-TMS ether of this compound is presented in Fig. 30 (top). The characteristic isotope clusters facilitated recognition of several prostaglandin metabolites, including 7α-hydroxy-5,11-diketotetranorprostanoic acid (a minor human metabolite of prostaglandin E$_1$) and 5β,7α-dihydroxy-11-ketotetra-

---

*The most common isotopic species in this labeled compound contained three deuterium atoms.

norprostanoic acid (a major metabolite of prostaglandin E_1 in the guinea pig), but none of the desired metabolite was observed.

A second rabbit experiment using improved dosing and isolation procedures resulted in our obtaining the desired labeled metabolite in a partially purified state suitable for use as the internal standard.

A partial mass spectrum of the metabolite (as its dimethyl ester-di-*O*-methyloxime-TMS ether) is presented in Fig. 30 (bottom). The most common isotopic clusters are those of *m/e* 367 and 457, as the most common isotopic species in this labeled compound contains two deuterium atoms (because of the position of labeling in the precursor, one of the deuterium atoms is lost during metabolism). The ion of *m/e* 367 was chosen for monitoring in our GC–MS assay for endogenous metabolite in human urine. The rabbit metabolite, in an underivatized state, was added to aliquots of urine from a normal adult male collected before and after his taking 12 aspirin tablets (over a two-day period). Comparison of the mass chromatograms in Fig. 31 shows that the *m/e* 365 (endogenous human metabolite) to 367 (internal standard) intensity ratio is diminished following administration of aspirin, probably reflecting suppression of PGE_1 and PGE_2 synthesis by the drug.

Mass spectrometry, while not the only method available, provides a highly accurate, precise, and sensitive method for determining isotope ratios and abundances. The method also permits location of the isotope sites in the molecule from the fragmentation patterns obtained. Also, the

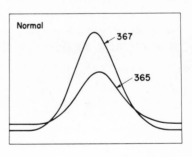

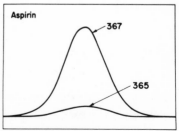

FIG. 31. Mass chromatograms resulting from monitoring the ions *m/e* 365 and 367 produced by derivatized prostaglandin metabolite isolates arising from the urine of a normal human male prior to (top) and following (bottom) oral administration of aspirin.

use of enriched samples has permitted greater sensitivity in quantitative studies.

3. Medicinal Chemistry

Mass spectrometry has played a significant role in the search for pharmaceutically active compounds. All of the major pharmaceutical companies have at least one mass spectrometer at their disposal; however, the medicinal chemical results obtained were rarely published. The obvious role of mass spectrometry in synthetic chemistry is the application of this technique (along with other instruments such as NMR, IR, and UV spectrometers) for the structural confirmation and identification of potential drugs. Previously, most of the work was done on pure analytical samples by using the probe technique. However, the advent of the GC–MS approach has reduced the need for extensive purification during the synthetic workup of the sample. This is illustrated by the reaction sequence presented in Fig. 32. In this example, the compound, when allowed to react with the reagents, produced three products as determined by GC analysis. Using the GC–MS system, the three products (A, B, and C) were identified without laborious purification. Analysis of aliquots of the reaction mixture as a function of time also provides information concerning completion of the reaction and formation of side products. In this manner mass spectrometry facilitates the development of ideal reaction conditions for a given reaction.

The usefulness of mass spectrometry in ascertaining the synthesis of intermediates in a long reaction sequence is illustrated below. If, after each reaction step, an aliquot of the reaction mixture is analyzed by direct-

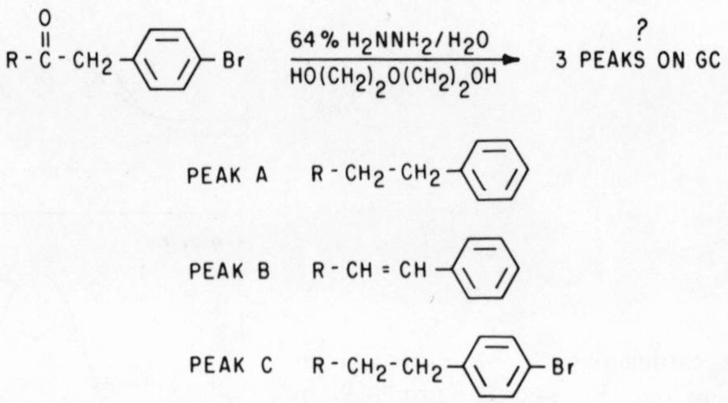

FIG. 32. A typical chemical reaction sequence. The products from this reaction afforded three gas chromatographic peaks.

probe mass spectrometry, an estimate of product formation can be obtained by examining the expected molecular ion region or diagnostic fragment ion. In this manner, if the desired compound is formed the next reaction can be initiated without extensive purification and characterization of the intermediate. Samples can be analyzed directly from crude reaction mixtures or after preliminary purification using TLC, CC, or solvent extraction procedures. In cases where TLC separation is employed, best results are obtained when the material is eluted from the adsorbent prior to analysis. In some instances, direct analysis of the silica gel scrapings has succeeded; however, we have found that better results are obtained when the TLC scrapings are shaken with water and an organic solvent (e.g., benzene or ethyl acetate). Subsequent mass spectrometric analysis is performed on the organic extract.

Peptide sequencing is rapidly becoming a rewarding application of mass spectrometry. The low volatility of peptides, which was previously a limiting factor, has been overcome by the use of new derivatization techniques. The advent of new ionization methods (e.g., field desorption) has added impetus to investigations into structural assignments of synthetic peptides (proteins).

4. Pharmaceutical Preparations

Very few papers have been published on the application of mass spectrometry to the possible identification of impurities or decomposition products in pharmaceutical preparations. This situation will undoubtedly change as a result of the high degree of specificity and sensitivity associated with this technique. Pharmaceutical chemists, in many instances, had not been made aware of the capabilities of a combined GC–MS system.

A number of pharmaceutical preparations are currently being examined for trace impurities or for potential decomposition products which may form upon sample storage. Possible interaction of drug with packaging material has been noted upon storage of certain samples. In one instance in our laboratories, several microscopic crystals were observed to form on vial stoppers upon storage. The crystals were collected and analyzed. The mass spectrum of the crystals (Fig. 33, top) was subsequently compared with that of the synthetic drug (Fig. 33, bottom). As is evident, the spectra were nearly identical except for the ions at m/e 48 and 64 due to SO and SO_2, respectively. The drug formed a sulfate complex with the material present in the stoppers. Analysis of head-space gases by using mass spectrometry has also provided insight into possible drug decomposition products.

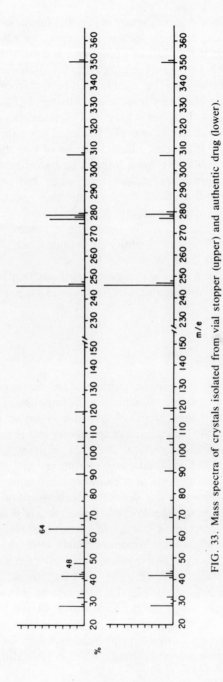

FIG. 33. Mass spectra of crystals isolated from vial stopper (upper) and authentic drug (lower).

5. Summary

We have tried to emphasize the value of mass spectrometric techniques in unraveling some of the problems associated with structural assignments in the areas of synthetic chemistry and especially drug metabolism. Examples were presented for the structural identification of the metabolites from a number of therapeutic agents. The use of combined GC–MS (both high- and low-resolution MS) and derivatization as an adjunct in structure assignment was demonstrated. The inherent high degree of specificity and sensitivity of the combined system has permitted the quantitation of drugs and their metabolites by use of SIM and repetitive-scanning techniques. These approaches have opened new fields of research in clinical pharmacology, safety assessment, and drug metabolism by permitting the measurement of plasma levels of drugs and their metabolites following therapeutic doses. The use of stable isotopes, as shown, has made easier ion fragment assignments as well as providing access to highly desirable compounds for use as internal standards in quantitative studies.

Mass spectrometry in its present state of development is not our ultimate technique but, when used in conjunction with other methods, it offers a powerful approach for the identification and quantitation of drugs and their metabolites.

6. Exercises

1. Methylation of MK-196 with diazomethane affords the methyl ester shown in Fig. 2 as compound 3b. Metabolism of some drugs proceeds via methylation. How could one distinguish between metabolic methylation and methylation which takes place during derivatization?

2. Several unsaturated drugs afford metabolites containing an epoxide moiety. Reaction of such metabolites with trimethylchlorosilane yields silylated chlorohydrins. What is the most distinctive feature of their mass spectra?

3. A urinary steroid fraction contains hundreds of natural hormones and their metabolites. How could one use GC–MS to (i) locate and (ii) identify metabolites of the drug Nilevar (17α-ethyl-17β-hydroxy-4-estren-3-one) in a urinary extract? The major metabolites are reduced in the A ring, and TMS derivatives are suitable for GC–MS.

4. The identities of drugs taken in over-dose can be determined by examination of gastric fluids. Compare the use of GC–MS and direct-probe CIMS for this task.

5. Caffeine is frequently encountered in the analysis of human body fluids. What is its origin? What other unexpected compounds might be found occasionally?

7. Suggested Reading

G. HORVÁTH, in *Progress in Drug Research* (E. Juker, ed.), Vol. 18, Birkhaüser Verlag, Basel, 1974, pp. 399–473.

B. J. MILLARD, in *Advances in Drug Research* (N. J. Harper and A. B. Simmonds, eds.), Vol. 6, Academic Press, New York, 1971, pp. 157–231.

C. -G. HAMMER, B. HOLMSTEDT, J. -E. LINDGREN, AND R. THAM, in *Advances in Pharmacology and Chemotherapy* (S. Garattini, A. Goldin, F. Hawkins, and I. J. Kapin, eds.), Vol. 7, Acadmeic Press, New York, 1969, pp. 53–89.

E. COSTA AND B. HOLMSTEDT (eds.), *Advances in Biochemical Psychopharmacology,* Vol. 7, Raven Press, New York, 1973.

D. J. JENDEN AND A. K. CHO, *Ann. Rev. Pharmacol.* **13,**371–390 (1973).

S. AHUJA, Derivatization in gas chromatography, *J. Pharm. Sci.* 65, 163–182 (1976).

W. J. A. VANDENHEUVEL AND A. G. ZACCHEI, Gas–liquid chromatography in drug analysis, in *Advances in Chromatography* (J. C. Giddings, ed.), Vol. 14, Marcel Dekker, Inc., New York, 1976, pp. 199–263.

Applications of Mass Spectrometry in the Petrochemical Industry

Thomas Aczel

1. Introduction

1.1. The Role of Mass Spectrometry in the Petrochemical Industry

The petrochemical industry is one of the essential industries in a modern economy. Its products range from gasoline to synthetic rubber, from simple solvents to complex chemicals. Research and quality control related to the literally thousands of products rely heavily on sophisticated analytical instrumentation.

Mass spectrometry plays a key role in the analysis of petroleum and petroleum products. Mass spectrometric methods are particularly well suited to the analyses of complex mixtures of relatively simple hydrocarbons which comprise the bulk of petroleum and its products.

This chapter will introduce the reader to the mass spectrometric methodology used by the most advanced laboratories in the petroleum industry. Subjects covered will include a general overview of the principal advantages and limitations of this methodology, a historical background, principles of qualitative and quantitative analyses, applications of high- and ultrahigh-resolution instruments, computerization, miscellaneous applications, and a listing of areas where further research is needed.

Mass spectrometric methods are used in the petrochemical industry for a variety of purposes. These can be classified as follows:

Thomas Aczel • Exxon Research and Engineering Company, Baytown, Texas 77520

(i) Identification of separated pure compounds with the aid of reference spectra.

(ii) Identification of separated pure compounds through the use of spectra–structure correlations.

(iii) Qualitative and quantitative analysis of unseparated or only partially separated complex mixtures. Most of the methodology particular to the petroleum industry is related to this problem. This will thus be the methodology emphasized in this chapter.

The principal advantages of mass spectrometric methods are high reproducibility of quantitative analyses, the possibility of obtaining detailed data on the individual components and/or carbon number homologs in complex mixtures, and the minimum sample size required for analysis. The ability of mass spectrometry to determine individual components in complex mixtures is unmatched by any modern analytical technique, with the exception of gas chromatography. Mass spectrometry is well suited to computerization.

Systems for data acquisition, data handling, and data retrieval are available and increasingly popular with users.

Disadvantages of the mass spectrometric approach are the limitation to organic materials that are volatile and stable at temperatures up to 300°C, the difficulty of separating positional isomers, and the expensive and complex instrumentation required. The sample is usually destroyed, but this is seldom a great disadvantage. If an analytical problem, however, requires examination of small amounts of sample by more than one instrument, mass spectrometry is usually the last technique used.

1.2. Historical Background

Early applications of analytical mass spectrometry were developed to a large extent by workers in the petroleum industry, during and immediately after World War II. These early developments included the analysis of gaseous mixtures of hydrocarbons, a heated inlet system which permitted vaporization and hence analysis of liquids boiling up to 500–600°C at atmospheric pressures, compound type analyses, and low-voltage techniques. Many of these methods are still in use and will be discussed in more detail in subsequent sections of this chapter.

The advent of high-resolution mass spectrometry in the early sixties represents another important milestone. The first application of this approach was carried out by John Beynon, then working at Imperial Chemical Industries, Ltd., in Manchester, England, on an MS8 prototype mass spectrometer. It was soon realized that the exploitation of the full potential of high-resolution technology required computerized data acquiition and handling systems, and work in this area started in the early and middle sixties. Further impetus to computerization was given by the development of GC–MS systems and by the general trend towards laboratory automation.

Alternate techniques for ionization such as chemical ionization were also introduced in the sixties, first in the petroleum industry, then for biomedical applications. Extension of mass spectrometry to the latter field has brought about a tremendous growth in instrumentation and applications in recent years, with benefits extending to all users of the technique.

Mass spectrometric technology is still growing. One of the newest developments, particularly important for the petrochemical industry, is ultrahigh-resolution instrumentation, which has been made commercially available since 1974.

2. Methods

2.1. Methods for Qualitative Analysis

Examination of the approaches used in the petrochemical industry for qualitative analysis requires a short introduction to the overall composition of petroleum and related materials.

Although there are great variations in composition among crude oils, and according to the processing techniques used for the various products, certain important features are common among all petroleum-derived materials.

In general, most petroleum samples are very complex mixtures, containing literally hundreds and even thousands of individual components.

These components have relatively simple structures. Most of them possess very similar chemical properties, and they form several homologous series of the same general formulas. Physical separation of individual members of these series (homologs) is very difficult, if not impossible in most cases, in particular for the higher boiling fractions. Separation of individual components (isomers) of the same formula is even more difficult.

The net effect of these features is the limitation of qualitative analysis to classes or types of components, or to individual homologs at the most, and the extreme difficulty of identifying individual isomers by mass spectrometry alone. This difficulty can be overcome to some extent by the GC–MS approach, discussed in Chapter 3. GC–MS is, however, limited by the resolving power of the gas chromatograph and the difficulties inherent in the similarities between the large number of components. In practice, the utility of GC–MS is greatly decreased for mixtures containing components with carbon numbers above C_{10}–C_{15}.

TABLE 1. Main Compound Types in Petroleum Steams

Class	Main compound types
Saturated hydrocarbons	Iso and normal paraffins, one- to eight-ring condensed and noncondensed cycloparaffins with alkyl side chains up to C_{40}.
Saturated heterocompounds	Alkyl sulfides
Aromatic hydrocarbons	One- to eight-ring condensed and noncondensed types, with alkyl side chains up to C_{40}.
Olefinic hydrocarbons	Present only in specially processed products, such as cracked materials, and pyrolysis products, such as shale oils. These yield both linear and cyclic olefins.
Aromatic heterocompounds	One- to eight-ring aromatic furans, phenols, thiophenes, pyrroles, pyridines, etc.

Carbon number range: 1–70+
Number of compound types: >200
Number of individual homologs: >5000
Number of individual components: >50,000

In spite of the above limitations, mass spectrometry does furnish useful and adequate information on the composition of petroleum streams, even if this information is not as exhaustive as one might desire. Structural similarities hindering identification of individual components make, in effect, identification by type or by homolog only more meaningful, as these similar structures behave similarly in most processing situations and knowledge of the individual isomeric distribution would add little to our understanding of the relationships between composition and processing parameters. The principal classes of compounds present in petroleum fractions boiling from 50 to 650°C (the range subject to mass spectrometric analysis with conventional volatilization techniques) are listed in Table 1. All of these types possess distinctive mass spectral features, which are summarized in the following paragraphs.

2.2. Spectral Features of Major Compound Types in Petroleum- and Coal-Derived Materials

2.2.1. Paraffins

Members of this homologous series have the general formula C_nH_{2n+2} (methane, CH_4; ethane, C_2H_6; propane, C_3H_8; . . .; hexadecane, $C_{16}H_{34}$; 2-methylpentane, C_6H_{14}; etc.). Molecular ions appear at corresponding *m/e* values: 16, 30, 44, 58, 86, 226, etc. The major fragment ions are due to cleavage of the C–C bonds and thus appear at *m/e* values of general formula C_nH_{2n+1}, such as 29, 43, 57, 71, 85, 99.

In the case of isoparaffins, the preferred fragmentation occurs at the highly branched carbon atoms. This fact allows one to identify the position of these carbon atoms in many cases. For example, 3-methylheptane has a large peak at m/e 85, due to the loss of an ethyl group, while 4-methylheptane yields a large ion at m/e 71, due to the loss of a propyl group.

Isoparaffins also give prominent peaks at even mass numbers, in the C_nH_{2n} series.

The molecular ions are not very stable in paraffins. In general their stability decreases with increasing degree of branching and increasing molecular weight. This feature is illustrated by the data listed in Table 2 in which the relative abundances of the molecular ions with respect to the base peaks, the strongest peaks in the spectrum, are listed for some typical paraffins.

The implications of these data in Table 2 are obvious. Note in particular the almost complete absence of the molecular ions in the spectra of compounds with quaternary carbon atoms.

Fragments which cannot be formed by breakage of a single C–C bond are usually very improbable, and correspond to low-intensity ions. Absence or low intensity of certain ions in a fragment ion series also furnishes the analyst with useful clues on the location of branching in isoparaffin structures. For example, in the spectrum of 4-methylheptane the location of the substituent methyl group is evident not only by the large fragment ion at m/e 71 but also by the relatively low intensity of the fragment ion at m/e 57, the formation of which requires cleavage of two C–C bonds.

The fragmentation rules discussed for paraffins are generally applicable to compounds with paraffinic side chains. The molecular ion of *n*-propylbenzene is, for example, more stable than that of isopropylbenzene or dodecylbenzene.

TABLE 2. Molecular Ion Intensities of Typical
Paraffins

Compound	Relative intensity of molecular ions
n-Octane, C_8H_{18}	6.74
n-Hexadecane, $C_{16}H_{34}$	4.36
n-Triacontane, $C_{30}H_{62}$	0.86
n-Octane	6.74
2-Methylheptane	4.91
2,3-Dimethylhexane	1.68
2,2-Dimethylhexane	0.03
2,2,4-Trimethylpentane	0.02
2,2,3,3-Tetramethylbutane	0.02

2.2.2. Naphthenes

The molecular ion series of single-ring C_5, C_6, C_7, etc., naphthenes (or cycloparaffins), appears at m/e values 70, 84, 98, etc., in the C_nH_{2n} series. The general formulas are identical for cyclopentanes and cyclohexanes. The major fragment ions appear at m/e 69, 83, 97, in the C_nH_{2n-1} series. Typical fragmentation occurs at bonds connecting the ring to the rest of the molecule (bond α to the ring). Cyclopentyl derivatives often yield high-intensity ions at m/e 56, 69, 70, while cyclohexyl derivatives give strong peaks at m/e 55, 84, 98, and thus there is some possibility of distinguishing these two types. The distinction, however, is not always feasible; gem-disubstituted cyclohexanes, for example, also give high-intensity ions at m/e 69.

Molecular ions of naphthenes are generally more abundant than those of paraffins. Higher-molecular-weight members of the series tend, however, to give intense rearrangement ions (formed by a C–C cleavage, followed by migration of a hydrogen atom) at m/e values corresponding to the molecular ions of lower-molecular-weight members of the series. This feature makes identification of the lower-molecular-weight members of a series very difficult in the presence of higher-molecular-weight members.

One-ring naphthenes have the same general formulas as olefins of the same carbon number. The fragmentation patterns are also quite similar. In general, naphthenes cannot be distinguished from olefins. The latter are usually eliminated by chemical treatment, or hydrogenated to the corresponding paraffins. Bromination and deuteriation of the olefinic double bonds are also used for this purpose.

Condensed-ring naphthenes possess spectral features generally similar to those of the single-ring naphthenes. Their molecular and main fragment ions are listed in Table 3.

The mass difference between the m/e values of the first members of homologous series differing by one ring in the structure is 40, corresponding to the addition of a cyclopentyl moiety, C_3H_4, to an existing ring structure, as shown in Table 3.

TABLE 3.   Molecular and Fragment Ions of Multiring Naphthenes

Number of naphthenic rings per molecule	Molecular ion series	Fragment ion series
2	110, 124, 138, . . . ,(C_nH_{2n-2})	109, 123, 137, . . . ,(C_nH_{2n-3})
3	150, 164, 178, . . . ,(C_nH_{2n-4})	149, 163, 177, . . . ,(C_nH_{2n-5})
4	190, 204, 218, . . . ,(C_nH_{2n-6})	189, 203, 217, . . . ,(C_nH_{2n-7})
5	230, 244, 258, . . . ,(C_nH_{2n-8})	229, 243, 257, . . . ,(C_nH_{2n-9})
6	270, 284, 298, . . . ,(C_nH_{2n-10})	269, 283, 297, . . . ,(C_nH_{2n-11})

2.2.3. Alkylbenzenes

These compounds, as generally all aromatics, are stabilized by the resonance structures of the aromatic ring(s) and possess intense molecular ions as a rule.

The molecular ion series of alkylbenzenes appears at m/e values 78, 92, 106, . . .,etc., in the C_nH_{2n-6} series. The most abundant fragment ions are tropylium ions formed by benzylic cleavages (p. 12). These are at m/e values 91, 106, 119, . . .,etc., in the C_nH_{2n-7} series.

The fragment ion at m/e 91 is an intense one in all alkylbenzenes, as well as in naphtheno-benzenes such as indanes and tetralins. Rearrangement ions of m/e 92 are also frequently observed.

Polymethylsubstituted benzenes, such as xylenes, yield ions due to α cleavage, but at least one of the methyl groups usually remains on the ring, giving again abundant ions of m/e 91.

2.2.4. Other Aromatic Hydrocarbons

Other aromatics also give abundant molecular ions. These stand out in complex mixtures and greatly facilitate the identifications of aromatic homologs. A list of the major aromatic and heteroaromatic compound types in petroleum- and coal-derived materials is given in Table 4. This list includes the general formulas of the various compound types, and the structures, molecular weights, and carbon numbers of the first members in each series. As several isomeric structures such as phenanthrenes and anthracenes can have the same general formulas, the structures given should be considered only as possibilities, rather than definitive and unique assignments. The data in Table 4 can be used nevertheless with success in identifying various aromatic compounds from their general formulas, for example, at m/e values in the C_nH_{2n-13} series for naphthalenes, whose general formula is C_nH_{2n-12}. This feature is due to the fact that the prevalent fragmentation is caused by cleavages of C–C bonds in the sidechains.

Aromatics also give significant amounts of doubly and triply charged ions, that appear at masses of ½ and ⅓ of the mass of the singly charged ions. Double charged ions of even mass species appear at even m/e values, for example, at m/e 64 for the doubly charged naphthalene molecular ion, and thus are not immediately distinguishable from other peaks in the spectrum. Doubly charged odd mass fragments appear at nonintegral nominal masses, for example, at 63.5 for the doubly charged $C_{10}H_7$ ion from naphthalene, and furnish easy clues to the identification of aromatic mixtures. Similar reasoning applies to triply charged ions.

The presence of heteroatoms in aromatic structures is due mainly to

TABLE 4. Major Aromatic and Heteroaromatic Compound Types in
Petroleum and Coal Liquefaction Products

		Features of unsubstituted nucleus	
General formula	Typical structure(s)	MW	Carbon number
C_nH_{2n-6}	Benzene	78	6
C_nH_{2n-8}	Indane	118	9
	Tetralin	132	10
C_nH_{2n-10}	Indene	116	9
	Dihydronaphthalene	130	10
	Tetrahydroacenaphthene	158	12
	Hexahydrofluorene	172	13
	Octahydrophenathrene	186	14
C_nH_{2n-12}	Naphthalene	128	10
	Decahydropyrene	212	16
C_nH_{2n-14}	Acenaphthene	154	12
	Biphenyl	154	12
	Tetrahydrophenanthrene	182	14
C_nH_{2n-16}	Acenaphthylene	152	12
	Fluorene	166	13
	Dihydrophenanthrene	180	14
	Hexahydropyrene	208	16
C_nH_{2n-18}	Phenanthrene	178	14
	Anthracene	178	14
	Tetrahydropyrene	206	16
C_nH_{2n-20}	Cyclopentanophenanthrene	190	15
	Dihydropyrene	204	16
C_nH_{2n-22}	Pyrene	202	16
	Fluoranthene	202	16
C_nH_{2n-24}	Chrysene	228	18
C_nH_{2n-26}	Benzo[*ghi*]fluoranthene	226	18
	Cholanthrene	240	19
C_nH_{2n-28}	Benzopyrene	252	20
C_nH_{2n-30}	Benzochrysene	278	22
C_nH_{2n-32}	Benzo[*ghi*]perylene	276	22
	Anthanthrene	276	22
C_nH_{2n-34}	Dibenzopyrene	302	24
C_nH_{2n-36}	Coronene	300	24
C_nH_{2n-38}	Dibenzo[*cd,lm*]perylene	326	26
C_nH_{2n-40}	Dibenzo[*a,o*]perylene	352	28
C_nH_{2n-42}	Benzocoronene	350	28
C_nH_{2n-44}	Tribenzo[*de,kl,rst*]perylene	376	30
$C_nH_{2n-4}S$	Thiophene	84	4
$C_nH_{2n-6}S$	Thiol	110	6
	Naphthenothiophene	124	7
$C_nH_{2n-10}S$	Benzothiophene	134	8
$C_nH_{2n-12}S$	Thionaphthol	160	10
	Naphthenobenzothiophene	174	11
$C_nH_{2n-14}S$	Indenothiophene	172	11

TABLE 4 (continued)

| General formula | Typical structure(s) | Features of unsubstituted nucleus | |
		MW	Carbon number
$C_nH_{2n-16}S$	Naphthothiophene	184	12
	Dibenzothiophene	184	12
$C_nH_{2n-18}S$	Acenaphthenothiophene	210	14
$C_nH_{2n-20}S$	Fluorenothiophene	222	15
$C_nH_{2n-22}S$	Phananthrenethiophene	234	16
$C_nH_{2n-24}S$	Naphtheophenanthrenothiophene	246	17
$C_nH_{2n-26}S$	Pyrenothiophene	258	18
$C_nH_{2n-28}S$	Chrysenothiophene	284	20
$C_nH_{2n-30}S$	Cholanthrenothiophene	296	21
$C_nH_{2n-32}S$	Benzopyrenothiophene	308	22
$C_nH_{2n-14}S_2$	Thiophenobenzothiophene	190	10
$C_nH_{2n-20}S_2$	Thiophenodibenzothiophene	240	14
$C_nH_{2n-28}S_2$	Thiophenophenanthrenothiophene	290	18
$C_nH_{2n-6}O$	Phenol	94	6
$C_nH_{2n-8}O$	Indanol	134	9
	Hydroxytetralin	148	10
$C_nH_{2n-10}O$	Benzofuran	118	9
$C_nH_{2n-12}O$	Naphthol	144	10
	Naphthenobenzofuran	158	11
$C_nH_2,yn_{14}O$	Hydroxyacenaphthene	170	12
$C_nH_{2n-16}O$	Dibenzofuran	168	12
	Hydroxyfluorene	182	13
$C_nH_{2n-18}O$	Hydroxyphenanthrene	194	14
	Naphthenonaphthofuran	194	14
$C_nH_{2n-20}O$	Hydroxycyclopentanophenanthrene	206	15
	Fluorenofuran	206	15
$C_nH_{2n-22}O$	Hydroxyphenanthrene	218	16
	Phenanthrenofuran	218	16
$C_nH_{2n-24}O$	Hydroxychrysene	244	18
$C_nH_{2n-26}O$	Pyrenofuran	242	18
	Hydroxychlolanthrene	256	19
$C_nH_{2n-28}O$	Hydroxybenzopyrene	266	20
$C_nH_{2n-30}O$	Hydroxybenzochrysene	294	22
$C_nH_{2n-32}O$	Hydroxyanthanthrene	292	22
$C_nH_{2n-34}O$	Hydroxydibenzopyrene	318	24
$C_nH_{2n-36}O$	Hydroxycoronene	316	24
$C_nH_{2n-6}O_2$	Dihydroxybenzene	110	6
$C_nH_{2n-8}O_2$	Dihydroxytetralin	164	10
$C_nH_{2n-14}O_2$	Dihydroxyacenaphthene	186	12
$C_nH_{2n-16}O_2$	Hydroxydibenzofuran	184	12
	Dihydroxyfluorene	198	13
$C_nH_{2n-18}O_2$	Dihydroxyphenanthrene	210	14
$C_nH_{2n-22}O_2$	Dihydroxypyrene	234	16
$C_nH_{2n-24}O_2$	Dihydroxychrysene	260	18

continued overleaf

TABLE 4 (continued)

General formula	Typical structure(s)	Features of unsubstituted nucleus	
		MW	Carbon number
$C_nH_{2n-26}O_2$	Furanophenanthrenofuran	258	18
$C_nH_{2n-3}N$	Pyrrole	67	4
$C_nH_{2n-5}N$	Pyridine	79	5
$C_nH_{2n-7}N$	Naphthenopyridine	119	8
$C_nH_{2n-9}N$	Indole	117	8
	Dinaphthenopyridine	159	11
$C_nH_{2n-11}N$	Quinoline	129	9
$C_nH_{2n-13}N$	Naphthenoquinoline	169	12
$C_nH_{2n-15}N$	Carbazole	167	12
$C_nH_{2n-17}N$	Acridine	179	13
$C_nH_{2n-19}N$	Dihydropyridine	205	15
$C_nH_{2n-21}N$	Pyrenide	203	15
	Benzcarbazole	217	16
$C_nH_{2n-22}N$	Chrysenide	229	17
$C_nH_{2n-25}N$	Cholanthrinide	241	18
$C_nH_{2n-27}N$	Benzopyrenide	253	19
	Dibenzcarbazole	267	20
$C_nH_{2n-29}N$	Benzochrysenide	279	21
$C_nH_{2n-31}N$	Benzo[ghi]perylenide	277	21
$C_nH_{2n-16}SO$	Hydroxydibenzothiophene	200	12
$C_nH_{2n-18}SO$	Hydroxyacenaphthenothiophene	226	14
$C_nH_{2n-20}SO$	Hydroxyfluorenothiophene	238	15
$C_nH_{2n-22}SO$	Hydroxyphenanthrenothiophene	250	16
$C_nH_{2n-26}SO$	Hydroxypyrenothiophene	274	18
$C_nH_{2n-9}NO$	Hydroxyindole	133	8
$C_nH_{2n-11}NO$	Hydroxyquinoline	145	9
$C_nH_{2n-15}NO$	Hydroxycarbazole	183	12
$C_nH_{2n-17}NO$	Hydroxyacridine	195	13
$C_nH_{2n-21}NO$	Hydroxybenzcarbazole	233	16
$C_nH_{2n-16}O_3$	Dihydroxybenzofuran	200	12
$C_nH_{2n-22}O_3$	Trihydroxypyrene	250	16
$C_nH_{2n-16}SO_2$	Dihydroxydibenzothiophene	216	12
$C_nH_{2n-15}NO_2$	Dihydroxycarbazole	199	12

furanic, phenolic, thiophenic, pyrrolic, and pyridinic groups. It is useful to remember the changes in general formulas related to the addition of such groups to aromatic structures (Table 5).

Naphthenoaromatic or hydroaromatic components, such as indanes, tetralins, tetrahydrophenanthrenes, and octahydrophenanthrenes, give mass spectra that combine both aromatic and naphthenic features. Alkyl chains on the saturate moieties of these molecules fragment predominantly

TABLE 5. Comparison of General Formulas of Aromatic
and Related Heteroaromatic Types[a]

Generalized expressions

Group added	General formula of aromatic before addition	General formula after addition
Hydroxy	C_nH_{2n-x}	$C_nH_{2n-x}O$
Furan	C_nH_{2n-x}	$C_nH_{2n-(x+4)}O$
Thiophene	C_nH_{2n-x}	$C_nH_{2n-(x+4)}S$
Pyridinic N	C_nH_{2n-x}	$C_nH_{2n-(x-1)}N$
Pyrrole	C_nH_{2n-x}	$C_nH_{2n-(x+3)}N$

Examples

Aromatic compound	General formula	Heteroaromatic compound	General formula
Benzene	C_nH_{2n-6}	Phenol	$C_nH_{2n-6}O$
Naphthalene	C_nH_{2n-12}	Dibenzofuran	$C_nH_{2n-16}O$
Benzene	C_nH_{2n-6}	Benzothiophene	$C_nH_{2n-10}S$
Naphthalene	C_nH_{2n-12}	Quinoline	$C_nH_{2n-11}N$
Benzene	C_nH_{2n-6}	Indole	$C_nH_{2n-9}N$

[a] *Note*: Addition of heteroatoms also changes the precise masses of the components.

through α cleavages; those on the aromatic moieties through β cleavages. This rule allows one to identify the position of the substituents on alkyl tetralins.

2.3. General Approach for Structure Identification

Exploitation of the simple relationship existing between the *m/e* values of the most abundant fragment ions and the general formulas of the homologous series can lead to qualitative identification of some of the compound types present. The scope of this approach can be greatly enhanced through the use of high-resolution techniques for identifying molecular formulas, as discussed in Section 3 of this chapter.

In addition to high resolution, formulas can also be determined by measuring the abundance of the ^{13}C, ^{18}O. ^{34}S, ^{37}Cl, and ^{81}Br isotopes given by the various ions, in particular the molecular ions in pure compounds, that are free from interferences from fragment ions. The intensity of [M+1] and [M+2] ions, where M is the *m/e* value of the molecular ion, is related to the natural abundance of the heavier isotopes and the number of the atoms of a given element in the formula (p. 4).

It is thus very helpful if one can immediately identify the molecular ion(s) in a mass spectrum. This is reasonably feasible by noting a few simple rules.

(1) *Chemical rules:*

    (a) All compounds containing any number of C, H, O, S, halogen, and P atoms and an even number of N atoms have even number molecular ions.

    (b) All compounds containing an odd number of N atoms have odd number molecular ions. These rules derive from the fact that N is the only element with even nominal atomic weight but odd valence state.

(2) *Mass spectrometric rules:*

    (a) The molecular ion and its isotopes appear at the highest *m/e* values in the spectrum of a pure compound.

    (b) In mixtures, the start of the molecular ion region can sometimes be identified by the greater abundances of the even mass ions with respect to those of the odd masses from the same series (odd mass vs. even mass in N compounds). Isotope-corrected molecular ions (the isotope correction procedure subtracts the contribution of the ^{13}C and other isotopes of a lower mass peak from the intensity of the higher mass peak) show often a bimodal distribution of intensities vs. carbon number. The first low carbon number "hump" in such a plot is usually due to fragment and rearrangement ions, the second to the parent peaks. The "minimum" between the humps signals, thus, the beginning of the molecular ion region for the series.

    (c) Molecular ions in the spectra of pure compounds or simple mixtures can be identified by the reasonable mass differences existing between parent and fragment ions. At high resolution, these differences have to correspond to exact formulas.

    (d) Decay in peak sensitivities due to diffusion through the leak in the inlet system ("rate of leak") is proportional to the square root of the molecular weight of the compound and is equal for all ions from that compound. Observation of the rate of leak might thus be helpful in pinpointing sets of molecular ions and the corresponding fragment ions in simple mixtures.

The strategy for the qualitative analysis of complex petrolim mixtures might thus include the following steps:

    (i) Identify the major compound types by their characteristic fragmentation features.

(ii) Identify the major molecular ions.

(iii) Determine precise formulas of both molecular and fragment ions if high resolution is available.

(iv) Identify major fragmentation processes.

(v) Compare mass spectrometric evidence with data from IR, NMR, GC, etc., if feasible.

Additional information can be gathered through restriction of the spectra to the molecular ions only, as will be shown in Section 2.5.2.

Separation into saturate, aromatic, and polar aromatic fractions is also helpful, in particular for the characterization of the saturate fractions.

Other separations, including distillations, are used occasionally. All have the objective of simplifying the materials to be analyzed by mass spectrometry.

The foregoing discussion dealt mainly with the qualitative analysis of complex mixtures such as petroleum fractions. Identification of pure compounds is carried out often in the petrochemical industry following general approaches common to all mass spectrometric approaches.

2.4. Principles of Quantitative Analysis

In the inlet system of a mass spectrometer, the partial pressures of the components of a mixture are proportional to their concentrations. The intensities of the molecular and fragment ions of a component are directly related both to its partial pressure and to its specific structure. These intensities can thus be used for developing analytical procedures for quantitative analyses.

The component of the ion intensity related to the structure is determined by calibration with a known amount of that compound and is usually expressed in arbitrary units (chart paper divisions, digitizer counts, etc.) per milligram of charge or microns of pressure. Quantitative determination can be based on one ion or group of ions per component.

A typical approach in setting up a quantitative procedure is to obtain the mass spectrum of each of the components to be analyzed for, select the ions to be used for the quantitative procedure, and to determine the specific calibration coefficient(s) or sensitivity. The most important part of this approach is the selection of the particular ions used. These have to be both as abundant and as specific as possible. In other words, they need to be intense in the spectrum of the component they represent and absent, or at least weak, in the spectra of the other components.

Mutual interferences, the general rule, are handled by setting up simultaneous equations, solved by matrix algebra, as shown below:

$$zA_1 + yA_2 = M_1 \qquad (1)$$

$$zB_2 + yB_1 = M_2 \qquad (2)$$

where z and y are the weight fractions of two components in a mixture; A_1, A_2, B_1, and B_2 are the sensitivities of characteristic peaks in the spectra of the two components, and M_1 and M_2 are the peak heights in a mixture of the two components.

For an accurate analysis, the intensity of the peak A should be stronger in the spectrum of the first component than in the second, and B should be more abundant in the second component.

Equations (1) and (2) can be solved as follows:

$$z = (M_1 B_1 - M_2 A_2)/(A_1 B_1 - B_2 A_2) \qquad (3)$$

$$y = (M_2 A_1 - M_1 B_2)/(A_1 B_1 - B_2 A_2) \qquad (4)$$

The procedure can be illustrated by an example.

Calibration for n-nonane shows a relative sensitivity of 900 divisions/mg at m/e 128 and a sensitivity of 14,200 divisions/mg at m/e 43. The sensitivities at the same m/e values are, respectively, 8,200 divisions/mg and 14 divisions/mg for naphthalene. In a 50/50 weight fraction mixture, theoretical intensities would be $7,100 + 7 = 7,107$ at m/e 43 and $450 + 4100 = 4550$ at m/e 128. Equations (1) and (2) in this case become

$$(0.5 \times 14,200) + (0.5 \times 14) = 7,107 \qquad (5)$$

$$(0.5 \times 900) + (0.5 \times 8,200) = 4,550 \qquad (6)$$

If the concentrations were unknown, one could calculate them from the measured intensities and the pure compound calibration data. Substituting the assumed values in equations (3) and (4) indeed gives the correct concentrations for x and y, as shown below:

$$z = \frac{(7,107 \times 8,200) - (4,550 \times 14)}{(14,200 \times 8,200) - (14 \times 900)} = 0.50 \qquad (7)$$

$$y = \frac{(4,550 \times 14,200) - (7,107 \times 900)}{(14,200 \times 8,200) - (14 \times 900)} = 0.50 \qquad (8)$$

It is customary to express z and y in percentages by multiplying the weight fractions by 100.0. In practical cases, the sum of the concentrations is slightly different from 100.0 but the data are normalized to this value. Large differences of the unnormalized sum of concentrations from 100.0 usually indicate difficulties, such as irreproducible sample charges, presence of unexpected components, or sudden changes in instrument sensitivity.

The procedure illustrated above can be extended to 30–50 components. Unnormalized analyses in which not all the components are determined are also used. In these cases one cannot compensate for slight inequalities of volumes or weights of sample and calibrant charged, and accuracy of the analysis is strictly related to the reproducibility of sample introduction. When using a capillary pipette for charging liquid samples, this reproducibility is approximately 1% of the amount charged.

2.5. Simplifying Approaches

The complexity of petroleum mixtures generally limits analysis of individual components to gases and light aromatic mixtures. Analysis of middle and heavy distillates requires simplification of the approach. This is essentially achieved in two ways. The first approach is to reduce the scope of the analysis by limiting it to the determination of groups of similar compounds. This approach is usually referred to as compound type analysis. The second approach is to simplify the spectra, preferably to the molecular ions. This can be achieved by using different ionization techniques, such as field ionization, field desorption, chemical ionization, or low-voltage ionization.

Another way of simplifying the complexity of petroleum mixtures is the use of very efficient separation techniques, such as GC–MS.

Approaches discussed in this chapter will be the compound type analysis and the low-voltage techniques. The other approaches are presented elsewhere.

2.5.1. Compound Type Analysis

This approach takes advantage of the similarity of the spectra in a homologous series. As mentioned earlier, all compound types of formulas C_nH_{2n-z} give strong, characteristic fragment peaks at m/e values corresponding to $C_nH_{2n-(z+1)}$. The most abundant of these can be used to determine a given compound type in a mixture without, of course, determining the individual homologs. In effect, each compound type is treated as a single component and mathematical treatment of the interferences between types is the same as the one used for individual components.

The number of ions used to represent a type can vary from 3 to 30 or 40. In general, the more complex a type is, the more ions are used to determine it. Four or five ions such as those of m/e 71, 85, 99, and 113 are deemed sufficient for paraffins, but a larger number is used for six-ring naphthenes which can include various combinations of condensed and noncondensed five- and six-member naphthenic rings. Two component methods for determining total aromatics and total saturates need to use a

large number of ions for each "component" in order to include characteristic fragment or molecular ions of the various saturate and aromatic types
expected.

Ions used in these methods can include the molecular ion series.
Values used for both calibration and analyses are the summations of the
abundances of the ions selected to represent each compound type.

Preparation of a compound type procedure is similar to that of a
quantitative procedure of a mixture of pure components. Selection of the
ions used is carried out according to the criteria discussed above, that is,
the ions preferred for a given type are those which are abundant in the
spectra of homologs in a given type, weak in the others. Spectral features
are determined either from the examination of the spectra of a large
number of individual components, or from physically separated fractions.
The latter approach is the preferred one, if feasible, as pure compounds
are not always available, and well-separated fractions have the same or
similar isomeric distribution as the expected samples. They offer, thus, a
high similarity between calibrant and sample that is essential for accurate
analyses. Separation techniques used include distillations, liquid–solid
chromatography, such as the separation of saturates and aromatics on
silica gel or the separation of aromatics of one, two, three, etc., rings on
alumina.

The actual calibration procedure is carried out either with blends of
pure compounds resembling the expected isomeric distribution or with the
separated fractions.

Compound type procedures generally involve spectra obtained at
high-voltage, low-resolution conditions. Their advantages are related to
this mode of operation. Reproducibility of the analyses is very high: 1–5%
of the amount of a given compound type. The detectability is also high;
detection of types at concentration levels as low as 0.1% is common.

Reproducibility is maintained by carefully standardizing instrumental
conditions. One of the most important of these is the temperature of the
ion source. Small variations in this temperature can cause large variations
in the spectra, in particular in the stability of the molecular ion and the
ratios between molecular and fragment ions. The ion source temperature
is usually monitored by the ratio between a parent and a fragment peak in
a pure compound, typically m/e 226/57 or m/e 226/127 in n-hexadecane.
This ratio is kept at a constant value, for example, 0.046 ± 0.002 for the
226/57 ratio. Instrument sensitivity (response to a given amount of a pure
compound) and linearity of response are also controlled closely.

The calculations required are easy, and are readily computerized. Use
of inverses allows manual solutions of 6–8 component analyses in a matter
of minutes.

Disadvantages of this approach are mainly related to calibration
difficulties. Accuracy is a function of the similarity between calibrant and

sample. Unexpected and unobserved dissimilarities may occur frequently. In general, any high-voltage method is restricted to materials that have a similar origin, process treatment history, and boiling range.

Quantitative procedures are also restricted to *a priori* selected components, and the presence of unexpected materials can cause difficulties.

High-voltage, low-resolution compound type analyses are particularly useful for comparison purposes, where the importance of reproducibility, and the need of detecting small variations and trends in sample composition in a series of processing experiments, outweighs the need for absolute accuracy.

The most common compound type methods in use today include the determination of individual saturate and aromatic types, such as paraffins, one- through six-ring naphthenes, benzenes, indanes, tetralins, naphthalenes, and more condensed aromatics, aromatic thiophenes, and hydroaromatics. Most methods require a minimum of preliminary separations, such as separation of total saturates or aromatics prior to determination of the individual saturate or aromatic types. Methods that do not require such preliminary separations are also available.

2.5.2. Low-Voltage Techniques

A very successful technique for limiting mass spectra to the molecular ions is the use of low ionizing voltages. The appraoch is based on the fact that the ionization potential of the molecular ions of aromatic and olefinic compounds is significantly lower than the appearance potentials of their fragments, as well as the fragments from saturate or polar aromatic components. There is thus a region, generally between 8 and 10 eV in which the only peaks observed are those corresponding to aromatic and olefinic molecular ions.

Operation at low-voltage conditions offers several important advantages:

(1) Aromatic and olefinic molecular ions can be easily distinguished from their fragment ions. The latter will disappear as one slowly lowers the voltage from high-voltage conditions (70 eV) to 10 eV. Sensitivity of the molecular ions also decreases, but at a much slower rate, so that in effect one observes a very considerable enhancement of the relative intensities of the molecular ions.

(2) Each component yields only its molecular ion (and the isotopes). There are no interferences between components in a mixture except for those that have the same nominal molecular weight or formula. Interferences among components of the same nominal molecular weight, but of different formula, can be resolved by high-resolution techniques. Isotopes do cause interferences, but these can be easily corrected.

(3) Calibration is easy in that only the molecular ions need to be

considered. The calibration set can be extended to components not available for instrumental calibration by extrapolation and by using theoretical considerations. Dependence on sample origin, history, and boiling range is essentially eliminated.

(4) All individual aromatic and heteroaromatic homologs in a mixture can be determined individually, with no *a priori* limitation of the number of components. The only limitation is the resolving power available to separate components of the same nominal molecular weight. In practice, the number of components that can be analyzed may be as high as 300 at a resolving power of 500 and as high as 3000 at a resolving power of 10,000. This number is being increased further by the advent of high-resolution instrumentation capable of dynamic resolving power of 40–80,000.

(5) Data on the individual homologs lead to a very detailed characterization of aromatic and heteroaromatic mixtures.

Disadvantages of the low-voltage approach include the following:

(1) It is restricted to aromatic and heteroaromatic components. Paraffins, naphthenes, alkylsulfides, and other important constituents of petroleum and coal liquefaction products cannot be detected by this technique.

(2) Sensitivity of the molecular ions is lower than at high voltages. This decrease in sensitivity amounts to a factor of 5–10 and makes the determination of components present in trace amounts more difficult at low voltages than at high voltages. This deficiency can be compensated in some cases by charging large amounts of sample (up to 10 or 15 mg).

(3) Small variations in the ionizing voltage cause significant changes in the sensitivities. This fact, coupled by the lower sensitivity, leads to the inherently lower reproducibility of low-voltage determinations. In general, these are no better than 1–5% of the amount present for the major components and 5–10% for the minor ones. Even this result can be achieved only by careful standardizations of all instrumental conditions. In most cases, the voltage used is not that shown by the meter on the instrument, but it is established by using criteria selected in the spectrum of a pure compound. Low-voltage operation is thus carried out at voltages corresponding to the disappearance of the molecular ion of a saturate compound, say, cyclohexane, or to a ratio of about 100 between molecular ion (m/e 106) and the major fragment ion (m/e 91) in the spectrum of *m*-xylene.

(4) Some rearrangement ions are still present in low-voltage spectra, in particular in those of olefins and one-ring aromatics. These can be mistaken for real molecular ions. The m/e values of these ions are, however, generally low, and thus can be distinguished from the real molecular ions in the higher m/e region.

Low-voltage sensitivities are closely related to the value of the ionization potential. The lower the value of the ionization potential, the higher is the sensitivity at a given low voltage value at approximately 2 or 3 V above the ionization potential. Ionization potentials of aromatic components range approximately from 6 to 9.5 eV; customary low-voltage conditions range from 10 to 12 eV. Ionization potentials in general decrease with the aromaticity of a component, and thus one can establish a reasonable correlation between the number of aromatic double bonds and low-voltage sensitivity.

Low-voltage sensitivities are, of course, limited by the vaporization characterizations of the compounds, and hence by their molecular weights. Low-voltage sensitivity can be set at zero for infinite molecular weight, and in general is inversely proportional to molecular weight for the higher members of a given homologous series.

These and other considerations allow extrapolation of calibration data obtained for 100–200 compounds to larger calibration sets containing data for thousands of components. The main rules used can be illustrated by the following examples.

(1) Low-voltage sensitivities increase with:

(a) aromaticity (phenanthrene > benzene > pyridine)
(pyrrole > benzene > pyridine)
(indole > naphthalene > quinoline)
(carbazole > phenanthrene > acridine)

(b) number of side chains (xylenes > toluene > benzene)

(2) Low-voltage sensitivities decrease with:

(a) length of side chains (*n*-propyl benzene < ethylbenzene < toluene)

(b) degree of branching of side chain (isopropyl benzene < *n*-propyl benzene).

3. Applications of High- and Ultrahigh-Resolution Instruments

The availability of high-resolution instrumentation extends the scope of mass spectrometry for both qualitative and quantitative applications and for all analytical approaches including high- and low-voltage techniques, chemical ionization, GC–MS, and field ionization and desorption, etc. The main advantage of high-resolution techniques is their advantage of separating fragment and molecular ions of the same nominal molecular weight but different elemental composition. In addition, most high-resolution instruments are capable of furnishing precise mass measurements, so that the formulas separated can be calculated from these measurements.

The advantages of high resolution are obvious for qualitative analysis, as illustrated by the following simple example. At low resolution, the *m/e* value of the molecular ion of a compound can be defined only as an integral number, say 106. At high resolution, one can determine whether the mass of this ion is either 106.0782 or 106.0418, and hence calculate whether the formula is C_8H_{10} or C_7H_6O. The knowledge of the elemental composition is obviously a great step forward in identifying a particular structure.

Advantages of high-resolution techniques for quantitative analysis are equally impressive. The number of components or compound types which can be determined in a mixture is increased by a large factor as individual components of the same nominal molecular weights but different formulas can now be separated.

High-resolution mass spectrometry is based on the fact that the atomic weights of the elements are, except for ^{12}C, nonintegral values, as shown in Table 1 in Chapter 1.

Different combinations of these elements thus possess formulas with slightly different precise masses. The number of possible combinations at a given nominal molecular weight increases with molecular weight and the number of elements considered. Examples of different combinations are shown in Table 6.

The resolving power required to separate ions of slightly different masses is generally defined as the ratio of the mass of the ions to the mass difference, $M/\Delta M$. For example, the resolving power required to separate the ion of formula $C_{16}H_{28}$ (mass 220.2191) from an ion of formula $C_{15}H_{24}O$ (mass 220.1827) is $220/(220.2191-220.1827)=220/0.0364 \simeq 6050$. The mass difference between the two formulas (0.0364) corresponds to the difference

TABLE 6. Examples of Different Combinations of Formulas

Formula	Mass	Isotopic ratios[a](%) M+1	M+2	Number of C, H, O, N formulas at nominal mass
CO	27.9949	1.12	0.200	4
CH_2N	28.0187	1.49	0.005	
C_2H_4	28.0313	2.23	0.013	
N_2	28.0071	0.76	0.035	
$C_{17}H_{16}$	220.1252	18.63	1.64	99
$C_{14}H_{20}S$	220.1286	16.21	5.33	
$C_{16}H_{28}$	220.2191	17.74	1.48	
$C_{15}H_{24}O$	220.1827	16.63	1.50	
$C_{14}H_{24}N_2$	220.1939	16.27	1.24	
$C_{14}H_{22}ON$	220.1701	15.90	1.38	
$C_{12}H_{16}N_2O_2$	220.1212	14.06	1.32	

[a] Differences in isotopic ratios are larger for the closer mass values, e.g., 220.1252 and 220.1212. Determination of precise isotopic ratios can thus help in resolving uncertainties derived from finding more than one formula within the accuracy limits of the mass measurement.

between CH_4 (16.0313) and O (15.9949). This is the value of the difference, commonly referred to as a mass doublet, between formulas differing by a CH_4 versus an O unit at any mass value. The resolving power required to separate formulas differing by the CH_4/O doublet increases with the mass of the ions and is, for example, approximately 12,100 at m/e 400.

Values of resolving power required to separate the doublets most frequently occurring in petroleum and coal liquefaction products are shown in Table 7. The formulas of the compound types distinguished are expressed in general terms with illustrative examples given at the bottom of the table. The table also gives the maximum m/e values at which the doublets can be resolved with a given resolving power. The resolving power values used (10,000, 80,000, 150,000) correspond, respectively, to the dynamic resolving power of an Associated Electric Industries model MS9 instrument, and the dynamic and static resolving power of a model MS50 instrument of the same manufacturer. Low-resolution instruments commonly in use have resolving powers ranging up to 800.

None of the doublets listed in Table 7 can be thus resolved with a low-resolution instrument. An instrument of capabilities equivalent to an MS9 can resolve only the H_{12}–C, C_2H_8–S, and some of the CH_4–O doublets in the 100–500 mass range that is generally of most interest to petroleum chemists. Resolution of most of the doublets listed in Table 7 requires either an ultrahigh-resolution instrument, such as the MS50, or special auxiliary procedures. Typical of the latter approach are the "separation" of the ^{13}C isotopes of hydrocarbons and oxygenated hydrocarbons from the interfering nitrogen and nitrogen–oxygen compounds and that of hydrocarbons from interfering sulfur compounds.

Nitrogen compounds can be determined by calculating and subtracting the isotopic contribution of the hydrocarbon peaks from the intensity of the nitrogen-containing peaks. The remainder intensities can be assumed to be due to the nitrogen compounds. A cue to the presence of unresolved nitrogen compounds is given by the precise mass measurements, which are the weighted average of the unresolved ^{13}C and N peaks and have lower mass values than that of an unicomponent isotope. An example of this approach is given in Table 8.

Separation of interfering sulfur and hydrocarbon components is more difficult. In the low-voltage approach, one can take advantage of the fact that the homologous series of the sulfur compounds starts generally at lower molecular weights than those of the hydrocarbons. For example, the first member of the benzothiophene series has a molecular weight of 134, while the first member of the interfering cyclopentanophenanthrene series has a molecular weight of 190. The first four homologs in the unresolvable $C_nH_{2n-10}S/C_nH_{2n-20}$ series can be thus unequivocally assigned to the $C_nH_{2n-10}S$ series, and an approximate separation of the higher-molecular-weight members can be achieved by extrapolation procedures

TABLE 7. Resolving Power Required to Separate the Most Common Multiples in Petroleum and Coal Liquids

Mass doublet	ΔM	General formulas of compound types separated	Resolving power required at m/e:			
			100	300	500	1000
H_{12}–C	0.0939	$C_nH_{2n-x}/C_nH_{2n-(x+14)}$	1,065	3,195	5,325	10,650
C_2H_8–S	0.0905	$C_nH_{2n-x}/C_nH_{2n-(x+4)}S$	1,105	3,315	5,525	11,050
CH_4–O	0.0364	$C_nH_{2n-x}/C_nH_{2n-(x+2)}O$	2,747	8,242	13,735	27,470
C_4–SO	0.0330	$C_nH_{2n-x}/C_nH_{2n-(x-8)}SO$	3,030	9,090	15,150	30,300
C_3–H_4O_2	0.0211	$C_nH_{2n-x}/C_nH_{2n-(x-10)}O_2$	4,739	14,218	23,695	47,390
S–O_2	0.0177	$C_nH_{2n-x}S/C_nH_{2n-x}O_2$	5,650	16,950	28,250	56,500
^{13}CH–N	0.0081	$C_{n-1}{}^{13}CH_{2n-x}/C_nH_{2n-(x-1)}N$	12,346	37,037	61,730	123,460
C_3–SH_4	0.0034	$C_nH_{2n-x}/C_nH_{2n-(x-10)}S$	29,412	88,235	147,060	294,120

Examples of general formulas	Examples of compound types	Maximum mass at which doublet can be separated with RP values of:		
		10,000	80,000	150,000
C_nH_{2n-6}/C_nH_{2n-20}	Benzenes/dihydropyrenes	940	7,500	14,000
$C_nH_{2n-6}/C_nH_{2n-10}S$	Benzenes/benzothiophenes	910	7,300	13,500
$C_nH_{2n-6}/C_nH_{2n-8}O$	Benzenes/hydroxytetralins	364	2,900	5,500
$C_nH_{2n-18}/C_nH_{2n-10}SO$	Phenanthrenes/hydroxybenzothiophenes	330	2,600	5,000
$C_nH_{2n-18}/C_nH_{2n-8}O_2$	Phenanthrenes/dihydroxytetralins	210	1,700	3,200
$C_nH_{2n-16}S/C_nH_{2n-16}O_2$	Dibenzothiophenes/dihydroxyfluorenes	117	1,400	2,600
$C_n{}^{13}C_1H_{2n-10}/C_nH_{2n-11}N$	^{13}C isotopes of naphthalenes/quinolines	81	650	1,200
$C_nH_{2n-20}/C_nH_{2n-10}S$	Cyclopentanophenanthrenes/ benzothiophenes	34	270	500

TABLE 8. Separation of Unresolved N and ^{13}C Peaks

Condition[a]	Measured values		Comments
	Precise mass	Intensity counts	
A	166.0782	3500	Isotope contribution of m/e 166.0782 $C_{13}H_{10}$, to m/e 167: $13 \times 1.1 = 14.3\%$
	167.0815	500	$3500 \times 0.143 = 500$. No residual after isotope correction
B	166.0782	3500	Residual after isotope correction: 500
	167.0758	1000	Measured mass if residual corresponds to nitrogen compound $C_{12}H_9N$ of mass 167.0701: $(0.5 \times 166.0782) + (0.5 \times 167.0701) = 167.0758$

[a] Condition A: No nitrogen compound present. Condition B: Presence of nitrogen compound suggested by mass measurement.

based on the carbon number distribution in the low-molecular-weight region.

Separation of compound types in a high-voltage procedure can also be based on the lower mass fragment peaks given by the sulfur components. The major fragment peaks in the $C_nH_{2n-11}S$ series are those at m/e 147, 161, 175, 191, while those in the C_nH_{2n-21} series appear at m/e values of 203, 217, 231, 245.

Although the auxiliary procedures discussed above are helpful, they are only a poor substitute for instrumental resolving power.

Full exploitation of high resolution requires the availability of precise mass measurement techniques. Mass measurement can be carried out instrumentally in a static mode by the method usually referred to as peak matching. This approach involves the focusing of two peaks, a known standard and the unknown peak to be measured, at a constant magnetic field but at different accelerating voltages. The ratio of the accelerating voltages multiplied by the mass of the known standard yields the precise mass of the unknown. This technique can be very precise, to fractional ppm of m/e values when the two peaks are alternately displayed on a storage oscilloscope and the ratio of the accelerating voltages is changed digitally and in very small increments until a perfect "match" of the two peaks is observed on the scope.

3.1. Computer Programs for Handling High-Resolution Mass Spectra

The large number of peaks to be measured in complex mixtures requires a more efficient approach. Mass measurements can be carried out in the dynamic mode while the mass spectrometer is scanning an entire spectrum, by using computerized data acquisition systems. The general functions of the computer programs developed for this purpose include the following:

(i) Measurement of peak intensities and positions (with respect to time or to space as on a photographic plate).
(ii) Recognition of the reference standard peaks of known m/e.
(iii) Precise mass measurement.
(iv) Assignment of formulas to major peaks.

Specific functions of programs developed for quantitative analyses have additional requirements:

(i) Assignment of unequivocal and unique formulas to all peaks.
(ii) Use of sample peaks as auxiliary reference standards.
(iii) Sorting the recognized formulas according to compound type and carbon number.
(iv) Quantitative analysis.

The details and the logic used in such programs are illustrated by the approach developed at Exxon Research and Engineering Company for the high-resolution, low-voltage analysis of complex aromatic and heteroaromatic mixtures. This approach also shows the great wealth of information that one can obtain from mass spectrometry. The equations and the logic given can also be used in manual calculations.

3.2. Measurements of Peak Intensities and Positions

Analog signals from the mass spectrometer can be converted continuously to digital values by an analog-to-digital converter at a rate corresponding to about 20 conversions per peak. The conversion rate, controlled by a crystal clock, is used as the time scale. If three or more consecutive signals have values above an *a priori* set baseline threshold, they are considered to constitute a peak. Areas of peaks are determined by summing (integrating) the signals; the baricenters are determined by a centroid calculation.

Peak positions with respect to time, and peak-height intensities can also be determined with electronic digitizers. These are less expensive than an analog-to-digital system, but are somewhat less accurate, in particular for smaller peaks that have irregular shapes.

Partially resolved peaks may be separated by mathematical treatment of the digital points describing the peaks (deconvolution). Such treatments include provisions for smoothing the contours of the peaks, criteria to recognize noise spikes, and criteria to recognize the *bona fide* maxima and minima.

3.3. Reference Standards

Mass measurement is carried out with the aid of reference standards. These are peaks of a known compound, or a blend of compounds, introduced into the mass spectrometer together with the sample, to provide the analyst or the computer program with a basis of comparison for mass measurement.

All types of compounds may be used for reference standards. Ideally they need to give a known peak at every 10–14 amu in the mass range of interest. Halogenated compounds are preferred in practice because they can be easily distinguished from hydrocarbons. The most widely used are perfluorokerosene, perfluorotriazines, and, for low-voltage work, blends of halogenated aromatics.

Reference peaks need to be recognized as such among the sample peaks. Criteria used for this prupose include the expected positions and sizes of the reference peaks in the spectrum as a whole, and their positions with respect to the other peaks in a narrow mass multiplet or cluster. Halogenated standards are located on the low-mass side of clusters of hydrocarbon peaks and this property is widely used in computerized recongition schemes.

Correct recognition is absolutely essential to carry out mass measurements. Erroneous recognition of a sample peak or of a noise spike as a reference ion leads to gross errors in mass measurements.

3.4. Mass Measurement

Precise masses of the sample peaks are determined by relating the known positions of the unknown peaks to the known masses and positions of the reference peaks. This relationship is described by equation (9):

$$\ln Mx = Ma + \frac{ta - tx}{\tau(a,b)} \tag{9}$$

$$\tau(a,b) = \frac{ta - tb}{\ln (Mb/Ma)} \tag{10}$$

where Mb and Ma are the known masses of the reference peaks bracketing a sample peak of mass Mx; ta, tb, tx are the respective occurrence times; and $\tau(a,b)$ is the time constant calculated between the reference peaks. Other methods for precise mass measurement include Lagrangian interpolation and other polynominal approximations.

The need for using equations including logarithmic expressions stems from the fact that the scanning function in most magnetic sector instruments is a logarithmic one.

Accuracy of mass measurement is usually better if the mass to be measured is near a reference peak.

As the number of available reference ions is generally limited, accuracy of mass measurement can be improved sometimes by using the already identified sample peaks as auxiliary reference peaks for the remainder of the sample peaks, in a "bootstrapping" type appraoch. In general, a 3-millimass-unit (mmu) accuracy is deemed sufficient.

3.5. Formula Assignment

Formulas are calculated from the precise mass measurements with simple algorithms based on the noninteger masses of the atomic weights of all elements and isotopes other than ^{12}C. Deviations from integer values of a measured mass can be used to calculate the number of hydrogen atoms and heteroatoms included in the formula corresponding to that measured mass. This concept is to be clarified by the examples given below.

In a hydrocarbon containing any number of ^{12}C and ^1H atoms, the deviation from integer masses will be entirely due to the ^1H atoms present, as ^{12}C and all its multiples have integer masses, multiples of the integer value 12.0000. Each ^1H atom in the formula has a mass of 1.0078246, and it thus contributes 0.0078246 to the fractional portion of the measured mass. In an ion containing ten hydrogen atoms, the value of this fractional mass will be $10 \times 0.0078246 = 0.0782460$. Thus the precise mass of $C_{10}H_{10}$ is 130.078426, that of $C_{20}H_{10}$ is 250.078426, etc. It is convenient conceptually to divide measurements as these in an integer portion, say 250, and a fractional portion, 0.078246. The number of hydrogen atoms in an ion can thus be calculated by simply dividing the fractional portion of the measured mass by the fractional mass of hydrogen, 0.0078246. The number of carbon atoms in the same ion is obtained by subtracting the number of hydrogen atoms found from the integer portion of the measured mass and by dividing the integer portion by 12.0000. For example, the formula corresponding to a mass of 212.250388 is $C_{15}H_{32}$ (0.250388 : 0.007824 = 32.0001; 212−32=180; 180 : 12.0000 = 15.0000).

Mass measurements are not always accurate as in the above example, and thus allowances have to be made for errors. It can be shown that a mass measurement error of 1 millimass unit will lead to a quotient differing by ± 0.128 from an integer number when dividing the fractional portion of the mass measurement by the hydrogen deficiency. When calculating the number of hydrogen atoms, one may thus accept fractional values differing from an integer by up to 0.128, or 0.256, or 0.384 for this quotient, according to whether the maximum acceptable mass measurement error is 1, 2, or 3 mmu. The number of hydrogen atoms is then given by the value of the closest integer.

In the above example, a mass measurement error of 1 mmu gives a mass of 212.251388; 0.251388 : 0.0078246 is equal to 32.128, indicating 32 H atoms in the formula. A measured mass of 212.249388 would give a quotient of 31.872, still indicating 32 H atoms.

There cannot be, however, any deviation from integer values when dividing the integer portion of the mass measurement (after subtracting the number of hydrogens) by 12.0000. A fractional value indicates that there is no C, H combination corresponding to the measurement within the error limit.

The presence of heteroatoms, including ^{13}C, O, S, N, P, is also a possibility. The above calculations are still valid, provided that the mass of the heteroatom is first subtracted from the measured mass. In practice, the type and number of heteroatoms to be considered in these calculations is specified *a priori*, and all combinations among these heteroatoms are subtracted, one by one, from the measurement before calculating the number of H and C atoms. If the quotients in these divisions satisfy the criteria set above (integer ±0.3840 for H, integer ±0.000 for C) a valid formula is found. This will contain the number of C and H atoms calculated, plus the particular combination of heteroatom(s) whose masses were subtracted from the measured mass in the given step.

All valid formulas are considered distinct possibilities and are printed out by the computer. There is no limit to the number of heteroatoms one can consider, but as the number of valid formulas increases drastically with the number of elements considered, it is advisable to limit the combinations to only those expected in the sample. For a typical petroleum or coal liquefaction product the combinations would include no more than two sulfur, two oxygen, and two nitrogen atoms.

The above considerations are clarified by the simple example given in Table 9.

The number of ^{12}C and ^1H atoms considered in these calculations is usually unlimited, except for some rudimental chemical considerations. For example, formulas containing more than the number of H atoms required for completely saturated structures are discarded.

An example of a computer output showing formulas identified in a small portion of the low-voltage mass spectrum of the aromatic fraction of a petroleum distillate is shown in Table 10.

The data shown were obtained on an MS9 operated at a resolving power of approximately 10,000, and contain some unresolved peaks. The potential of ultrahigh-resolution mass spectrometry is illustrated by examples of mass multiplets identified in a complex coal liquefaction product (Table 11). A typical MS9 run on a same sample would separate only four of the five peaks at *m/e* 320 and only two of the five at *m/e* 321. A low-resolution spectrum would show, of course, only one peak at *m/e* 320 and one at *m/e* 321. Aromatic and heteroaromatic compound types determined

TABLE 9.   Example of Formula Calculation

Mass measurement	: 296.1218
Error limit	: 0.0030 (3 mmu)
Heteroatoms considered	: $^{32}S(31.9721)$, $^{16}O(15.9949)$
Number of heteroatoms	: sulfur: 1; oxygen: 2

Calculations and Comments

1. C, H

   0.1218:0.007825=15.5655   Not a valid formula; quotient differs from integer by more than 0.384 (3 × 0.128)

2. C, H, S (31.9721)
   (a) 296.1218-31.9721=264.1497   Subtraction of heteroatom mass
   (b) 0.1497:0.007825=19.1310   19 is a possibility for number of H atoms
   (c) 264-19=245   Subtraction of number of H atoms
   (d) 245:12=20.42   Not a valid formula; number of carbons calculated is not an integer.

3. C, H, O (15.9949)
   (a) 296.1218-15.9949=280.1269
   (b) 0.1269:0.007825=16.2173   16 is a possibility for number of H atoms
   (c) 280-16=264 $C_{22}H_{16}O$ is a valid formula. The mass
   (d) 264:12=22.00   measurement error is +1.7 mmu. This can be calculated either by comparing the measured mass with the theoretical one, 296.1201, or by dividing 0.2173 by 0.128.

4. C, H, O_2 (31.9890)   17 is a possibility for the number of H
   (a) 296.1218-31.9898=264.1320   atoms
   (b) 0.1320:0.007825=16.869
   (c) 264-17=247   Not a valid formula
   (d) 247:12=20.58

5. C, H, SO (47.9670)
   (a) 296.1218-47.9670=248.1548   20 is a possibility for the number of H atoms
   (b) 0.1548:0.007825=19.783
   (c) 248-20=228 $C_{19}H_{20}SO$ is a valid formula; mass measurement error is −1.7 mmu.
   (d) 228:12=19

6. C, H, SO_2 (63.9619)
   (a) 296.1218-63.9619=232.1599   Not a valid formula
   (b) 0.1599:0.007825=20.4345

The valid formulas within the mass measurement error allowed are thus $C_{22}H_{16}O$ and $C_{19}H_{20}SO$. Either or both compounds could be present. A decision between these possibilities can be based on auxiliary factors, such as isotopic ratios, elemental analysis, and carbon number distribution of the series.

BLE 10.   Portion of a Typical Computer Output of a High-Resolution, Low-Voltage Spectrum[a]

sity	Measured m/e	mmu error	C	H	O	S	N	General formula
)20	284.0840	+0.3	20	12	2	0	0	C(N)H(2N−28)O(2)
ˈ50	284.1061	COMP NOT FOUND						(Noise?)
?50	284.1189	−1.2	21	16	1	0	0	C(N)H(2N−26)O
ˈ10	284.1423	COMP NOT FOUND						(Isotope of N comp.)
?90	284.1557	−0.7	22	20	0	0	0	C(N)H(2N−20)
ˈ80	285.1142	−1.1	20	15	1	0	1	C(N)H(2N−25)NO
ˈ70	285.1240	ISOTOPE						
ˈ70	285.1504	−1.3	21	19	0	0	1	C(N)H(2N−23)N
ˈ20	285.1583	ISOTOPE						
ˈ40	285.8803	COMP NOT FOUND						(Reference not used)
ˈ70	286.0800	−1.6	20	14	0	1	0	C(N)H(2N−26)S
ˈ10	286.0979	−1.5	20	14	2	0	0	C(N)H(2N−26)O(2)
ˈ30	286.1184	COMP NOT FOUND						(Isotope of N comp)
ˈ20	286.1345	−1.2	21	18	1	0	0	C(N)H(2N−24)O(2)
ˈ20	286.1541	COMP NOT FOUND						(Isotope of N comp.)
ˈ70	286.1709	−1.2	22	22	0	0	0	C(N)H(2N−22)
ˈ60	286.6275	COMP NOT FOUND						(Noise)
.30	287.1302	−0.8	20	17	1	0	1	C(N)H(2N−23)NO
ˈ60	287.1378	ISOTOPE						
ˈ30	287.1675	+0.1	21	21	0	0	1	
.40	287.1724	ISOTOPE						

a obtained on an AEI model MS50 mass spectrometer resolving power set to ~40,000 for dynamic scan.

TABLE 11.   Multiplets Separated by MS50 in Polar Fraction of a Coal Liquefication Product[a]

Formula	Intensity	Precise mass	Mass measurement error	RP required
$C_{24}H_{16}O$	2,340	320.1201	+0.0010	
$C_{23}H_{17}^{13}CN$	60	320.1395	+0.0007	16,500
$C_{25}H_{20}$	279	320.1565	+0.0010	18,800
$C_{24}H_{24}O_2$	676	320.1776	+0.0000	15,200
$C_{23}H_{28}O$	236	320.2140	+0.0012	8,800
$C_{23}H_{15}NO$	178	321.1153	+0.0005	
$C_{23}H_{16}^{13}CO$	452	321.1235	+0.0022	39,200
$C_{24}H_{19}N$	2,078	321.1517	+0.0002	11,380
$C_{21}H_{23}NS$	154	321.1551	+0.0005	94,400
$C_{24}H_{20}^{13}C$	37	321.1599	+0.0012	66,900

a Experimental data obtained by Elliott, Evans, and Hazelby of AEI, Ltd., Manchester, England.

by the computerized high-resolution, low-voltage approach include all those listed in Table 4 and a large number of additional types with up to 30–40 homologs per compound type. The direct mass spectral identifications are restricted, however, to the formulas of the molecular ions. Identification of the major corresponding structures requires further effort.

Structures of the major homologs can be postulated by observing the carbon number distribution of a series in a large number of samples. The lowest-molecular-weight homolog will correspond to the unsubstituted aromatic nucleus. Additional information on the isomeric structures of the major components can be obtained by combining mass spectrometric observations with those from NMR, UV, IR, hydrogenation, dehydrogenation, separations, etc.

3.6. Determination of Average Sample Properties

High-resolution, low-voltage techniques yield values for the concentrations of up to 3000 aromatic and heteroaromatic components per sample. These data may be used to calculate sample properties related to composition, including elemental analysis, average molecular weight, average carbon number, average condensation, distribution of carbon in aromatic, naphthenic, and paraffinic (side-chain) carbon, and boiling point distribution.

These calculations are particularly useful when the amount of sample available for characterization is limited. In these cases, mass spectrometry can furnish information on physical properties that could not be obtained otherwise.

The calculations involved are based mainly on simple, stoichiometric relationships. Elemental analysis for C, H, N, O, and S is obtained simply by calculating the percent C, H, N, O, and S in each identified formula, multiplying these values by the concentration of the corresponding homolog expressed as a weight fraction of the sample, and adding the results. The average molecular weight, average carbon number, and the average condensation or hydrogen deficiency (z number) are calculated similarly. Calculations related to the various forms of carbon require some assumptions on the structures of the unsubstituted aromatic nuclei. Values obtained for these parameters will be different, for example, whether one assumes tetralin or indane as the prevalent structure in C_nH_{2n-8} series. Experience with a large number of samples and comparison of the average calculated with NMR data can furnish, however, reasonable assumptions.

Calculation of the boiling point distribution requires the knowledge of the individual boiling points of all expected homologs. These can be obtained from the literature or by extrapolation. Concentrations of the aromatic homologs are then sorted according to their boiling points, and summed within successive boiling point ranges of 5°F intervals.

In theory, these calculations can be carried out manually. In practice they can be obtained only through computerized data handling. The computer is particularly well suited to carry out this type of work, which consists in a very large number of simple operations. A small portion of a typical computer output, obtained by using proprietary computer programs developed at Exxon Research and Engineering Company, is shown in Table 12.

4. Established Procedures

The general approaches discussed in the section on quantitative analysis (Section 2.4) led to the development of a large number of quantitative, standardized analytical methods which are in common use throughout the petroleum industry. These include the following:
(1) High-voltage, low-resolution methods for up to 40 individual

TABLE 12. Typical Summary Data Calculated from Aromatic Concentrations

Elemental analysis by MS, wt. %		Characteristic averages	
Carbon	89.04	Molecular weight	318.658
Hydrogen	10.17	Carbon number	26.057
Sulfur	0.67	z Number	15.070
Oxygen	0.12	C atoms in side chains	11.555
Total	100.000		
Atomic H/C ratio	1.360		

Number of aromatic rings	Weight %			
	Hydrocarbons	Sulfur compounds	Oxygen compounds	Totals
1	26.273	1.172	0.121	27.565
2	35.220	4.127	0.785	40.131
3	17.149	1.248	0.526	18.923
4	9.586	0.138	0.518	10.242
5	1.893	0.023	0.125	2.041
6	0.786		0.022	0.808
7+	0.293			0.293
Totals	91.201	6.707	2.097	100.005

TABLE 13   List of ASTM Methods Using MS

Aromatic types in aromatic fractions of gas oils	D 3239
C number distribution of C_6–C_9 aromatics in naphthas	D 1658
Gas analysis (26 components)	D 2650
Saturate types in saturate fractions of gas oils (8 types)	D 2786
Hydrocarbon types in naphthas (6 types)	D 2789
Hydrocarbon types in middle distillates (12 types)	D 2425
Hydrocarbon types in propylene polymers	D 2424
(6 types, 16 olefin homologs)	
Isomer distribution of straight-chain detergent	
[alkylate (1–9 phenyl isomers in	
C_{15}–C_{24} Range)]	D 2498
Low voltage analysis of propylene tetramer	
(4 types, C_9–C_{18} olefin homologs)	D 2601
Molecular distribution for monoalkylbenzenes (C_{12}–C_{42})	D 2567

components. Applications range from the analyses of gases and light hydrocarbons to the analysis of complex oxygenated aromatics.

(2) Compound type analyses at high voltages. These are mainly carried out at low resolution, although there is a definitive trend of occasionally using high-resolution data. The most widespread applications are the analyses of saturate types in saturate fractions, the analysis of aromatic types in aromatic fractions, and the analysis of both saturate and aromatic types in unseparated gasolines and in middle distillates. Compound type analyses have been recently extended to the typical aromatic and hydroaromatic components of coal liquefaction products. Types included in the above methods range from 7 to about 40. Most methods in routine use belong to this category.

(3) High-resolution, low-voltage methods, as discussed previously.

(4) High-resolution, high-voltage procedures. A recent example of the potential of this approach is the analysis of heterocompounds in heavy petroleum residues.

Many of the quantitative methods used throughout industry are standardized through the American Society for Testing and Materials (ASTM). A list of these standard methods is given in Table 13.

5. Miscellaneous Applications

The number of mass spectrometric applications throughout the petroleum industry is a very large one. Each laboratory has its specific needs and its specific methods to satisfy them. A few interesting applications are involved in the analysis of polymers and polymer additives.

Trace amounts of solvents in polymers can be determined at concentration levels as low as parts per billion. The approach consists of charging a large amount of polymer, up to 300 mg, in the solid inlet system of a mass spectrometer and heating it. For most polymers, there is a temperature, usually slightly above the melting point, at which all the light components, such as entrapped solvents and additives, will volatilize without a simultaneous volatilization or decomposition of the polymer. This temperature can be determined empirically for each polymer type. The large amount of polymer charged assures usable spectra for the detection of components in trace amounts.

These analyses are usually on an unnormalized basis. Volatile additives can be determined by analogous procedures.

Another interesting application is the identification of fire-retardant materials. These usually contain molecules with up to ten chlorine and bromine atoms. The resulting spectra show clearly the expected isotopic distribution pattern.

5. Future Research

Although analytical mass spectrometry in the petroleum industry is in its "middle years," there are a large number of areas in need of further research. Others will no doubt appear in the future.

One of the most pressing needs is to extend the applicability of quantitative mass spectrometric methods for materials boiling higher than 1000–1200°F, such as heavy residues. Alternate ionization techniques such as field desorption might provide a partial solution.

There is also a need to evaluate the applicability of methods developed primarily for petroleum streams to coal liquefaction products. Some of this work is already in progress, but further methods are needed to cope with the analytical problems that will arise from the widespread application of coal liquefaction technology.

Further quantitative methods are also needed to handle routinely the large number of heteroaromatic and, in general, heteroatomic components in the higher boiling petroleum fractions.

Increased computerization of all phases of data acquisition and data handling is also a constant challenge.

Continued application of GC–MS and alternative ionization techniques, such as chemical ionization, will also require considerable research effort. Particularly promising are the expected applications of the chemical ionization approach. The specific problems related to these needs are, however, outside the scope of this chapter.

7. Exercises

1. The partial mass spectra of three petroleum-derived samples are given in Table 14. Observe the major fragment ions and determine whether each spectrum corresponds to prevalently paraffinic, cycloparafinnic, or aromatic material.

TABLE 14

| | Peak heights (digitizer counts) | | |
m/e	A	B	C
69	902	1410	3352
70	348	1048	795
71	575	4736	1131
72	74	311	67
73	148	96	—
75.5	124	—	—
76	182	—	35
77	1727	44	1155
78	783	19	372
79	669	56	1811
80	205	28	811
81	1020	162	5123
82	914	231	3332
83	936	919	3241
84	244	592	493
85	343	3149	626
86	393	236	53
87	583	43	13
88	432	—	10
89	910	—	46
90	211	—	30
91	2221	42	867
92	397	17	175
93	61	21	944
94	224	—	686
95	819	111	4115
96	571	136	2441
97	596	602	2496
98	385	417	351
99	259	718	133
105	1030	30	219
106	316	—	87
107	910	18	481
108	448	—	415
109	385	56	1501
110	286	67	1010
111	298	276	919
112	108	279	195
113	288	500	113

. Select the most suitable peaks to use for the quantitative analysis of the components in the following blends:

(a) Benzene/cyclohexane

(b) Benzene/toluene

(c) Benzene/ethylbenzene

(d) Benzene/toluene/ethylbenzene

(e) Benzene/toluene/ethylbenzene/Cyclohexane

(f) Benzene/p-xylene

(g) p-Xylene/ethylbenzene

(h) p-xylene/m-xylene

Use the partial spectra given in Table 15.

TABLE 15

m/e	Benzene	Cyclohexane	Toluene	Ethylbenzene	p-Xylene	m-Xylene
39	14.2	28.0	18.7	9.7	15.7	18.6
51	18.6	2.7	9.6	13.1	15.9	15.4
56	0.0	100.0	0.0	0.0	0.0	0.0
69	0.0	22.0	0.0	0.0	0.0	0.0
77	14.4	0.8	0.4	7.8	13.5	13.2
78	100.0	0.2	0.1	6.9	7.2	7.4
79	6.4	0.6	0.0	3.5	6.3	6.4
83	—	4.2	0.0	0.0	0.0	0.0
84	—	70.5	0.2	0.0	0.0	0.1
85	—	4.2	0.6	0.1	0.2	0.2
91	—	—	100.0	100.0	100.0	100.0
92	—	—	75.6	7.5	7.5	7.6
93	—	—	5.2	0.2	0.2	0.2
105	—	—	—	5.6	29.6	28.2
106	—	—	—	30.6	61.5	62.7
107	—	—	—	2.6	4.7	5.2

TABLE 16

m/e	Intensity	m/e	Intensity	m/e	Intensity
120	11	148	21	166	129
128	413	154	211	168	193
132	146	156	127	170	53
134	25	158	24	172	20
142	226	160	35	174	17
146	58	162	13	178	365

3. Examine the partial low-voltage spectrum given in Table 16 and identify the carbor number and compound type of the individual aromatic homologs. For example, the homolog at *m/e* 120 should be identified as a C_9 alkylbenzene. Consult Table 4 for the molecular weights of the unsubstituted aromatic nuclei.

4. Determine the resolving power required for separating the doublets listed below.

(a) $^{12}C-12H$ at *m/e* 450   (d) C_3-SH_4 at *m/e* 276

(b) $^{13}C-^{12}CH$ at *m/e* 136   (e) $C_3-H_4O_2$ at *m/e* 284

(c) CH_4-O at *m/e* 520   (f) $^{13}CH-N$ at *m/e* 167

5. Calculate the formula(s) corresponding to the mass measurements listed below.

(a) 254,1095   (e) 301,1832

(b) 274,2659   (f) 306,1620

(c) 288,2437   (g) 290,0770

(d) 278,1120   (h) 248,0655

Calculate the mass measurement errors for each formula. Consider that in addition to ^{12}C and H, the formulas contain at the maximum one sulfur, two oxygen, and one nitrogen atom. Set the error limit to 3 mmu.

8. Suggested Reading

M. J. O'NEAL, T. P. WIER, *Anal. Chem.* **23**, 830 (1951).

R. A. BROWN, *Anal Chem.* **23** 430 (1951).

H. E. LUMPKIN, *Anal. Chem.* **30**, 321 (1958).

J. H. BEYNON, *Mass Spectrometry and Its Applications to Organic Chemistry,* Elsevier Publishing Co., Amsterdam, 1960.

D. DESIDERIO, P. BOMMER, AND K. BIEMANN, *Tetrahedron Lett.,* 1725 (1964).

W. J. MCMURRAY, B. N. GREENE, AND S. R. LIPSKY, *Anal. Chem.* **38**, 1194 (1966).

A. L. BURLINGAME, D. H. SMITH, AND R. W. OLSEN, *Anal. Chem.* **40**, 136 (1968).

T. ACZEL, D. E. ALLAN, J. H. HARDING, AND E. A. KNIPP, *Anal. Chem.* **42**, 341 (1970).

M.S.B. MUNSON, *Adv. Mass Spectrom.* **5**, 233 (1971).

Mass Spectral Data, American Petroleum Institute, Research Project 44.

B. H. JOHNSON AND T. ACZEL, *Anal. Chem.* **39**, 682 (1967).

T. ACZEL, *Rev. Anal. Chem.* **1**, 226–261 (1971).

T. ACZEL AND H. E. LUMPKIN, in *Proceedings of the 18th Annual Conference on Mass Spectrometry,* American Society for Mass Spectrometry, p. B37, 1970.

H. E. LUMPKIN AND D. E. NICHOLSON, *Anal. Chem.* **32**, 74 (1960).

A. HOOD AND M. J. O'NEAL, *Advances in Mass Spectrometry* Pergamon Press, London, 1959, pp. 175–190.

H. E. LUMPKIN, *Anal. Chem.* **28,** 1946 (1956).

American Society for Testing Materials, *Annual Books of Standards*, parts 23, 24.

C. J. ROBINSON AND G. C. COOK, *Anal. Chem.* **41,** 1548 (1969).

H. T. BEST AND A. B. MARCUS, in *Proceedings of the 14th Annual Conference on Mass Spectrometry*, American Society for Mass Spectrometry (1976).

A. W. PETERS AND J. G. BENDORAITIS, *Anal. Chem.* **48,** 968 (1976).

H. G. HILL, *Introduction to Mass Spectrometry*, Heyden Co., London, 1972.

K. BIEMANN, *Mass Spectrometry, Organic Chemical Applications*, McGraw-Hill, New York, 1962.

H. BUDZIEKIEWICZ, C. DJERASSI, AND D. H. WILLIAMS, *Interpretation of Mass Spectra of Organic Compounds*, Holden-Day Inc., San Francisco, 1967.

F. W. McLAFFERTY, *Mass Spectrometry of Organic Ions*, Academic Press, London, 1963.

13

Cosmochemical and Geochemical Applications of Mass Spectrometry

J. Oró and Daryl Nooner

1. Introduction

Organic cosmochemistry considers processes of synthesis of compounds containing C, N, O, H, etc., which take place or may have occurred on the primitive Earth surface and atmosphere, on the various planets and bodies of the solar system, in interstellar clouds, in nebulae, or in other cosmic bodies. Experimental approaches to this type of study include:

1. Analysis of extraterrestrial material such as lunar samples from the Apollo and Luna missions and meteorites.
2. Simulation studies in which attempts are made to reproduce cosmic processes in the laboratory.
3. Examination of ancient terrestrial sediments.
4. Remote analysis of planetary atmospheres and surfaces.

Studies in the field of organic cosmochemistry were initiated in the 1950's by a few scientists. This activity is now worldwide and supported by highly sophisticated instrumentation. The most useful instrument for the examination of organic matter of geochemical or extraterrestrial origin has been the mass spectrometer used alone or with a gas chromatograph.

J. Oró and Daryl Nooner • Department of Biophysical Sciences, University of Houston, Houston, Texas 77004. Daryl Nooner's present address: Spectrix Corporation, Houston, Texas 77054

This is a noncomprehensive survey of the use of mass spectrometry in this work. Most of the specific examples presented are from work carried out in our laboratory.

2. Meteorites

The extraterrestrial material available in largest amount and variety is represented by meteorites. Meteorites are objects of great age that fall to the Earth rather frequently. They are classified on the basis of chemical comparison and mineral content as iron, stoney iron, and stoney meteorites. The greatest interest, from an organic cosmochemical point of view has been in a special subgroup of the stoney meteorites, the carbonaceous meteorites, or carbonaceous chondrites. A portion of a carbonaceous chondrite, the Murchison meteorite, that fell in Australia in 1969, is shown in Fig. 1. The extensive literature on carbonaceous meteorites has been recently cataloged in great detail in a reference volume by B. Nagy.

Carbonaceous meteorites have long (since 1806) been known to contain organic compounds. However, the separation and identification of these compounds, present in very small amounts, had to await the development of modern analytical methods like gas chromatography and mass spectrometry. Interest in the hydrocarbons in meteorites became acute in 1960 when it was reported that the hydrocarbons in the Orgueil carbonaceous meteorite appeared to be of biological origin. This conclusion was based on mass spectrometric analyses of a pyrolyzate collected from the meteorite.

Because pristane and phytane were thought to be indicators of biological activity, they were sought in the hydrocarbons in meteorites. These and other isoprenoid hydrocarbons were detected almost simulta-

FIG. 1. The Murchison, Australia, carbonaceous meteorite.

neously in mid-1960 by J. Oró and D. W. Nooner at the University of Houston and by W. G. Meinschein of Esso Research Laboratories.

The analysis of the Murray meteorite is given as an example of our work with numerous carbonaceous meteorites.

Figure 2 shows a chromatogram of the aliphatic hydrocarbons, i.e., normal and isoprenoid alkanes, extracted from the meteorite. As previously mentioned, the isoprenoid hydrocarbons are of particular interest because they are thought to result from living systems. Some of the mass spectra used to identify the isoprenoid hydrocarbons from the Murray meteorite are shown in Fig. 3. The ubiquity of isoprenoid and other hydrocarbons in carbonaceous meteorites is indicated in Table 1, which summarizes results from the Mokoia, Murray, Orgueil, and Vigarano meteorites.

The significance of the presence of isoprenoid hydrocarbons in meteorites, if truly indigenous, is so great that special efforts have been made to determine their origin. Although their source is not definitely established, the isoprenoid hydrocarbons, expecially those of 14 carbon atoms and larger, are believed to be terrestrial contaminants. Part of the evidence indicating contamination comes from experiments with graphite–troilite nodules in iron meteorites. These nodules have had a history of high temperatures necessary to bring out the melting of the nickel-iron alloy, which should preclude the presence of hydrocarbons of the type observed in Murray and other carbonaceous meteorites.

Since these nodules are buried in a nickel-iron matrix, they are made accessible by sawing, which is a potentially contaminating process. Figure 4 shows how various nodules may appear in the matrix and the portions usually analyzed. Data from a nodule of the type diagrammed in Fig. 4c, presented in Fig. 5, show that a concentration gradient exists, i.e., the amounts of hydrocarbons decrease as we move into the nodule. Mass spectra indicated that isoprenoids and other hydrocarbons of the type found in Murray and other carbonaceous chondrites are also present in the nodule (see peak nos. in parentheses, Table 1).

Along with the aliphatic hydrocarbons, a variety of aromatic hydrocarbons have been found in meteorites. These compounds are generally separated by gas chromatography and identified by the interfaced mass spectrometer. We have identified, by this procedure, the following aromatic hydrocarbons in the Murchison meteorite: naphthalene, methyl and dimethyl naphthalenes, acenaphthene, phenanthrene, anthracene, fluoranthene, and pyrene.

In addition to the hydrocarbons, one of the most important classes of organic compounds that have been found in meteorites are the amino acids. Relatively large amounts have been found in the Murchison meteorite. Figure 6 illustrates our initial analysis of a water extract of this meteorite. The amino acids identified by mass spectrometry include

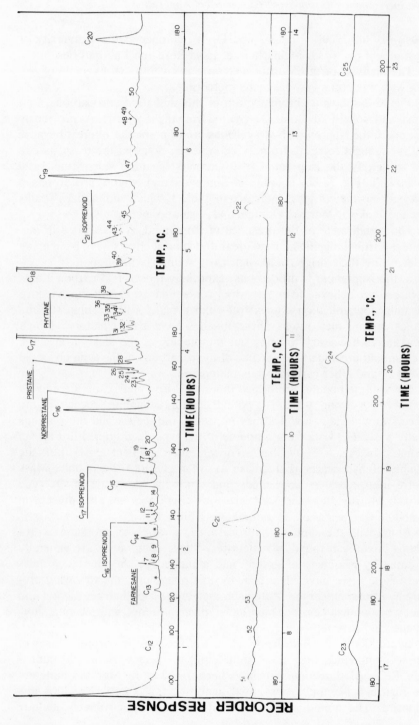

FIG. 2. Gas chromatogram showing aliphatic hydrocarbons extracted from the Murray carbonaceous meteorite. Separated on Polysev.

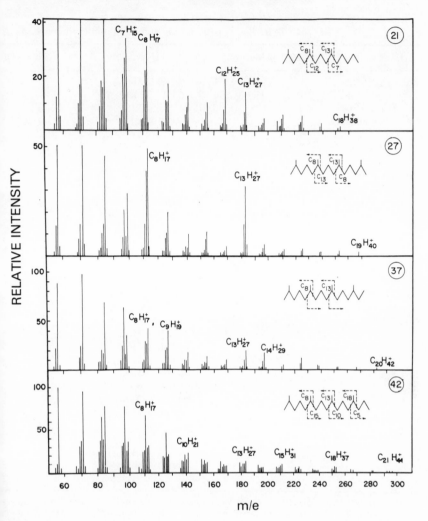

FIG. 3. Mass spectra of isoprenoid hydrocarbons extracted from the Murray carbonaceous meteorite. Numbers refer to peaks in Fig. 2.

alanine, glycine, valine, and glutamic acid. Also indicated on the figure are two nonprotein amino acids, α-aminobutyric acid and β-alanine. Mass spectra of the isopropyl N-trifluoroacetyl derivatives of the nonprotein amino acids are shown in Fig. 7. These compounds did not give molecular ions.

The development of techniques for separating D and L amino acids by gas chromatography provides a means for determining whether the amino acids are indigenous. If the amino acids are contaminants, they should be

TABLE 1. Isoprenoids and Aliphatic Hydrocarbons in

Hydrocarbon	Mo.	Mu.	O.	V.	Hydrocarbon	E.	Gr
nC_{11}				1	(19) 2-MeC$_{15}$	1	
nC_{12}	0-1	1		1	(20) 3-MeC$_{15}$	1	
1 Decahydro-	1			1	nC_{16}	1	1
naphthalene					(21) 2,6,10-	1	1
2 2,6,10-TriMeC$_{11}$		1		0-1	TriMeC$_{15}$		
3 2-MeC$_{12}$		1		1	(23) 6-MeC$_{16}$		
4 3-MeC$_{12}$		1		1	24 5-MeC$_{16}$		
5 4,8-DiMeC$_{12}$		1			(25) 4-MeC$_{16}$		
nC_{13}	1	1		1	(26) 2-MeC$_{16}$	1	1
6 2,6,10-TriMeC$_{12}$	1	1		1	(27) 2,6,10,14-	2	1
7 2-MeC$_{13}$	1	1		1	TetraMeC$_{15}$		
8 3-MeC$_{13}$	1	1		1	(28) 3-MeC$_{15}$	1	
9 4,8-DiMeC$_{13}$		1		1	nC_{17}	3	1
nC_{14}	1	1		3	(30) 2,6,10-		
10 2,6,10-TriMeC$_{13}$	1	1		1	TriMeC$_{16}$		
11 4-MeC$_{14}$		1			(31) 4,7,11-		
(12) 2-MeC$_{14}$	1	1		3	TriMeC$_{16}$		
(13) 3-MeC$_{14}$	1	1		3	32 Decylcyclo-		1
14 4,8-DiMeC$_{14}$		1			hexane		
nC_{15}	2	2	1	4	(33) 6-MeC$_{17}$		
(16) 2,6,10-TriMeC$_{14}$	1	1		0-1	(34) 5-MeC$_{17}$		
17 5-MeC$_{15}$		1		1	(35) 4-MeC$_{17}$	1	
18 4-MeC$_{15}$	1	1			(36) 2-MeC$_{17}$	1	1
					(37) 2,6,10,14-	3	1
					TetraMeC$_{16}$		
					(38) 3-MeC$_{17}$	1	1
					nC_{18}	6	1
					(39) Dodecyl-		
					cyclopentane		
					(40) Undecyl-		
					cyclohexane		

[a] E. (Essebi); Gr. (Grosnaja); Mo. (Mokoia); Mu. (Murray); O. (Orgueil); V. (Vigarano). The amounts given (in ppm) have been obtained by rounding the actual values to the next higher number. Blanks in the columns of the table do not necessarily indicate the absence of certain alkanes, but rather that no

predominantly of L configuration. Such separations have been carried out in our laboratory and at Arizona State University. However, the most extensive work has been done at the NASA-Ames Research Center at Moffett Field in California. The amino acids identified by gas chromatography–mass spectrometry in extracts from the Murchison meteorite are shown in Table 2. The compounds produced during peak discharge experiments recently carried out by S. L. Miller and his colleagues and analyzed by gas chromatography–mass spectrometry are also shown. The D–L distribution of amino acids in Murchison and the overall resemblance

Meteorites: GC-MS Identification[a]

Mo.	Mu.	O.	V.	Hydrocarbon	E.	Gr.	Mo.	Mu.	O.	V.
1	1	1	1	41 5-MeC$_{18}$			1			1
1	1		1	(42) 2,6,10,14-	1	0-1	1	1		1
4	6	1	4	TetraMeC$_{17}$						
2	2		2	(43) 4-MeC$_{18}$			1	1		
				(44) 2-MeC$_{18}$	1		2	1		1
1	1			(45) 3-MeC$_{18}$	1		1	1		1
0-1	1		1	nC$_{19}$	5	2	6	8	2	3
1	1			(46) Tridecyl-						0-1
1	1	1	1	cyclopentane						
6	4	1	3	(47) Dodecyl-			1	1		1
				cyclohexane						
1	1		1	48 4-MeC$_{19}$			1	1		1
8	14	3	8	(49) 2-MeC$_{19}$			1	1		1
1			1	(50) 3-MeC$_{19}$			1	1		1
				nC$_{20}$	1	2	3	5	1	2
0-1	1		1	(51) Tridecyl-			1	1		1
				cyclohexane						
1	1			(53) 3-MeC$_{20}$				1		
				nC$_{21}$	1	2	1	6	1	1
				54 Tetradecyl-				1		
	2		1	cyclohexane						
1	1			55 2-MeC$_{21}$				1		
1	1			nC$_{22}$			1	4	1	1
2	1	1	1	nC$_{23}$				2		1
4	3	1	2	nC$_{24}$				2		1
				nC$_{25}$						1
1	1	1	1	nC$_{26}$						1
7	9	3	5							
	1		0-1							
3	1		1							

[a] useful mass spectra were obtained. Compounds other than *n*-alkanes have an identification number corresponding to that given in the gas chromatograms. When this number is placed in parentheses, it indicates that the compound has also been identified in graphite nodules from iron meteorites.

to the spark discharge results indicate that the amino acids in the meteorite were synthesized by abiotic processes.

These few examples of the applications of mass spectrometry to the analysis of meteorites are illustrative only and are by no means comprehensive. Aliphatic fatty acids, both normal and branched, have been extracted from carbonaceous meteorites and identified by gas chromatography–mass spectrometry. Nitrogenous heterocyclic compounds have been extracted from the Murchison meteorite, derivatized, and identified in a similar manner.

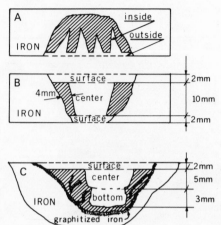

FIG. 4. Diagram showing how graphitic nodules often appear in the matrix of iron meteorites and the portions usually analyzed.

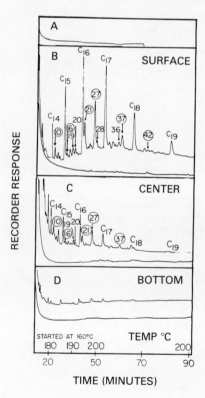

FIG. 5. Gas chromatograms showing aliphatic hydrocarbons extracted from different portions of a graphitic nodule from the Canyon Diablo iron meteorite. Separated on Polysev.

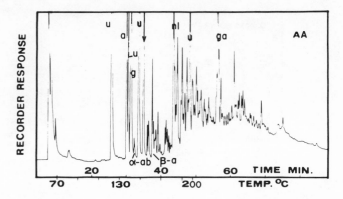

FIG. 6. Gas chromatogram showing amino acids (as *N*-trifluoroacetyl isopropyl esters) extracted from the Murchison carbonaceous meteorite. Separated on SF-96. The peaks are: U = unknown or not identified; a = alanine; g = glycine; v = valine; α-ab = α-aminobutyric acid; β-a = β-alanine; nl = norleucine (standard); ga = glutamic acid.

Mass spectrometry and gas chromatography–mass spectrometry have been used to identify compounds removed from meteorites by heat. The hydrocarbons and other organic compounds in the meteorites have been analyzed by heating samples in the ion source of a high-resolution mass spectrometer. Also, volatiles from samples of meteorites have been trapped, then separated by gas chromatography and introduced into the mass spectrometer. Pyrolysis–gas chromatography–mass spectrometry appears to be a useful procedure for studying the organic polymeric material in meteorites as well as the more simple compounds.

3. Abiotic Synthesis

Laboratory studies in which efforts were made to model conditions on the primitive earth or in cosmic bodies have been carried out for about 25 years. Application of gas chromatography–mass spectrometry to the identification of the products produced has been much more recent. One of the earlier applications of this technique was made in our laboratory by H. B. Skewes. Simulating a mild stellar environment, he heated ammonia, methane, and water in a graphite resistance reactor at 1500°C. The reactor effluent was quenched, heated at 90°C in water for 24 hr, and dried. The product was treated to convert any amino acids present to the *N*-trifluoroacetyl butyl esters and analyzed by gas chromatography–mass spectrometry. The gas chromatogram is shown in Fig. 8. This experiment demonstrated that significant amounts of amino acids were produced under

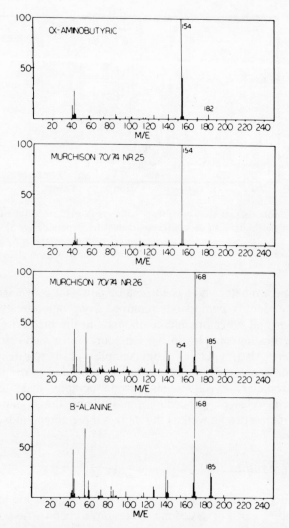

FIG. 7. Mass spectra of N-trifluoroacetyl isopropyl esters of two nonprotein amino acids extracted from the Murchison carbonaceous meteorite.

these unusual conditions. All of the amino acids shown were identified by mass spectrometry. The mass spectrum of the aspartic acid derivative from this study is given in Fig. 9.

The gas chromatographic–mass spectrometric analysis of derivatized amino acids is now generally standard procedure in our laboratory and others. Identifications so obtained are more acceptable than those based only on chromatographic techniques. One of our most recent applications

TABLE 2.   Relative Abundances of Amino Acids in
the Murchison Meteorite and an Electric-Discharge
Synthesis

| | Mole ratio to glycine (=100) | |
Amino acid	Murchison meteorite	Electric discharge
Glycine	>50	>50
Alanine	>50	>50
α-Amino-*n*-butyric acid	5–50	>50
α-Aminoisobutyric acid	>50	0.5–5
Valine	5–50	0.5–5
Norvaline	5–50	5–50
Isovaline	0.5–5	0.5–5
Proline	5–50	0.05–0.5
Pipecolic acid	0.05–0.5	0.05–0.5
Aspartic acid	5–50	5–50
Glutamic acid	5–50	0.5–5
β-Alanine	0.5–5	0.5–5
β-Amino-*n*-butyric acid	0.05–0.5	0.05–0.5
β-Aminoisobutyric acid	0.05–0.5	0.05–0.5
δ-Aminobutyric acid	0.05–0.5	0.5–5
Sarcosine	0.5–5	5–50
N-Ethyglycine	0.5–5	5–50
N-Methylalanine	0.5–5	0.5–5

has been the analysis of amino acids produced from synthetic mixtures similar in composition to the gases released from lunar samples.

Once amino acids were formed, attention was turned to their polymerization since polypeptides were necessarily involved in the evolution of life. Gas chromatography–mass spectrometry is now used in these investigations. The example presented here is from a study made in our laboratory. A suitable mixture (Table 3) was heated below 100°C in the

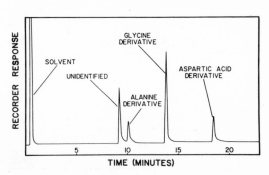

FIG. 8. Gas chromatogram showing *N*-trifluoroacetyl *n*-butyl esters of amino acids most abundant in ammonia–methane–water pyrolysis at 1500°C. Separated on STAP + polyvinylpyrolidone.

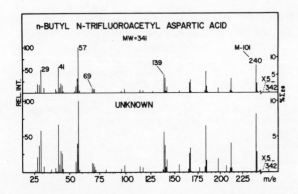

FIG. 9. Mass spectra of the N-trifluoroacetyl n-butyl ester of aspartic acid produced in ammonia–methane–water pyrolysis at 1500°C.

presence of simple condensing agents presumed to have existed on the primitive Earth, and the products separated by thin-layer chromatography. The recovered material was derivatized (trifluoroacetic anhydride and isopropyl alcohol–hydrochloric acid) and analyzed by gas chromatography–mass spectrometry. The gas chromatogram of a product in which the dipeptide of L-phenylalanine was detected is shown in Fig. 10. The identification of this product by mass spectrometry is shown in Fig. 11. This experiment demonstrates the formation of peptides under possible primitive Earth conditions.

As previously noted, carbonaceous meteorites contain a variety of aliphatic and aromatic hydrocarbons of unknown origin. Much effort has been directed toward synthesizing hydrocarbon mixtures that resemble those recovered from meteorites. This work, carried out primarily in our laboratory and by E. Anders' group at the University of Chicago, has utilized Fischer–Tropsch-type processes and has been very successful in preparing aliphatic hydrocarbons other than the high-molecular-weight isoprenoid hydrocarbons. The failure to synthesize these hydrocarbons,

TABLE 3.   Reaction Mixtures for Polymerizing Amino Acids

Reactants	Unlabeled reaction (moles)	[14]C amino acid reaction (moles)
Amino acid (gly, L-ile, L-phe)	7.5×10^{-5}	9.4×10^{-6}
Adenosine-5'-triphosphate	7.5×10^{-5}	9.4×10^{-6}
4-Amino-5-imidazole carboxamide	2.5×10^{-4}	3.1×10^{-5}
Cyanamide	2.5×10^{-4}	3.1×10^{-5}

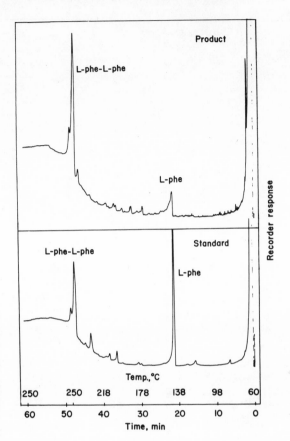

FIG. 10. Gas chromatogram showing *N*-trifluoroacetyl isopropyl ester of the dipeptide of L-phenylalanine. Separated on SP-2250.

especially phytane and pristane, which on Earth are decomposition prod-ucts of chlorophyll, lends credence to the belief that, when these isopren-oids are found in meteorites, they are probably terrestrial contaminants. The Fischer–Tropsch synthesis also produces aromatic hydrocarbon mix-tures that approach the composition of those observed in meteorites. The correspondence equals or exceeds that observed for aliphatic hydrocar-bons. The distribution of an array of aromatic hydrocarbons produced by Fischer–Tropsch synthesis is shown in Fig. 12. All of the labeled and most of the numbered peaks have been identified by mass spectrometry. Aromatic hydrocarbons are also produced when methane or other hydro-carbons are pyrolyzed.

Catalyst is well known to be a dominating factor in Fisher–Tropsch synthesis. Figure 13 shows the gas chromatograms of two aliphatic

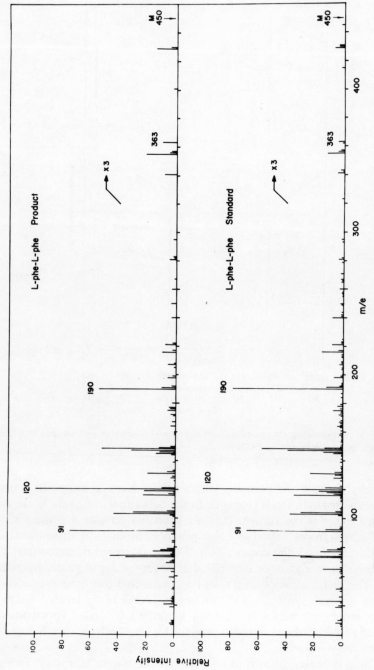

FIG. 11. Mass spectra of the N-trifluoroacetyl isopropyl ester of the dipeptide of L-phenylalanine.

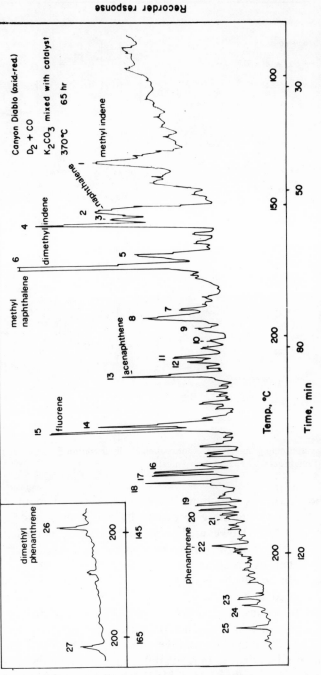

FIG. 12. Total ion current trace of perdeuterio aromatic hydrocarbons synthesized by a Fischer–Tropsch-type process. Separated on Polysev.

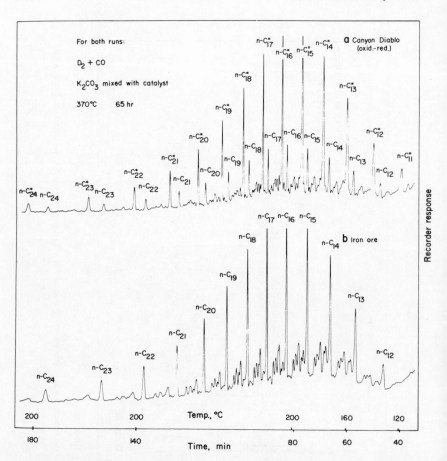

FIG. 13. Gas chromatogram of perdeuterio aliphatic hydrocarbons synthesized by Fischer–Tropsch-type process. Separated on Polysev.

hydrocarbon mixtures which we have synthesized. Without potassium carbonate the two catalysts give essentially the same saturated aliphatic hydrocarbon pattern. However, if carbonate is added, monoolefins dominate the distribution in which oxidized–reduced Canyon Diablo is the catalyst, but there is no effect upon the distribution in which iron ore is the catalyst. These results point to a possible difference in mechanism between reduced and oxidized catalyst. Mass spectra confirmed the assigned identifications. Mass spectra of one of the deuteriated monoolefins and a perhydromonolefin from the API series (No. 1013) are compared in Table 4. The difference in relative intensity is believed due to the instruments used and not to intrinsic differences between the two types of hydrocarbons. This is shown in Table 5, where the ions in the spectra of a perhydro and a perdeuterio hydrocarbon, both obtained on the LKB 9000, have about the same relative abundances.

Fatty acids are a class of compounds necessary for the development and maintenance of life and, according to S. L. Miller and L. E. Orgel, no generally satisfactory procedure for their prebiotic synthesis has been reported. We have prepared small amounts of fatty acids (10:0–16:0 range) by Fischer–Tropsch synthesis under conditions and using catalysts that may make the process plausible. The results, presented as the chromatogram in Fig. 14, show the perdeuterio fatty acids synthesized in the reaction and the perhydro fatty acids that are contaminants. Partial mass spectra which provide a comparison of the 16:0 fatty acids are given in Table 6. The fact that a perhydro methyl group is attached to a perdeuterio fatty acid provides a check on the fragmentation scheme. We have continued the study of fatty acids synthesis by Fischer–Tropsch-like processes and have obtained yields comparable to those for hydrocarbons. Furthermore, we have shown that reduced catalyst (meteoritic iron) mixed with potassium carbonate produced fatty acids, while oxidized catalyst (meteoritic iron) also mixed with potassium carbonate does not. These results indicate that Fischer–Tropsch processes could have been involved in the prebiotic synthesis of fatty acids.

TABLE 4.   Partial Mass Spectra of Perhydro and Perdeuterio Hexadecene

C_n	m/e	$C_{16}H_{32}$ Relative intensity, %[a]	m/e	$C_{16}D_{32}$ Relative intensity, %[b]
	111	38.8	126	95.5
C_8	112	16.0	128	38.4
	125	12.3	142	42.9
C_9	126	8.5	144	24.1
	139	4.0	158	17.9
C_{10}	140	5.2	160	15.2
	153	1.9	174	8.9
C_{11}	154	3.3	176	10.7
	167	0.9	190	5.4
C_{12}	168	1.5	192	8.0
	181	0.5	206	
C_{13}	182	1.1	208	6.3
C_{14}	196	1.9	224	8.9
	224 (M)[c]	2.4	256 (M)[c]	19.6
C_{16}	225	0.5	258	

[a] API No. 1013; base peak is m/e 83.
[b] Base peak is m/e 94.
[c] M stands for molecular ion.

TABLE 5.   Partial Mass Spectra of Perhydro and Perdeuterio
Hexadecane

C_n	$C_{16}H_{34}$		$C_{16}D_{34}$	
	m/e	Relative intensity, %	m/e	Relative intensity, %
C_4	57	100.0	66	100.0
C_5	71	75.1	82	71.4
C_6	85	48.5	98	51.0
C_7	99	15.6	114	19.5
C_8	113	9.4	130	10.9
C_9	127	6.7	146	7.1
C_{10}	141	5.3	162	5.7
C_{11}	155	4.4	178	4.3
C_{12}	169	3.8	194	3.8
C_{13}	183	2.3	210	1.9
C_{14}	197	2.0	226	1.4
C_{16}	226(M)[a]	5.0	260 (M)[a]	5.2

[a] M stands for molecular ion.

4. Lunar Samples

Besides the meteorites, the only extraterrestrial material available for
analysis are lunar samples from the Apollo missions and the unmanned
Soviet Luna 16 and 20 probes. In contrast to the meteorites, precise
control and monitoring of contamination were maintained for the lunar
samples. It was hoped at one time that lunar samples would help to answer
questions about the organic compounds in carbonaceous chondrites. For
example, if a carbonaceous chondrite could have been found on the Moon,
some of the questions regarding terrestrial contamination could have been

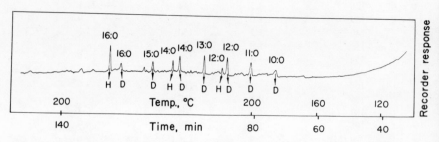

FIG. 14. Gas chromatogram showing the methyl esters of perdeuterio fatty acids synthesized by a Fischer–Tropsch-type process. Separated on Polysev. Canyon Diablo (oxidized–reduced); D_2 + 2CO; K_2CO_3 mixed with catalyst; 370°C 50 hr. D stands for
perdeuterio fatty acids; H stands for perhydro fatty acids.

answered. However, no carbonaceous chondrite was found although they make a measurable (1–3%) contribution to the lunar regolith. Apparently, carbonaceous chondrites striking the lunar surface at high velocities without the mediating effect of an atmosphere leave no macro-sized particles or high-molecular-weight organic compounds.

Studies of the carbon chemistry of lunar samples have involved several laboratories, including those of J. Oró, University of Houston; K. Biemann, Massachusetts Institute of Technology; A. L. Burlingame, University of California at Berkeley; G. Eglinton, University of Bristol; C. B. Moore, Arizona State University; B. Nagy, University of Arizona; and the NASA-Ames Research Center. Mass Spectrometry or gas chromatography–mass spectrometry have been the methods of choice for analyses of compounds released, usually by hydrolysis with selected acids or by thermal extraction.

In our laboratory at the University of Houston we have analyzed lunar samples by solvent extraction, acidolysis, and thermal treatment. Figure 15 shows the analysis of a benzene–methanol (3:1 v/v) extract of an Apollo 11 sample. The absence of organic compounds is obvious. The large peak following the solvent peak is a contaminant introduced by the York meshes used as an aid in packaging the Apollo samples. This contaminant was identified as diisopropyl disulfide by mass spectrometry.

The results obtained by the acidolysis and thermal treatment of lunar samples are more promising than those obtained by analyzing solvent extracts. Acid treatment released H_2, N_2, CO, CO_2, methane, ethane, propylene, propane, and H_2S from all the samples of Apollo 11, 12, and 14 tested. DC1 was used to distinguish between hydrocarbons generated by acid treatment and those initially present in the samples; the use of reconstructed mass chromatograms permitted separation of deuteriated

TABLE 6. Partial Mass Spectra of Methyl Esters of Perhydro and Perdeuterio Fatty Acids (16:0)

$C_{16}H_{31}O_2CH_3$		$C_{16}D_{31}O_2CH_3$	
m/e	Relative intensity, %	*m/e*	Relative intensity, %
74	100	77	100
87	52	91	53
129	7	139	6
143	17	155	14
213	3	235	6
227	10	251	4
239	7	270	3
270 (M)[a]	21	301 (M)[a]	11

[a] M stands for molecular ion.

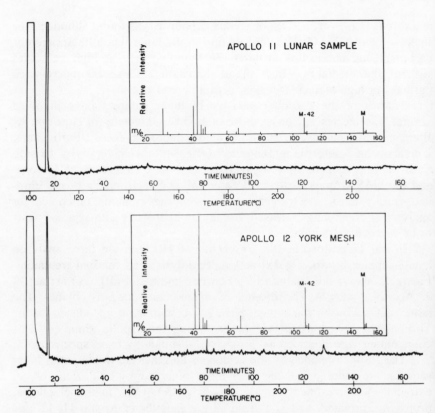

FIG. 15. Gas chromatograms showing compounds extracted from a lunar sample and York mesh. Separated on Polysev. Insets: mass spectra of extracted compounds.

and nondeuteriated hydrocarbons in the DC1 products. Figure 16 presents such a separation in the CH_4–CD_4 region of a gas chromatogram from DC1 treatment of an Apollo 14 sample. As shown, all possible deuteriated species were obtained. These results could mean that partially hydrogenated carbon species were present. Thermal treatments also released H_2, N_2, CO, CO_2, methane, ethane, propylene, and propane. However, H_2S was not released. These results were obtained by using the LKB 9000 gas chromatograph–mass spectrometer equipped with a Porapak Q column.

Quadrupole mass spectrometry was used to monitor the release of H_2O, CO_2, CO, and N_2. In these experiments the samples were volatilized directly into the quadrupole mass spectrometer. The figures were prepared by plotting peak intensity of the particular molecular ions vs. temperature. Since N_2 and CO have the same molecular ion at low resolution their relative contribution was determined by solving simultaneous equations using the m/e 12, 14, 16, and 28 intensities obtained from fragmentation patterns of the pure gases. The plots for H_2O and CO from an Apollo 14 sample are shown in Fig. 17. Most of the water is released below 500°C and a substantial amount of this water is terrestrial contamination. The high-temperature water probably results from solar wind proton irradiation. Carbon dioxide exhibits two peaks. The lower-temperature peak probably represents terrestrial contamination, while the peak at higher temperature (~475°C) indicates the possible presence of indigenous carbonates. The plot of quadrupole mass spectrometric data for CO and N_2 from Apollo 14 samples is shown in Fig. 18. The origin of the CO, which shows several peaks, is uncertain. It may result from iron, nickel, and cobalt carbides but not from calcium carbide. Whatever the source, it probably represents indigenous material. Nitrogen has more than one peak and is released at

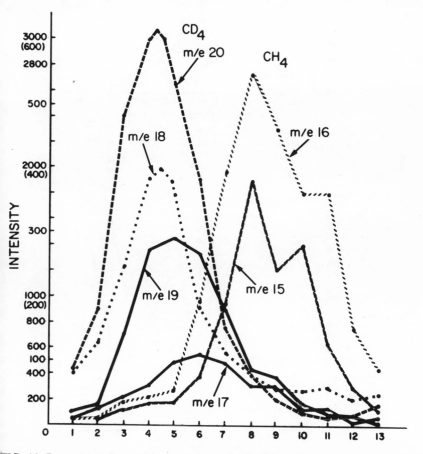

FIG. 16. Reconstructed mass chromatograms of CH_4–CD_4 region of gas chromatogram from DCl treatment of lunar sample.

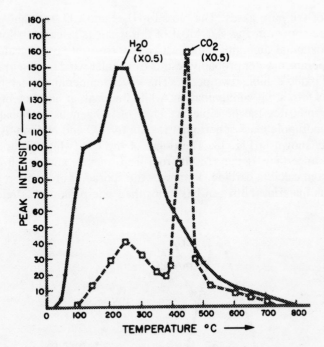

FIG. 17. Temperature release data for CO_2 and H_2O in a lunar sample.

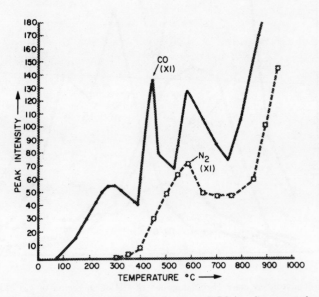

FIG. 18. Temperature release data for N_2 and CO in a lunar sample.

high temperature. This behavior could be due to rupture of traps in the matrix and/or to the presence of nitrides stable up to these high temperatures.

Samples from the Luna 16 and Luna 20 space probes were analyzed by A. L. Burlingame and co-workers at the University of California (Berkeley) using vacuum pyrolysis to 1400°C. As expected, the results are quite similar to those from the Apollo samples. The major gaseous components are H_2O, CO, CO_2, N_2, and CH_4, while the minor components include NH_3, HCN, NO, SO_2, H_2S, and C_2–C_3 hydrocarbons. As in the Apollo samples, several of the compounds were thought to be solar wind implanted or derived species. The release of CO and N_2 at 700°C and above was bimodal, the same as for the Apollo 14 and 15 samples.

5. Ancient Sediments

Calvin has called the insight gained by the study of ancient sediments and co-relative materials, such as recent sediments, contemporary bacteria, and algae, the "view from the present to the past." The development of this view entails in part the examination of ancient sediments for morphological remains of organisms and for molecular fossils (i.e., molecular species derived from deposited organisms). In addition the analysis of recent sediments, or sediments now forming, provides a basis for interpreting data from ancient sediments, and comparative studies of geologically significant molecules (e.g., hydrocarbons) in present day microorganisms morphologically similar to ancient organisms may help correlate molecular fossils with particular biota. Although notable contributions in this field have been made at several laboratories, this report, with few exceptions, will be restricted to selections from our experience at the University of Houston.

We have found hydrocarbons which may be molecular fossils in Gunflint chert (1.9×10^9 years, Precambrian of North America) and Fig Tree shale (3.2×10^9 years, Precambrian of Africa). The hydrocarbon distribution for Fig Tree shale is shown in Fig. 19. Mass spectra of three isoprenoid hydrocarbons extracted from this shale are given in Fig. 20. We consider these hydrocarbons as being probably molecular fossils although it is not possible to unequivocally prove that extracted hydrocarbons are syngenetic with sediment deposition. Probably the best indication that a molecule is syngenetic would be to derive it in the laboratory from the kerogen in the sample, as B. Nagy has for some cyclic ethers from Transvaal kerogen (2.3×10^9 years).

The organic matter in many Precambrian sediments is believed to be derived from bacteria and algae, particularly blue-green algae. Thus the

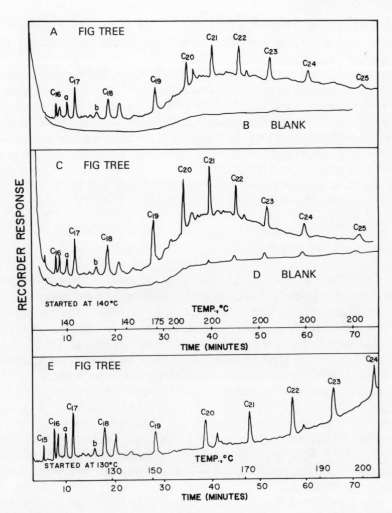

FIG. 19. Gas chromatograms showing aliphatic hydrocarbons extracted from Fig Tree shale. Separated on Polysev.

hydrocarbons found in present day organisms may have some relevance to those found in ancient sediments. Several bacteria have been analyzed in our laboratory and found to contain hydrocarbons. In particular, *Sarcina lutea* was shown to contain relatively high-molecular-weight (C_{22}–C_{29}) branched, unsaturated aliphatic hydrocarbons. A comparative analysis made in our laboratory of several microscopic algae, which are believed to be contemporary representatives of algae found in ancient sediments, was also important from a paleobiological standpoint. The results are summarized in Tables 7 and 8.

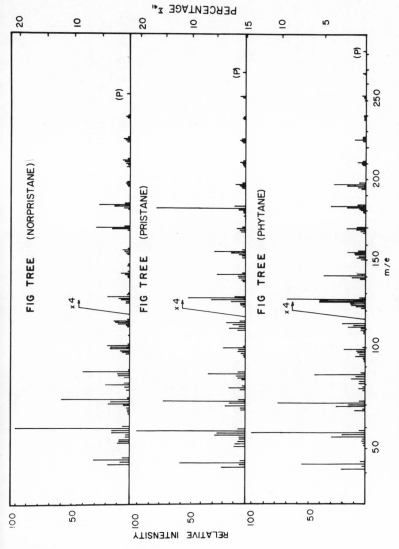

FIG. 20. Mass spectra of isoprenoid hydrocarbons extracted from Fig Tree shale.

TABLE 7. Hydrocarbons from Contemporary Representatives of Algae Found in Sediments

Division[a]	Organism	Sediment	Alkanes	
			Range	Major
Chlorophycophyta	*Coelastrum microsporum*	Green River	C_{17}	C_{17}
	Chlorella pyrenoidosa	Bitter Springs	C_{17}	ΔC_{17}
	Scenedesmus quadricauda	—	C_{17}–ΔC_{27}	$C_{17},\Delta C_{27}$
	Tetraedron sp.	Green River	C_{15}–C_{27}	C_{17},C_{23}
Cyanophycophyta	*Anacystis cyanea*	Bitter Springs	C_{17}, 7-MeC_{17}, 8-MeC_{17}	C_{17}
	Anacystis nidulans	—		C_{17}
	Spirulina platenis	Bitter Springs	C_{15}–C_{18}	C_{17}
	Lyngbya aestuarii	Bitter Springs	C_{15}–C_{17}	C_{17}
	Nostic sp.	—	C_{15}–C_{18}^{b}	C_{17}
	Chroococcus turgidus	Green River	C_{15}–C_{18}^{b}	C_{17}^{b}
	Anacystis montana	Bitter Springs	C_{16}–C_{19}^{b}	$\Delta C_{25},\Delta C_{27}$
			C_{17}–C_{29}	
Chrysophycophyta	*Botryococcus braunii*[c]	Tertiary	ΔC_{17}–$2\Delta C_{33}$	$2\Delta C_{29}, 2\Delta C_{31}$

[a] Chlorophycophyta: green algae. Cyanophycophyta: blue-green algae. Chrysophycophyta: golden brown algae.
[b] Compound with mass spectrometric fragmentation pattern corresponding to a polycyclic triterpenoid with empirical formula $C_{30}H_{50}$.
[c] Some authors classify this organism in the Chlorophycophyta.

| | Chlorophycophyta | | | | Cyanophycophyta | | | | | | | phycophyta |
Hydrocarbon	Coel. micro.	Chlor. pyr.	Scen. quad.	Tetr. sp.	Ana. cyan.	Ana. nid.	Spir. plat.	Lyng. aest.	Nost. sp.	Chroo. turg.	Ana. mont.	Boty. brau.
C_{15}				1		23	10	2	3	3		
C_{16}						8	20	6	4			
6-Me + 7-MeC_{16}										1.0		
C_{17}	100	18.5	26	30	87	44	70	35	48	32		
ΔC_{17}		76.9	0.6									
7-Me + 8-MeC_{17}											11.5	1.5
C_{18}					13	20	<1	38	27	22		
2-MeC_{18}						2		1.4		3		
C_{19}			6.8							<1		
2Δ–C_{21}										2		
C_{23}				40							8.9	0.14
ΔC_{23}											8.0	0.65
C_{24}				2.6							14.6	
C_{25}				20.0							3.8	
Δ–C_{25}											34.7	11.1
ΔC_{26}											2.8	50.4
C_{27}				5.9								5.5
Δ–C_{27}												27.9
$2\Delta C_{27}$			43.2									2.0
2Δ–C_{29}												
3Δ–C_{29}									16	10	38	
2Δ–C_{31}												
2Δ–C_{33}												
c											d	
Squalene[d]												d

[a] Differences to 100 made up by unidentified compounds.
[b] Organisms in columns 2 through 13: *Coel. micro.*—*Coelastrum microsporum*; *Chlor. pyr*—*Chlorella pyrenodosa*; *Scen. quad.*—*Scenedesmus quadricauda*; Tetr. sp.—*Tetraedron* sp.; *Ana. cyan.*—*Anacystis cyanea*; *Ana nid.*—*Anacystis nidulans*; *Spir. plat.*—*Spirulina platenus*; *Lyng. aest.*—*Lyngbya aestuarii*; *Nost* sp.—*Nostoc* sp.; *Chroo. turg.*—*Chroococcus turgidus*; *Ana. mont.*—*Anacystis montana*; *Boty. brau.*—*Botyroccus braunii*.
[c] Compound with mass spectrometric fragmentation pattern corresponding to a polycyclic triterpenoid structure with empirical formula $C_{30}H_{50}$.
[d] In benzene fraction.

The most salient feature common to all the algae is the high concentration of n-heptadecane. In sediments where algae significantly contribute to the organic matter the n-heptadecane content should be relatively high. This is indeed observed in most oil shales (see below). The data presented show that algae of different taxonomic classifications produce significant amounts of paraffins and olefins, and some contain methyl-branched hydrocarbons as well as acyclic and cyclic triterpanes. The mass spectra of squalene and "compound X" (probably a hopene) are shown in Fig. 21 and are typical of the results obtained that implicate these algae with ancient sediments.

Many ancient oil shales are very rich in both soluble and insoluble organic matter. In most cases contamination by postdepositional migration does not complicate interpretations of results based on extractable organics. These formations may be young in relation to the Precambrian but are often quite old (up to $250-300 \times 10^6$ years). We analyzed one such shale from the Irati formation (Permian of Brazil) and, on the basis of mass spectrometric identification of the extracted hydrocarbons, were able to show that the organic matter in the shale was derived predominantly from

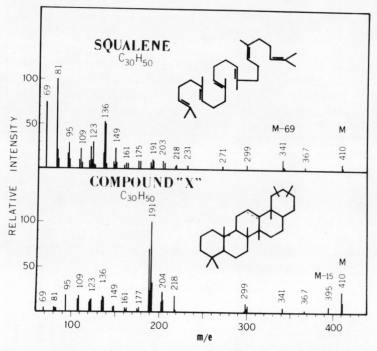

FIG. 21. Mass spectra of squalene and "compound X" (probably a hopene).

algae. Other oil shales that have been extensively examined by gas chromatographic and mass spectrometric techniques are the Green River (Western United States) and Messel (West Germany) shales. The results from these shales also correlate with the biotic residues found in the shale.

6. Viking Mission to Mars

The Viking mission to Mars probably represents the ultimate, to date, in the space application of mass spectrometry and gas chromatography–mass spectrometry. The miniaturized system on board the Viking landers consists of a magnetic sector double-focusing mass spectrometer to which a pyrolysis oven, a gas chromatograph, a palladium molecular separator, a small computer, and several other auxilliary systems have been fitted. This complex instrument package weighing only some 20 kg is capable of analyzing (1) volatile and/or pyrolytic products from the Martial soil and on-board experiments, and (2) the Martian atmosphere.

The laboratory support for the Viking mission has consisted in part of analyzing terrestrial analogs or simulations of Martian soils and atmosphere. In particular, soil from the desert valleys of Antarctica has been analyzed. Pyrolysis data from various compounds and assemblages of inorganic and organic matter has been gathered to help interpret data received from Mars. Examples of reference data are the pyrogram of an amino acid, shown in Fig. 22 and the mass spectrum of a selected peak of the pyrogram given in Fig. 23.

The landers of both Viking 1 and Viking 2 are now on Mars. Figure 24 shows the view from the Viking 1 lander on *Chryse Planitia*. Various experiments have been carried out and a plethora of data returned to Earth. The mass spectrometric capability of the two Vikings included also two additional mass spectrometers, each mounted on the heat shields of the spacecrafts, to analyze neutral gases in the upper atmosphere during the entry. This instrument indicated that the upper atmosphere was predominantly CO_2 with traces of N_2, Ar, O, O_2, and CO. The lander mass spectrometer have also sampled the atmosphere directly at the surface and indicate that the atmosphere is 95% CO_2, with the remainder being N_2 (2–3%), ^{40}Ar (1–2%), CO, and O_2 (0.3–4%). The relative abundances of noble gases (Ar, Kr, Xe) were similar to those in the Earth's atmosphere although their absolute abundances were 100 times smaller. The isotopic ratios $^{15}N/^{14}N$, $^{40}Ar/^{36}Ar$, and $^{129}Xe/^{132}Xe$ were also different from the terrestrial values, indicating that Mars had lost part of its atmosphere in the past.

Several biological experiments carried out on the landers indicated the possibility that life processes may be occurring on Mars. The analysis

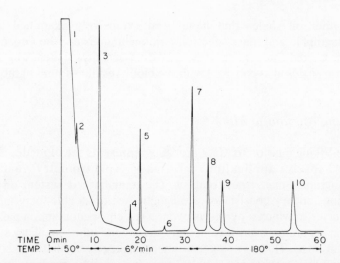

FIG. 22. Gas chromatogram of pyrolytic products from phenylalanine. These data were obtained from a miniaturized prototype instrument ("science breadboard") at JPL. The peaks are: (1) carbon dioxide, water, acetonitrile; (2) benzene; (3) toluene; (4) ethylbenzene; (5) styrene; (6) benzonitrile; (7) phenylacetronitrile; (8) 4-phenyl-1-butene; (9) isoquinoline; (10) 1,2-diphenylethane.

of Martial soil by pyrolysis–gas chromatography–mass spectrometry would be expected to indicate the presence of organic compounds in the soil if life, in fact, existed on Mars. No organic compounds have, to date, been detected in any Martian soil sample when heated up to 500°C. However, substantial amounts of water and carbon dioxide were detected. These results do not rule out life but indicate its probable absence. Life at least is not flourishing at the lander locations. Several reports on the results

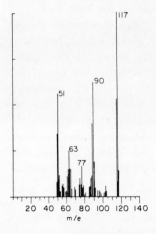

FIG. 23. Mass spectrum of peak 7 (phenylacetonitrile) of chromatogram shown in Fig. 22.

FIG. 24. View from Viking 1 lander: *Chryse Planitia* region of Mars.

obtained by the Viking gas chromatograph–mass spectrometers have been published, two of which are listed at the end of this chapter.

The study of Mars continues. Explanations, alternative to life, to reconcile the biological experiment results with the gas chromatograph–mass spectrometer analyses of the soil samples have been and are being devised.

7. Conclusion

The preceding section presents a brief look at the application of mass spectrometry and gas chromatography–mass spectrometry to the problems of cosmochemistry. The presentation has been problem- rather than technique-oriented. This approach is necessary because the problem or purpose of the experiments is the distinguishing feature of cosmochemistry. Many of the specific techniques are used in other fields. Several applications were not included, e.g., (1) work to establish the origin of enigmatic bitumen and carbon deposits; (2) stable isotope (D/H and $^{13}C/^{12}C$) distribution in terrestrial and extraterrestrial samples; and (3) use of mass spectrometry in absolute age determinations.

8. Exercises

1. A planetary atmosphere affords a mass spectrum with ions at m/e 12, 14, 16, 17, 18, 20, 28, 32, 40, and 44. Ascribe origins to these ions, and identify the constituents of the atmosphere.

2. Mass spectrometers sent to the Moon and Mars, as well as those used in interplanetary space, are much smaller than conventional mass spectrometers. Why is this?

3. What particular problems face the designers of instruments to be sent to investigate the composition of the atmospheres of (a) Mercury, (b) Venus, (c) Jupiter, (d) Saturn.

9. Suggested Reading

G. EGLINTON AND M. CALVIN, Chemical fossils, *Sci. Am.* **216**, 32–43 (1967).

J. ORÓ, Synthesis of organic molecules, *J. Brit. Interplanetary Soc.* **21**, 12–25 (1968).

M. CALVIN, *Chemical Evolution,* Oxford University Press, New York, 1969.

J. ORÓ AND D. W. NOONER, in *Advances in Organic Geochemistry,* 1966 (G. D. Hobson and G. C. Speers, eds.), Pergamon Press, New York, 1970, pp. 493–506.

M. H. STUDIER, R. HAYATSU, AND E. ANDERS, Origin of organic matter in early solar system-V. Further studies of meteoritic hydrocarbons and a discussion of their origin, *Geochim. Cosmochim. Acta* **36**, 189–215 (1972).

D. A. FLORY, S. WIKSTRÖM, S. GUPTA, J. M. GIBERT, AND J. ORÓ, Analysis of organic compounds in Apollo 11, 12 and 14 lunar samples, *Proc. Third Lunar Sci. Conf., Geochim. Cosmochim. Acta,* Suppl. 3, Vol. 2, MIT Press, Boston, 1972, pp. 2091–2108.

G. EGLINTON, J. R. MAXWELL, AND C. T. PILLINGER, The carbon chemistry of the Moon, *Sci. Am.* **216**, 32–43 (1967).

D. M. ANDERSON, K. BIEMANN, L. E. ORGEL, J. ORÓ, T. OWEN, G. P. SHULMAN, P. TOULMIN III, AND H. C. UREY, Mass spectrometric analysis of organic compounds, water and volatile constituents in the atmosphere and surface of mars: The viking mars lander, *Icarus* **16**, 111–138 (1972).

S. L. MILLER AND L. E. ORGEL, *The Origins of Life on the Earth*, Prentice-Hall, Englewood Cliffs, New Jersey, 1974.

J. ORÓ, S. L. MILLER, C. PONNAMPERUMA, AND R. S. YOUNG (eds.), *Cosmochemical Evolution and the Origins of Life*, Vols. I and II, Reidel, Dordrecht, 1974.

B. NAGY, *Carbonaceous Meteorites*, Elsevier, New York, 1975.

D. W. NOONER, J. M. GIBERT, E. GELPÍ, AND J. ORÓ, Closed system Fischer–Tropsch synthesis over meteoritic iron, iron ore and nickel–iron alloy, *Geochim. Cosmochim. Acta* **40**, 915–924 (1976).

R. S. YOUNG, Viking on Mars: A preliminary survey, *Am. Sci.* **64**, 620–627 (1976).

K. BIEMANN, J. ORÓ, P. TOULMIN III, L. E. ORGEL, A. O. NIER, D. M. ANDERSON, P. G. SIMMONDS, D. FLORY, A. V. DIAZ, D. R. RUSHNECK, AND J. A. BILLER, Search for organic and volatile inorganic compounds in two surface samples from the *Chryse Planitia* region of Mars, *Science* **194**, 72–78 (1976).

Solutions to Exercises

Chapter 1

1. (a) Vinyl chlorine has the formula H_2CCHCl. The nominal molecular weight is 62. If the contribution from ^{13}C is ignored, ions of m/e 62 and 64 will be observed in the ratio 75.5:24.5 (or 100:32.5). This is approximately 3:1 (Table 2).

   (b) Iodoform has the formula CHI_3. The nominal molecular weight is 394. Ions of m/e 394 and 395 will be observed in the ratio 98.9:1.1 (or 100:1.1). Such a low [M+1] abundance at this high mass is unusual, leading one to conclude immediately that there is only one carbon atom per molecule and to infer the presence of monoisotopic elements.

   (c) Chlorobromomethane has the formula CH_2BrCl. The nominal molecular weight is 128. If the contribution from ^{13}C is ignored, ions of m/e 128, 130, and 132 will be observed in the ratio 75.5:98.5:24.0 (or 76.6:100:24.4). This is approximately 3:4:1 (Table 2).

2. The empirical formula can be determined by consulting tables or using a suitably programmed computer (these are usually available to mass spectroscopists). With a compound of such a low molecular weight, trial-and-error soon leads to a formula of COS. A good approach would be to start by considering elements of high molecular weight. Br and I are too high. Subtraction of 34.9689 (for ^{35}Cl) affords 24.9981. The only reasonable combination of H, C, N, O, and F which would add up to a nominal mass of 25 and still leave room for a chlorine atom in the molecule is C_2H, but this is 25.0078. The presence of chlorine is therefore excluded. Next, try sulfur. Subtract 31.9721 from the molecular weight to leave 27.9949. Since ^{16}O is 15.9949, it is easily concluded that the other constituents of the molecule are carbon and sulfur, and that the identity of the compound is carbonyl sulfide. If the contribution of ^{13}C is ignored, ions of m/e 60 and 62 will be observed in the ratio 95.0:4.22 (or 100:4.44).

3. n-Butyraldehyde can undergo a McLafferty rearrangement to afford an ion of m/e 44, whereas isobutyraldehyde (with no γ-hydrogen atom) cannot.

4. The m/e value of the fragment ion is 100. From the formula $m^* = m_2^2/m_1$, we can determine that the apparent m/e value of the "metastable" peak is $(100)^2/115$ or 86.9.

5. $M + C_2H_5^+ \rightarrow [M+H]^+ + C_2H_4$

                           $M + C_2H_5^+ \rightarrow [M-H]^+ + C_2H_6$

These reactions will take place with molecules containing a suitable proton accepting and/or proton donating moiety.

Chapter 2

1. If there are $100n$ molecules (or atoms, in the case of He) of carrier gas for every 100 molecules of sample leaving the column, the concentration of sample will be $1/n$. After passing through the separator, n molecules of carrier gas and 75 molecules of sample will remain. The new concentration will be $75/n$. The enrichment factor, therefore, will be 75. [LKB claims enrichment factors of 50–200 under practical conditions.]

2. There is a linear relation between accelerating voltage and m/e. The m/e ratio is 388/386 = 1.0052, and the new accelerating voltage is 3500/1.0052 = 3481.9, a difference of 18.1 V.

3. For typical commercial instruments, the magnetic analyzer offers the following advantages:

   (i) greater mass range,
   (ii) little problem with high-mass discrimination, and
   (iii) metastable ions can be observed.

   Advantages of the quadrupole analyzer include:

   (i) greater versatility for SIM,
   (ii) small size, and
   (iii) lower price.

   The most significant factors in making a choice between these types of instrument are usually mass range and price. The mass spectroscopist who is interested in elucidating fragmentation pathways may wish to use an instrument which produces metastable ions, but he would probably also need a high-resolution instrument. If there is a need for extreme sensitivity, specifications of individual instruments should be compared to the extent that this is possible.

4. Mass range can be increased by increasing the magnet current. The limiting factor here is the amount of heat generated in the magnet coils when a high current is flowing. For a general-purpose instrument, it is uneconomic to install a large magnet and magnet current supply or cooling system.

   Decreasing the accelerating voltage increases the mass range, but the sensitivity eventually becomes too low. Ions of high molecular weight usually fragment readily so that such ions are usually present in low relative abundance, compounding this problem. The physical dimensions of the instrument cannot be changed without rebuilding the instrument. A smaller sector angle could lead to a wider m/e range, but resolution might be decreased to an unacceptable level. The ultimate limitation is that of volatility and/or thermal stability of the sample molecules.

5. (a) The precise m/e values of the molecular ions are:

$$CO = 27.9949$$
$$C_2H_4 = 28.0313$$

The difference is 0.0364. A resolution of $28/0.0364 = 769$ is required. These ions can usually be resolved using a low-resolution instrument.

(b) A resolution of $170/0.0364 = 4670$ is required. A moderately good high-resolution instrument is needed.

(c) A resolution of $310/0.0364 = 8516$ is required. This is well within the capabilities of most high-resolution instruments. The problem here is more one of distinguishing each of these ions from others with nominal m/e 310.

Chapter 3

1. Degree of column "bleed." In most gas chromatographs, "bleed" leads only to an elevated baseline which may limit sensitivity or, in the case of temperature programming, afford a rising baseline. In GC–MS, this bleed enters the ion source and may contaminate it, leading to defocusing or arcing across insulators. Silicones are among the most stable stationary phases, but they give rise to silica deposits on the ion source components. If a well-conditioned column is used below 250°C, ion source cleanings will be required only at infrequent intervals.

2. See text.

3. See text.

4. The trace amounts of radioisotopes frequently used are diluted too much by the natural isotopes to be noticed in most spectra. If radioisotopes of high specific activity are employed, they can be measured by mass spectrometry. Since the pump effluent is vented into the laboratory, any use of radioisotopes is inadvisable.

5. There is none.

Chapter 4

1. If all of the instrument parmeters are equal, so are the sensitivities of these techniques. The resolution of the instrument may be decreased for single ion monitoring, leading to a potentially higher sensitivity.

2. There are a number of instances where this is useful. Greater sensitivity can be obtained by monitoring the full range of molecular ions, for example, of compounds containing multi-isotopic elements such as selenium and mercury. This approach is also used for polychlorinated or polybrominated compounds.

3. The internal standard should behave (chemically and physically) in a manner as similar as possible to the compound being determined. Isotope-labeled analogs are best.

Homologs are sometimes an acceptable substitute. Unrelated compounds rank a poor third. Some analysts claim that ^{13}C, ^{15}N, and ^{18}O analogs are better than those containing ^2H. The choice frequently depends upon availability, ease of synthesis, or cost of the various candidates. A large excess of isotope-labeled analog may act as both a carrier and an internal standard.

4. Lack of interference by ions from column bleed is a major concern. These m/e values should be avoided. Ions in the spectra of other components of a sample should not be used, either. Almost all compounds other than those with an odd number of nitrogen atoms per molecule have an even molecular weight. Fragment ions formed by simple cleavage of most molecular ions, therefore, have odd m/e values. Thus one can expect less interference by performing SIM at even m/e values. There is an even lower risk of interference at the higher m/e values. The ion selected should also be an ion of relatively high abundance in the spectrum of the compound of interest. In summary, the best sensitivity can be expected with compounds affording a base peak of even m/e value in the upper portion of the mass range of the instrument, when there is no interference from column bleed or other compounds.

5. (a) They all undergo a McLafferty rearrangement to afford ions of m/e 74 as the base peaks of the spectra.

   (b) The molecular ion (m/e 298) is relatively weak, but is one of the few ions distinguishing this compound from other saturated straight chain methyl esters.

Chapter 5

1. Use a second trap in series with the first. If the second trap does not collect anything, the first is not overloaded.

2. The internal standard is usually used to compensate for losses in recovery or variation in injection volume. For most compounds, all of the sample is trapped and all of it is subsequently desorbed. An external standard can therefore be used for calibration of the gas chromatograph. Problems can sometimes be encountered if the sample is obtained by purging an aqueous solution. Acrolein is an example of a compound which cannot be purged from water at room temperature.

3. (a) Perform an appropriate bioassay on the sample after desorption from a Tenax tube to ascertain whether the activity has been changed.

   (b) Use an effluent splitter, collect the entire sample, and perform a bioassay.

Chapter 6

1. No. But it can work a lot faster. It used to take many hours to manually determine the mass scale of a spectrum, measure the relative abundance of each ion, draw a line

diagram, and search a library of spectra. It can now be done in seconds, freeing the analyst for more creative work (or writing more ingenious programs!) Very few people reconstruct SIM profiles without a computer, also.

2. Yes. If the analyst sees only plotted line diagrams, he may be unaware of a deterioration in resolution or increase in noise. Moreover, many data systems are programmed to ignore important features such as multiply charged ions or "metastable" ions.

3. The advantages of using the gas chromatograph and mass spectrometer are discussed in Chapter 3. See also exercise 1 in this chapter. It is clearly seen that the physical connection of the three units is far superior to using them separately.

4. The output from the mass spectrometer is comprised of a series of mass spectral peaks together with electrical "noise." The output from the gas chromatograph is comprised of a series of gas-chromatographic peaks together with carrier gas. Both the MS–COM and the GC–MS interface should permit the passage of undistorted peaks. However, the MS–COM interface should reject as much electrical "noise" as possible, while the GC–MS interface removes most of the carrier gas from the sample stream.

Chapter 7

1. Don't forget!

Chapter 10

1. A water sample might be extracted with a liter of solvent, the extract reduced in volume to 10 μl, and half of this quantity injected into the GC–MS instrument. If 50 ng of a contaminant is sufficient to interfere with your analysis, the original solvent should not contain more than 100 ng liter^{-1} (0.1 ppb) of this compound. This is an illustration of the "amplification factor" referred to on p. 226.

2. Chloroform is stabilized with 0.25–2.0% ethanol to minimize phosgene formation. Antioxidants are added to diethyl ether to prevent the formation of explosive peroxides. There are other examples: it is not always easy to find out what the additives are.

3. Everywhere. They are present in large quantities in many plastics. They may be introduced to the laboratory air by air conditioning filters. Be particularly careful of any solvents which are not distilled from glass and stored in glass bottles with Teflon-lined caps. Deionized water is usually not free of organic contaminants. If you use detergent from plastic bottles it will contaminate your glassware. The mass spectra of most phthalates are dominated by an ion of m/e 149. The ease with which they can be recognized is a help in tracking down their source.

4. Elemental sulfur. It is present in most petroleum-contaminated samples. Like most other inorganic substances, it is not detected by the FID detector. The spectrum is that of the S_8 molecule, with fragment ions corresponding to 1–7 sulfur atoms.

5. The criteria in Table 2 should be satisfied. Also, it should be noted that many pairs of isomers have identical GC retention times and mass spectra.

Chapter 11

1. One could examine the metabolite prior to derivatization if it were sufficiently stable. One could use a CD_3 derivative instead of a CH_3 derivative. One could also use other derivatives such as ethyl or trimethylsilyl, if appropriate.

2. The chlorine isotope ratio in the molecular ion and chlorine-containing fragment ions.

3. TMS derivatives of 17β-hydroxy steroids afford fragment ions of m/e 129 (p. 97). With a 17α-ethyl group, this ion appears at m/e 157. One can therefore monitor at m/e 157 throughout a GC–MS run to locate potential metabolites and then scan spectra at the appropriate times when the analysis is repeated. If a data system is available which stores repetitive scans, only one run is required. For further details, see: C. J. W. Brooks, A. R. Thawley, P. Rocher, B. S. Middleditch, G. M. Anthony, and W. G. Stillwell, Characterization of steroidal drug metabolites by combined gas chromatography–mass spectrometry, *J. Chromatogr. Sci.* **9**, 35–43 (1971).

4. GC–MS will afford both gas chromatographic and mass spectral data. However, derivatization may be required, and the analysis time is relatively long. Direct-probe CIMS is much more rapid, although less data are obtained. Since speed is usually the controlling factor in such cases, the latter procedure is more appropriate.

5. Caffeine is found in tea, coffee, and cola nuts, so beverages made from ingredients are an obvious source. It is also contained in a number of prescription and nonprescription drugs. The list of other "unexpected" compounds is endless. Phosphoric acid is present in many soft drinks, and nicotine in tobacco smoke. Both compounds are frequently encountered during GC–MS analyses of body fluids.

Chapter 12

1. Sample A is prevalently aromatic, B is paraffinic, and C is naphthenic. The most intense fragment ion series are, respectively, 77, 91, 105; 71, 85, 99, 113; and 69, 83, 97, 111, 81, 95, 109. Ions characteristic of the prevalent type are always the strongest ones within a carbon number (C_6: 73–86; C_7: 87–100; C_8: 101–114). An additional clue to the aromatic nature of sample A is furnished by the presence of doubly charged ions, as the one listed at m/e 75.5.

2. (a) Benzene: 78; cyclohexane:84

   (b) Benzene: 78; toluene: 91 or 92.

   (c) Benzene: 78; ethylbenzene: 106 or 91.

   (d) Benzene: 78; toluene: 92; ethylbenzene: 106. Note the restriction of options for toluene and ethylbenzene.

   (e) Benzene: 78; toluene: 92; ethylbenzene: 100; cyclohexane: 84.

   (f) Benzene: 78; *p*-xylene: 106, 91, or 92.

   (g) *p*-Xylene: 105; ethylbenzene: 106. Because of the similarity of the spectra, the selection is rather difficult. Accuracy of the quantitative data should be generally worse than in the previous samples.

(h) Quantitative analysis is impossible in this case. The spectra are too similar. This example illustrates the difficulty of mass spectrometry to distinguish among positional hydrocarbon isomers.

3. The start of a homologous series is indicated by the appearance of intense peaks, such as at m/e 128 for naphthalene (C_nH_{2n-12}), m/e 132 for tetralins (C_nH_{2n-8}), m/e 158 for tetrahydroacenaphthenes or, in general dinaphthenobenzenes (C_nH_{2n-10}), m/e 154 for acenaphthenes and biphenyls (C_nH_{2n-14}), m/e 166 for fluorenes, (C_nH_{2n-16}), and m/e 178 for phenanthrenes (C_nH_{2n-18}). (See Table 16.) A reasonable procedure is to identify at first the members in each homologous series. These are those at 14 mass units apart. The m/e value of the first member of the series, that is, the one with the lowest m/e value (lowest molecular weight) can then be compared with the values given in Table 4. The homologs present in the mixture are thus the following:

C_9–C_{12} in the C_nH_{2n-6} series (alkylbenzenes)
C_{10}–C_{13} in the C_nH_{2n-8} series (tetralins and indanes)
C_{12}–C_{13} in the C_nH_{2n-10} series (naphthenonaphthalenes)
C_{10}–C_{13} in the C_nH_{2n-12} series (naphthalenes)
C_{12}–C_{13} in the C_nH_{2n-14} series (acenaphthene and biphenyls)
C_{13} in the C_nH_{2n-16} series (fluorene)
C_{14} in the C_nH_{2n-18} series (phenanthrene or anthracene)

There is no information on the substitution pattern of the individual isomers of the same formula.

4. (a) ~ 4,800   (c) ~14,300   (e) ~13,500

    (b) ~30,200   (d) ~81,200   (f) ~20,600

5. The answers are given in the table below:

Measured m/e	Formula(s)	Mass measurement error	Measured m/e	Formula(s)	Mass measurement error
254.1095	$C_{20}H_{14}$	0.0	301.1832	$C_{22}H_{23}N$	+0.2
274.2659	$C_{20}H_{34}$	−1.0	306.1620	$C_{21}H_{22}O_2$	0.0
288.2437	$C_{20}H_{32}O$	−1.6	290.0770	$C_{19}H_{18}SO$	+0.5
278.1120	$C_{19}H_{18}S$	−0.9	248.0655	$C_{17}H_{12}S$	−0.5
	$C_{22}H_{14}$	+2.5		$C_{20}H_8$	+2.9

Note that there are two valid formulas at m/e 278 and at m/e 248. The mass measurement errors are lower in both cases for the sulfur-containing formulas, favoring the selection of these as the "correct" identifications. If these formulas correspond to molecular ions of pure compounds, the presence of sulfur can be further checked by considering the size of the [M+2] isotope, which will be much higher in a sulfur-containing formula. If the formulas correspond to molecular ions in a complex petroleum or coal liquid mixture (as they indeed are) the hydrocarbon formula at m/e 248 would be discarded as its molecular weight is below that of the reasonable structures in the C_nH_{2n-32} series (276). On the other hand, the sulfur-containing formula corresponds to a reasonable structure in the $C_nH_{2n-22}S$ series, namely, a methylphenanthrenothiophene. This consideration also applies for a pure compound spectrum. A decision on whether the sulfur or hydrocarbon formula is the correct one at m/e 278 is more difficult, because both correspond to reasonable structures, respectively, in the $C_nH_{2n-20}S$ and C_nH_{2n-30} series (Table 4). Both could be present in a petroleum sample. An unequivocal selection would require a resolving power of 82,000, such as provided by an instrument with the capabilities of the Kratos-AEI MS50. A semiquantitative decision could be made by examining the spectrum at m/e 208, 222, 236, and 264 for the presence of lower-molecular-weight homologs in the $C_nH_{2n-20}S$ series. If these are absent or very small with respect to the intensity of the peak at m/e 278, one would favor the hydrocarbon; one would select the sulfur formula if the opposite were true.

Chapter 13

1. Components of the atmosphere are:

    (i)   Carbon dioxide: $44(CO_2)$, $28(CO)$, $16(O)$, 12 (C)

    (ii) Argon: $40(Ar^{+\cdot})$, $20(Ar^{2+})$

    (iii) Oxygen: $32(O_2^{+\cdot})$, $16(O_2^{2+}$ or O)

    (iv) Nitrogen: $28(N_2^{+\cdot})$, $14(N_2^{2+}$ or N)

    (v)   Water: $18(H_2O)$, $17(OH)$, $16(O)$

    These are the major components of the Earth's atmosphere. It also contains small amounts of Ne, He, CH_4, Kr, and N_2O.

2. Payload weight limitations are a major factor. However, the need for large pumps for maintaining low ion source pressures is reduced or eliminated. Moreover, much of the control electronics, amplifiers, and recorders are left behind on Earth.

3. (a) High temperature on solar side, low temperature on dark side.

    (b) Corrosive atmosphere, high temperature.

    (c) High pressure, low temperature.

    (d) High pressure, low temperature.

    It may prove impossible to perform mass spectral analyses of these atmospheres. However, several of the satellites of the giant planets may have less hostile atmospheres.

Index